Peter Lucas, José A. Gámez, Antonio Salmerón (Eds.)

Advances in Probabilistic Graphical Models

Studies in Fuzziness and Soft Computing, Volume 213

Editor-in-chief
Prof. Janusz Kacprzyk
Systems Research Institute
Polish Academy of Sciences
ul. Newelska 6
01-447 Warsaw
Poland
E-mail: kacprzyk@ibspan.waw.pl

Further volumes of this series can be found on our homepage: springer.com

Vol. 197. Enrique Herrera-Viedma, Gabriella Pasi, Fabio Crestani (Eds.)
Soft Computing in Web Information Retrieval, 2006
ISBN 978-3-540-31588-9

Vol. 198. Hung T. Nguyen, Berlin Wu
Fundamentals of Statistics with Fuzzy Data, 2006
ISBN 978-3-540-31695-4

Vol. 199. Zhong Li
Fuzzy Chaotic Systems, 2006
ISBN 978-3-540-33220-6

Vol. 200. Kai Michels, Frank Klawonn, Rudolf Kruse, Andreas Nürnberger
Fuzzy Control, 2006
ISBN 978-3-540-31765-4

Vol. 201. Cengiz Kahraman (Ed.)
Fuzzy Applications in Industrial Engineering, 2006
ISBN 978-3-540-33516-0

Vol. 202. Patrick Doherty, Witold Łukaszewicz, Andrzej Skowron, Andrzej Szałas
Knowledge Representation Techniques: A Rough Set Approach, 2006
ISBN 978-3-540-33518-4

Vol. 203. Gloria Bordogna, Giuseppe Psaila (Eds.)
Flexible Databases Supporting Imprecision and Uncertainty, 2006
ISBN 978-3-540-33288-6

Vol. 204. Zongmin Ma (Ed.)
Soft Computing in Ontologies and Semantic Web, 2006
ISBN 978-3-540-33472-9

Vol. 205. Mika Sato-Ilic, Lakhmi C. Jain
Innovations in Fuzzy Clustering, 2006
ISBN 978-3-540-34356-1

Vol. 206. A. Sengupta (Ed.)
Chaos, Nonlinearity, Complexity, 2006
ISBN 978-3-540-31756-2

Vol. 207. Isabelle Guyon, Steve Gunn, Masoud Nikravesh, Lotfi A. Zadeh (Eds.)
Feature Extraction, 2006
ISBN 978-3-540-35487-1

Vol. 208. Oscar Castillo, Patricia Melin, Janusz Kacprzyk, Witold Pedrycz (Eds.)
Hybrid Intelligent Systems, 2007
ISBN 978-3-540-37419-0

Vol. 209. Alexander Mehler, Reinhard Köhler
Aspects of Automatic Text Analysis, 2007
ISBN 978-3-540-37520-3

Vol. 210. Mike Nachtegael, Dietrich Van der Weken, Etienne E. Kerre, Wilfried Philips (Eds.)
Soft Computing in Image Processing, 2007
ISBN 978-3-540-38232-4

Vol. 211. Alexander Gegov
Complexity Management in Fuzzy Systems, 2007
ISBN 978-3-540-38883-8

Vol. 212. Elisabeth Rakus-Andersson
Fuzzy and Rough Techniques in Medical Diagnosis and Medication, 2007
ISBN 978-3-540-49707-3

Vol. 213. Peter Lucas, José A. Gámez, Antonio Salmerón (Eds.)
Advances in Probabilistic Graphical Models, 2007
ISBN 3-540-68994-X

Peter Lucas
José A. Gámez
Antonio Salmerón
(Eds.)

Advances in Probabilistic Graphical Models

Springer

Dr. Peter Lucas
Institute for Computing
and Information Sciences
Radboud University Nijmegen
Toernooiveld 1
6525 ED Nijmegen
The Netherlands
E-mail: peterl@cs.ru.nl

Dr. Antonio Salmerón
Department of Statistics
and Applied Mathematics
The University of Almería
La Cañada de San Urbano s/n
04120 Almería
Spain
E-mail: Antonio.Salmeron@ual.es

Dr. José A. Gámez
Department of Computer Science
and Artificial Intelligence
University of Granada
c/. Daniel Saucedo Aranda, s/n
18071 Granada
Spain
E-mail: jgamez@info-ab.uclm.es

Library of Congress Control Number: 2006939264

ISSN print edition: 1434-9922
ISSN electronic edition: 1860-0808
ISBN-10 3-540-68994-X Springer Berlin Heidelberg New York
ISBN-13 978-3-540-68994-2 Springer Berlin Heidelberg New York

This work is subject to copyright. All rights are reserved, whether the whole or part of the material is concerned, specifically the rights of translation, reprinting, reuse of illustrations, recitation, broadcasting, reproduction on microfilm or in any other way, and storage in data banks. Duplication of this publication or parts thereof is permitted only under the provisions of the German Copyright Law of September 9, 1965, in its current version, and permission for use must always be obtained from Springer. Violations are liable for prosecution under the German Copyright Law.

Springer is a part of Springer Science+Business Media
springer.com
© Springer-Verlag Berlin Heidelberg 2007

The use of general descriptive names, registered names, trademarks, etc. in this publication does not imply, even in the absence of a specific statement, that such names are exempt from the relevant protective laws and regulations and therefore free for general use.

Typesetting: by the author and techbooks using a Springer LaTeX macro package
Cover design: Erich Kirchner, Heidelberg

Printed on acid-free paper   SPIN: 11778097   89/techbooks   5 4 3 2 1 0

Preface

Probabilistic graphical models, such as Bayesian networks and Markov networks, have been around for awhile by now, and have seen a remarkable rise in their popularity within the scientific community during the past decade. This community is strikingly broad and includes computer scientists, statisticians, mathematicians, physicists, and, to an increasing extent, researchers from various fields of application, such as psychology, biomedicine and finance.

With the increase of their popularity, the range of graphical models being investigated and used has also expanded. Bayesian networks, previously also called belief networks and probabilistic networks – terms used nowadays less commonly – remain popular as ever. They have the advantage that joint probability distributions can be specified locally, in terms of variables corresponding to the vertices in the associated acyclic directed graph and the variables corresponding to the parents of these vertices. This facilitates the development of real-world probabilistic models. Furthermore, prior and posterior univariate marginal probability distributions can be computed quickly if the associated acyclic directed graph is sparse, and easily visualised as plots associated with the vertices, and this has offerred a convenient basis for the development of intuitive, easy to understand user interfaces. Much of the success of the formalisms can be attributed to these features. Despite of this, there is an increasing tendency of researchers to go beyond Bayesian networks as we know them. A number of these extensions are discussed in this book. The most important ones are now briefly reviewed.

Whereas the graphical part of a Bayesian network has the form of an acyclic directed graph, which has the advantage of supporting an interpretation of the graph in terms of cause-effect relationships, a disadvantage is that many of the arcs in the graph can be reversed without affecting the meaning of the graph in terms of encoded independence information. Taking this into account has an impact on the number of different graphical representations of independence information to be considered, and, as a consequence, fewer graphs need to be visited when exploring the search space of possibilities when learning graphical representations from data. Chain graphs allow for encoding such information, by representing arcs that can be reversed as lines; arcs that cannot be revered without changing the meaning of the graph are kept unchanged. Thus, chain graphs are mixed graphs that offer a clear picture of the independence information that is represented graphically.

The assessment and representation of probabilistic information is another important issue in the area of probabilistic graphical models. Whereas in early work, there was an emphasis on discrete probability distributions, in recent years, various ways of specifying and reasoning with continuous probability distributions and mixtures of discrete and continuous distributions have attracted a great deal of attention.

Bayesian networks and other probabilistic graphical models are suitable for computing conditional and marginal (joint) probability distributions. However, in order to use probabilistic models for decision making some additional information needs to be available, such as ways to represent decisions or actions and ways to distinguish between the possible outcomes of the decisions or actions. Originally, influence diagrams were the formalism of choice, as they could be viewed as Bayesian networks extended with a sequence of decision vertices and a utility vertex. A range of techniques is available nowadays to deal with such decision problems, all offering special features supporting the development of specific decision-making models. The techniques are either extensions to the original influence diagram formalism, or based on so-called Markov decision processes, which is a versatile technique to model decision making over time.

Issues of representation lie at the heart of research in the area of probabilistic graphical models. However, any representation runs the risk of being useless if not properly supported by algorithm for its manipulation. This, therefore, also holds for probabilistic graphical models. The design of sophisticated inference algorithms for probabilistic graphical models remains an important research area.

Finally, whereas the research in probabilistic graphical models was originally done by a small group of researchers within their own community, there are now several research groups outside the probabilistic graphical models community that wish to exploit these models for problem solving in their area and this is also reflected by the content of the present book.

The aim of the present book is twofold. On the one hand, the reader is given an overview of a number of important topics in probabilistic graphical models. This has been accomplished by inviting researchers in the area of probabilistic graphical models to write a survey chapter on a topic on which they have special expertise. On the other hand, the book includes a number research papers on a variety of topics, based on earlier papers presented at the European Workshop on Probabilistic Graphical Models (PGM) 2004, in Leiden, the Netherlands. Based on reviews of papers for PGM 2004, revised and extended versions of the papers were requested from some of the authors.

The book starts with a part on foundations of probabilistic graphical models. This is followed by parts on probabilistic inference, on the learning of probabilistic graphical models, and on representation and algorithmic issue of decision processes. The book is rounded off by a part where various applications of probabilistic graphical models, in particular Bayesian networks, are described.

This book on Probabilistic Graphical will, therefore, give a good, representative impression of the state of the art of the research in probabilistic graphical models and we hope the reader will appreciate our effort in getting the contributions to this book together.

Nijmegen, *Peter J.F. Lucas*
Albacete, *José A. Gámez*
Almería, *Antonio Salmerón*
May 2006

Contents

Part I Foundations

Markov Equivalence in Bayesian Networks
Ildikó Flesch, Peter J.F. Lucas 3

A Causal Algebra for Dynamic Flow Networks
James Q. Smith, Liliana J. Figueroa 39

Graphical and Algebraic Representatives of Conditional Independence Models
Jiří Vomlel, Milan Studený... 55

Bayesian Network Models with Discrete and Continuous Variables
Barry R. Cobb, Rafael Rumí, Antonio Salmerón 81

Sensitivity Analysis of Probabilistic Networks
Linda C. van der Gaag, Silja Renooij, Veerle M.H. Coupé............. 103

Part II Inference

A Review on Distinct Methods and Approaches to Perform Triangulation for Bayesian Networks
M. Julia Flores, José A. Gámez 127

Decisiveness in Loopy Propagation
Janneke H. Bolt, Linda C. van der Gaag............................ 153

Lazy Inference in Multiply Sectioned Bayesian Networks Using Linked Junction Forests
Yang Xiang, Xiaoyun Chen 175

Part III Learning

A Study on the Evolution of Bayesian Network Graph Structures
Jorge Muruzábal, Carlos Cotta 193

Learning Bayesian Networks with an Approximated MDL Score
Josep Roure Alcobé 215

Learning of Latent Class Models by Splitting and Merging Components
Gytis Karčiauskas 235

Part IV Decision Processes

An Efficient Exhaustive Anytime Sampling Algorithm for Influence Diagrams
Daniel Garcia-Sanchez, Marek J. Druzdzel 255

Multi-currency Influence Diagrams
Søren Holbech Nielsen, Thomas D. Nielsen, Finn V. Jensen 275

Parallel Markov Decision Processes
L. Enrique Sucar 295

Part V Applications

Applications of HUGIN to Diagnosis and Control of Autonomous Vehicles
Anders L. Madsen, Uffe B. Kjærulff 313

Biomedical Applications of Bayesian Networks
Peter J.F. Lucas 333

Learning and Validating Bayesian Network Models of Gene Networks
Jose M. Peña, Johan Björkegren, Jesper Tegnér 359

The Role of Background Knowledge in Bayesian Classification
Marcel van Gerven, Peter J.F. Lucas 377

Part I

Foundations

Markov Equivalence in Bayesian Networks

Ildikó Flesch and Peter J.F. Lucas

Institute for Computing and Information Sciences, Radboud University Nijmegen,
Toernooiveld 1, 6525 ED Nijmegen, The Netherlands,
{ildiko,peterl}@cs.ru.nl

Summary. Probabilistic graphical models, such as Bayesian networks, allow representing conditional independence information of random variables. These relations are graphically represented by the presence and absence of arcs and edges between vertices. Probabilistic graphical models are nonunique representations of the independence information of a joint probability distribution. However, the concept of Markov equivalence of probabilistic graphical models is able to offer unique representations, called essential graphs. In this survey paper the theory underlying these concepts is reviewed.

1 Introduction

During the past decade Bayesian-network structure learning has become an important area of research in the field of Uncertainty in Artificial Intelligence (UAI). In the early years of Bayesian-network research at the end of the 1980s and during the 1990s, there was considerable interest in the process of manually constructing Bayesian networks with the help of domain experts. In recent years, with the increasing availability of data and the associated rise of the field of data-mining, Bayesian networks are now seen by many researchers as promising tools for data analysis and statistical model building. As a consequence, a large number of papers discussing structure learning related topics have been published during the last couple of years, rendering it hard for the novice to the field to appreciate the relative importance of the various research contributions, and to develop a balanced view on the various aspects of the field. The present paper was written in an attempt to provide a survey of the issue lying at the very heart of Bayesian-network structure learning: (statistical) independence and its representation in graphical form.

The goal of structure learning is to find a good generative structure relative to the data and to derive the generative joint probability distribution over the random variables of this distribution. Nowadays, graphical models are widely used to represent generative joint probability distributions.

A *graphical model* is a knowledge-representation formalism providing a graph representation of structural properties of uncertain knowledge. Bayesian networks are special cases of graphical models; they offer a representation that is both powerful and easy to understand, which might explain their current high level of popularity among UAI, machine-learning and data-mining researchers.

Structure learning consists of two main, interrelated, problems:

(i) the evaluation problem, and
(ii) the identification problem.

The *evaluation problem* amounts to finding a suitable way of judging the quality of generated network structures. Using a scoring criterion we can investigate how well a structure with its associated constraints fits the data. Note that a scoring criterion allows comparing structures with each other in such way that a structure-ordering becomes possible. As we compare Bayesian networks comprising the same vertices, scoring criteria are based only on relationships modelled by means of arcs. One expects that the better the independence relations implied by the graph representation match the knowledge hidden in the data the higher the score obtained by a scoring criterion, and this should be taken as one of a set of requirements when designing a scoring criterion.

The *identification problem* concentrates on finding efficient methods to identify at least one, maybe more, network structures given a scoring criterion. The total number of possible graph representations for a problem grows superexponentially with the total number of random variables [11]. As a consequence, the application of brute-force algorithms is computationally speaking infeasible. Thus practical methods for learning Bayesian networks use heuristic search techniques to find a graph with a high score in the space of all possible network structures.

In this survey paper we do not go into details of network scoring criteria; rather we focus on another, however closely related, important element in all modern research regarding the learning of Bayesian networks: the identification of Bayesian networks that represent the same joint probability distribution, i.e. they are Markov equivalent. Exploiting the notion of Markov equivalence can yield computational savings by making the search space that must be explored more compact [4]. There are various proposals in the literature to represent Markov equivalent Bayesian networks. One of them, a proposal by Andersson et al, [1], uses a special type of graph, called an *essential graph*, to act as a class representative for Bayesian networks that encodes the same probabilistic independence information. Markov independence is therefore a key issue in learning Bayesian networks. This paper summarises the theory underlying equivalence of graphical models in terms of the underlying independence relationship.

The paper is organised as follows. In the next section, basic notions from graph theory and the logical notion of (statistical) independence are introduced. These act as the basis for the graph representation of independence information as described in Section 3. Equivalence of Bayesian networks is studied in depth in Section 4, where we are also concerned with the properties of essential graphs. The paper is rounded off with remarks with respect to the consequences of the theory summarised in this paper for the area of Bayesian-network structure learning.

2 Preliminaries

We start by introducing some elementary notions from graph theory in Section 2.1. Next, we review the foundation of the stochastic (or statistical) independence relation as defined in probability theory in Section 2.2. We assume that the reader has access to a basic textbook on probability theory (cf. [6]).

2.1 Basic Concepts from Graph Theory

This subsection introduces some notions from graph theory based on Ref. [5]; it can be skipped by readers familiar with these notions.

Sets of objects will be denoted by bold, upright uppercase letters, e.g. V. For singleton sets $\{v\}$, we will often only write the element v instead of the set $\{v\}$. A *graph* is defined as a pair $G = (V, E)$, with V a finite set of *vertices*, where a vertex is denoted by an lowercase letter such as v, u and w, and $E \subseteq V \times V$ is a finite set of *edges*. A graph $G' = (V', E')$ is called an *induced subgraph* of graph $G = (V, E)$ if $V' \subseteq V$ and $E' = E \cap (V' \times V')$. A graph $G = (V, E)$ for which it holds that for each $(u, v) \in E$: $(v, u) \in E$, $u \neq v$, is called an *undirected graph* (UG). An edge $(u, v) \in E$ in an undirected graph is also denoted by $u - v$ and called an *undirected edge*. However, we will usually refer to undirected edges simply as edges if this will not give rise to confusion. A graph $G = (V, A)$ is called a *directed graph* if it comprises a finite set of vertices V, but, in contrast to an undirected graph, contains a finite set of *arcs*, by some authors called *arrows* or *directed edges*, $A \subseteq V \times V$ for which it holds that for each $(u, v) \in A$: $(v, u) \notin A$. An arc $(u, v) \in A$ is also denoted by $u \rightarrow v$ in the following.

A *route* in a graph G is a sequence $v_1, v_2, \ldots, v_k$ of vertices in V, where either $v_i \rightarrow v_{i+1}$, or $v_i \leftarrow v_{i+1}$, and possibly $v_i - v_{i+1}$, for $i = 1, \ldots, k-1, k \geq 1$; k is the *length* of the route. Note that a vertex may appear more than once on a route. A *section* of a route $v_1, v_2, \ldots, v_k$ is a maximal undirected subroute $v_i - \cdots - v_j$, $1 \leq i \leq j \leq k, k \geq 1$, where v_i resp. v_k is called a *tail terminal* if $v_{i-1} \leftarrow v_i$ resp. $v_k \rightarrow v_{k-1}$, and v_i resp. v_k is called a *head terminal* if $v_{i-1} \rightarrow v_i$ resp. $v_k \leftarrow v_{k-1}$. A *path* in a graph is a route, where vertices v_i and v_{i+1} are connected either by an arc $v_i \rightarrow v_{i+1}$ or by an edge $v_i - v_{i+1}$. A path is a *directed path*, if it contains at least one arc. A *trail* in a graph is a route where each arc appears at most once and, in addition, the vertices in each section of the trail appear at most once in the section. A *slide* is a special directed path with $v_1 \to v_2$ and $v_i - v_{i+1}$ for all $2 \leq i \leq k-1$. A graph has a *directed cycle* if it contains a directed path, which begins and ends at the same vertex, i.e. $v_1 = v_k$.

A graph $G = (V, E)$ is called a *chain graph* if it contains no directed cycles. An *acyclic directed graph* (ADG) is a chain graph that is a directed graph. Note that undirected graphs are special cases of chain graphs as well. Due to the acyclicity property of chain graphs, the vertex set of a chain graph can be partitioned into subsets $V_1 \cup V_2 \cup \cdots \cup V_L, L \geq 1$, such that each partition only consists of edges and if $u \to v$, then $u \in V_i$ and $v \in V_j$, $i \neq j$. Based on this we can define a total order $\leq$ on vertices in a chain graph, such that if $u \in V_i$ and $v \in V_j$, with $i < j$, then $u < v$, and if $i = j$, then $u = v$ (i.e. they are in the same V_i). This order can be generalised to sets such that $U \leq V$, if for each $u \in U$ and $v \in V$ we have that $u \leq v$. Subsets $V_1, V_2, \ldots, V_L$ are called the *chain components* of the graph. A set of *concurrent variables* of V_l is defined as $C_l = V_1 \cup V_2 \cup \cdots \cup V_l, 1 \leq l \leq L$. Any vertex u in an ADG that is connected by a directed path to a vertex v is called a *predecessor* of v; the set of predecessors of v is denoted by $pr(u)$.

We say that the vertex $u \in V$ is a *parent* of $v \in V$ if $u \to v \in A$; the set of parents of v is denoted by $\pi(v)$. Furthermore, v is then called u's *child*; the set of children of vertex u is denoted by $ch(u)$. Two vertices $u, v \in V$ are *neighbours*, if there is an edge between these two vertices. The *boundary* of vertex $u \in V$, denoted by $bd(u)$, is the set of parents and neighbours of u, while the *closure* of u, denoted by $cl(u)$, is defined as $cl(u) = bd(u) \cup \{u\}$. Note that the boundary of a vertex u in an undirected graph is equal to its set of neighbours. The set of *ancestors* of a vertex u is the set of vertices $\alpha(u) \subseteq V$ where there exists a path from each $v \in \alpha(u)$ to u, but there exists no path from u to v, whereas the set of *descendants* of u, denoted by $\delta(u)$, is the set of vertices $\delta(u) \subseteq V$, where there exists a path from u to each $v \in \delta(u)$, but no path from v to u. The set of *non-descendants* of u, denoted by $\bar{\delta}(u)$, is equal to $V \setminus (\delta(u) \cup \{u\})$. If for some $W \subseteq V$ it holds that $bd(u) \subseteq W$, for each $u \in W$, then W is called an *ancestral* set. By $an(W)$ is denoted the *smallest* ancestral set containing W.

From the chain graph G we can derive the *moral graph* G^m by the following procedure, called *moralisation*:

(i) add edges to all non-adjacent vertices, which have children in a common chain component, and
(ii) replace each arc with an edge in the resulting graph.

A moral graph is therefore an undirected graph.

A *chord* is an edge or arc between two non-adjacent vertices of a path. A graph is called *chordal* if every cycle of length $k \geq 4$ has a chord.

As mentioned above, two vertices can be connected by an arc or an edge. If two distinct vertices $u, v \in V$ are connected but it is unknown whether by an edge or arc, we write $u \cdots v$, where the symbol $\cdots$ denotes this connection.

2.2 Axiomatic Basis of the Independence Relation

Let V be a finite set and let X_v be a *discrete* random variable corresponding to $v \in V$; thus v acts as *index* to X. Define $X_W = (X_v)_{v \in W}$ for any subset $W \subseteq V$; in particular $X = X_V = (X_v)_{v \in V}$. Let P denote a joint probability distribution, or JPD for short, of X. We continue by providing the basic definition of (conditional) independence that underlies almost all theory presented in this paper. The idea that conditional independence is a unifying notion of relationships among components of many mathematical structures was first expressed by Dawid [2].

Definition 1. *(conditional independence) Let X_V be a set of random variables with $U, W, Z \subseteq V$ disjoint sets of vertices, and let P be a joint probability distribution defined on X, then X_U is said to be* conditionally independent *of X_W given X_Z, denoted by $X_U \perp\!\!\!\perp_P X_W \mid X_Z$, if*

$$P(X_U \mid X_W, X_Z) = P(X_U \mid X_Z). \tag{1}$$

Conditional independence can be also interpreted as follows: learning about X_W has no effect on our knowledge concerning X_U given our beliefs concerning X_Z, and vice versa. If Definition 1 does not hold, then X_U and X_W are said to be *conditionally dependent* given X_Z, which is written as follows:

$$X_U \not\perp\!\!\!\perp_P X_W \mid X_Z. \tag{2}$$

As an abbreviation, conditional independence $X_U \perp\!\!\!\perp_P X_W \mid X_Z$ and conditional dependence $X_U \not\perp\!\!\!\perp_P X_W \mid X_Z$ will also be denoted by $U \perp\!\!\!\perp_P W \mid Z$ and $U \not\perp\!\!\!\perp_P W \mid Z$, respectively. We will often write $X \perp\!\!\!\perp_P Y \mid Z$ instead of $\{X\} \perp\!\!\!\perp_P \{Y\} \mid \{Z\}$.

Next, we will introduce the five most familiar axioms, called the *independence axioms* or *independence properties*, which the independence relation $\perp\!\!\!\perp_P$ satisfies. An example is provided for each of the axioms, freely following Ref. [3]. As the independence axioms are valid for many different mathematical structures, and we are concerned in this paper with independence properties represented by graphs–examples of such mathematical structures–we will use graphs to illustrate the various axioms. However, a discussion on how such graphs should be interpreted in the context of probability theory is postponed to the next section. In the example graphs of this paper, the first set X_U in the triple $X_U \perp\!\!\!\perp_P X_W \mid X_Z$ is coloured lightly grey, the second set X_W is coloured medium grey and the set X_Z dark grey. If a vertex in the graph does not participate in an independence property illustrated by the example, it is left unshaded.

The independence relation $\perp\!\!\!\perp_P$ satisfies the following independence axioms or independence properties:[1]

---

[1] Here and in the following we will always assume that the sets participating in the various independence relations $\perp\!\!\!\perp$ are disjoint, despite the fact that this is not

- **P1**: *Symmetry*. Let X_V be a set of random variables and $U, W, Z \subseteq V$ be disjoint sets of vertices, then

$$U \perp\!\!\!\perp_P W \mid Z \iff W \perp\!\!\!\perp_P U \mid Z.$$

If knowing U is irrelevant to our knowledge about W given that we believe Z, then the reverse also holds. An example is given in Figure 1(a).

- **P2**: *Decomposition*. Let X_V be a set of random variables and $U, W, Z, Q \subseteq V$ be disjoint sets of vertices, then

$$U \perp\!\!\!\perp_P W \cup Q \mid Z \Rightarrow U \perp\!\!\!\perp_P W \mid Z \wedge U \perp\!\!\!\perp_P Q \mid Z.$$

This property states that if both W and Q are irrelevant with regard to our knowledge of U assuming that we believe Z, then they are also irrelevant separately. See the example in Figure 1(b).

- **P3**: *Weak union*. Let X_V be a set of random variables and $U, W, Z, Q \subseteq V$ be disjoint sets of vertices, then

$$U \perp\!\!\!\perp_P W \cup Q \mid Z \Rightarrow U \perp\!\!\!\perp_P Q \mid W \cup Z \wedge U \perp\!\!\!\perp_P W \mid Z \cup Q.$$

It expresses that when learning about W and Q is irrelevant with respect to our knowledge about U given our beliefs about Z, then Q will remain irrelevant when our beliefs do not only include Z but also W (the same holds for Q). The weak union relation is illustrated by Figure 1(c).

- **P4**: *Contraction*. Let X_V be a set of random variables and $U, W, Z, Q \subseteq V$ be disjoint sets of vertices, then

$$U \perp\!\!\!\perp_P W \mid Z \wedge U \perp\!\!\!\perp_P Q \mid W \cup Z \Rightarrow U \perp\!\!\!\perp_P Q \cup W \mid Z.$$

Contraction expresses the idea that if learning about W is irrelevant to our knowledge about U given that we believe Z and in addition learning about Q does not change our knowledge with respect to U either, then the irrelevance of Q with respect to U is not dependent on our knowledge of W, but only on Z. The notion of contraction is illustrated by Figure 1(d).

- **P5**: *Intersection*. Let X_V be a set of random variables and $U, V, Z, Q \subseteq V$ be disjoint sets of vertices, then

$$U \perp\!\!\!\perp_P W \mid Z \cup Q \wedge U \perp\!\!\!\perp_P Q \mid Z \cup W \Rightarrow U \perp\!\!\!\perp_P W \cup Q \mid Z.$$

The intersection property states that if learning about W has no effect on our knowledge about U assuming that we believe Z and Q, knowing, in addition, that our knowledge of Q does not affect our knowledge concerning U if we also know W, then learning about W and Q together has also no

---

strictly necessary. However, as disjoint and non-disjoint sets bear a completely different meaning, and it does not appear to be a good idea to lump these two meanings together, we have decided to restrict our treatment to disjoint sets, as this seems to offer the most natural interpretation of (in)dependence.

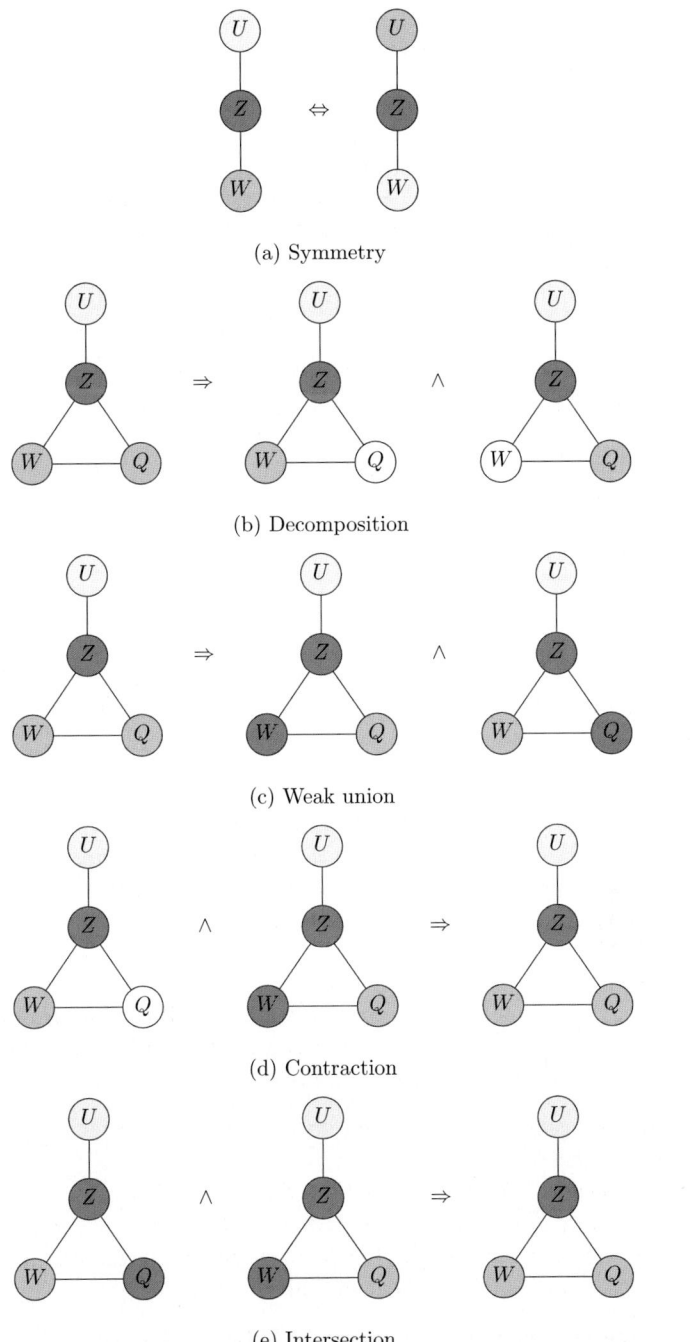

Fig. 1. Example graphs illustrating the following independence axioms: (a) Symmetry, (b) Decomposition, (c) Weak union, (d) Contraction and (e) Intersection.

effect on U. This property only holds for *strictly positive* joint probability distributions. An example of the intersection property is shown in Figure 1(e).

Any model satisfying the independence axioms **P1** to **P4** is called a *semi-graphoid*, whereas any model of the axioms **P1** to **P5** is called a *graphoid*. Any joint probability distribution P satisfies axioms **P1** to **P4**, while a joint probability distribution only satisfies **P5** if its co-domain is restricted to the open interval $(0,1)$, i.e. it is a joint probability distribution that does not represent *logical relationships*. A counterexample of the intersection property is shown in Table 1. Here, each random variable can take values a or b. There are only four possibilities, each having a probability equal to $\frac{1}{4}$, and the other possibilities have probability equal to 0. It holds that $U \perp\!\!\!\perp_P Q \mid Z \cup W$ and $U \perp\!\!\!\perp_P W \mid Z \cup Q$, however $U \not\!\perp\!\!\!\perp_P W \cup Q \mid Z$.

Table 1. Counterexample to the intersection property.

U	W	Q	Z
a	a	a	a
a	a	a	a
b	a	a	a
b	b	b	a

The independence axioms **P1** to **P4** were first introduced by Pearl (cf. [10]); he claimed that they offered a finite characterisation of the independence relation (Pearl's famous "completeness conjecture"). This statement, however, was shown to be incorrect by Studený after he discovered an axiom which indeed appeared to be a property of the independence relation, yet could not be deduced from axioms **P1** to **P4** [12]. Subsequently, Studený proved that no finite axiomatisation of the independence relation exists [13].

The five axioms mentioned above are well known by researchers in probabilistic graphical models; however, there are a number of other axioms which are also worth mentioning. We mention four of these axioms:

- **P6: Strong union.** Let X_V be a set of random variables and $U, W, Z, Q \subseteq V$ be disjoint sets of vertices, then

$$U \perp\!\!\!\perp_P W \mid Z \Rightarrow U \perp\!\!\!\perp_P W \mid Z \cup Q.$$

This property says that if learning about W does no convey any knowledge with regard to U given our beliefs concerning Z, then this knowledge concerning W remains irrelevant if our beliefs also include Q. An example is shown in Figure 2(a).

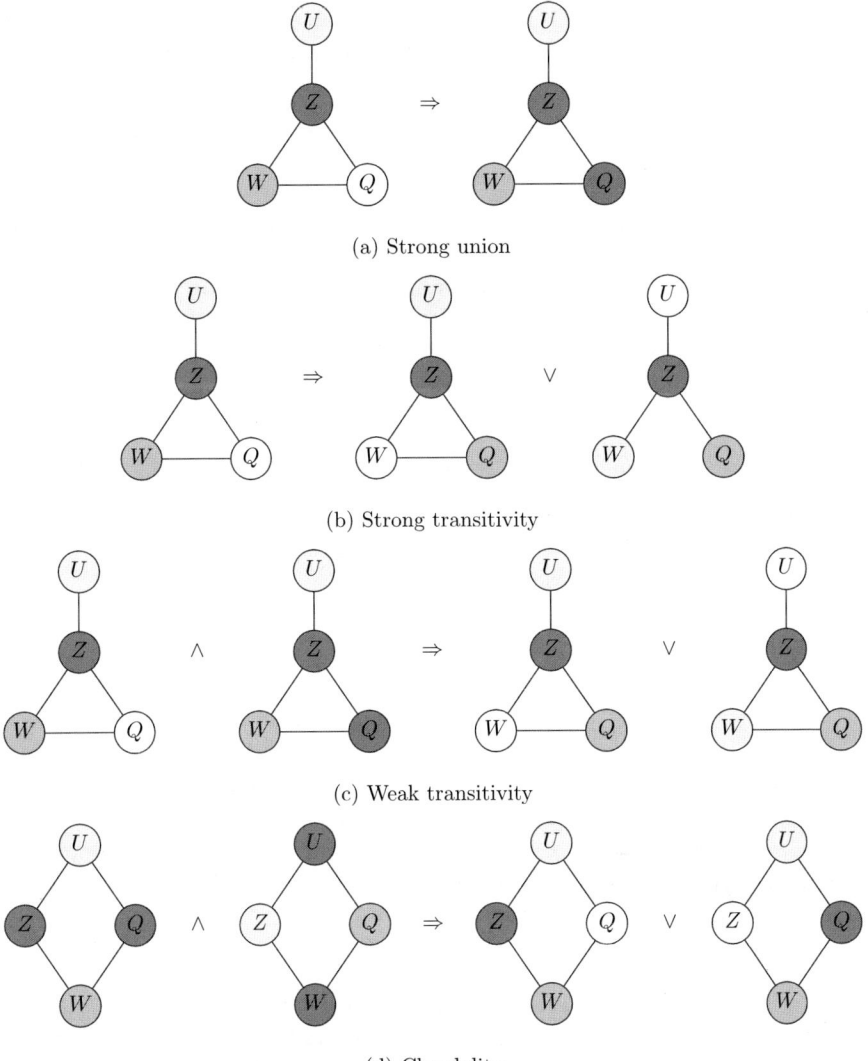

Fig. 2. Example graphs illustrating the following independence axioms: (a) Strong union, (b) Strong transitivity, (c) Weak transitivity and (d) Chordality.

It holds that strong union implies weak union [3].

- **P7**: *Strong transitivity*. Let X_V be a set of random variables and $U, W, Z, Q \subseteq V$ be disjoint sets of vertices, then

$$U \not\!\perp_P Q \mid Z \land W \not\!\perp_P Q \mid Z \Rightarrow U \not\!\perp_P W \mid Z.$$

This property says that if based on our beliefs concerning Z, observing Q will learn us something about both U and W, then our beliefs concerning

Z already made U and W relevant to each other. Applying the equivalence $a \Rightarrow b \equiv \neg b \Rightarrow \neg a$, strong transitivity can be rewritten to

$$U \perp\!\!\!\perp_P W \mid Z \Rightarrow U \perp\!\!\!\perp_P Q \mid Z \vee W \perp\!\!\!\perp_P Q \mid Z.$$

For an example see Figure 2(b).

- **P8**: *Weak transitivity.* Let X_V be a set of random variables and $U, W, Z, Q \subseteq V$ be disjoint sets of vertices, then

$$U \not\perp\!\!\!\perp_P Q \mid Z \wedge W \not\perp\!\!\!\perp_P Q \mid Z \Rightarrow U \not\perp\!\!\!\perp_P W \mid Z \vee U \not\perp\!\!\!\perp_P W \mid Z \cup Q.$$

Weak transitivity is an extension of strong transitivity and states that if U and W are separately dependent of Q given our beliefs about Z, then it holds that knowledge exchange between U and W is accomplished via Z or Z and Q. Applying the equivalence $a \Rightarrow b \equiv \neg b \Rightarrow \neg a$ the above mentioned dependence relation can also be written as

$$U \perp\!\!\!\perp_P W \mid Z \wedge U \perp\!\!\!\perp_P W \mid Z \cup Q \Rightarrow U \perp\!\!\!\perp_P Q \mid Z \vee W \perp\!\!\!\perp_P Q \mid Z.$$

This property is illustrated in Figure 2(c).
It holds that strong transitivity implies weak transitivity [3].

- **P9**: *Chordality.* Let X_V be a set of random variables and $U, W, Z, Q \subseteq V$ be disjoint sets of vertices, then

$$U \not\perp\!\!\!\perp_P W \mid Z \wedge U \not\perp\!\!\!\perp_P W \mid Q \Rightarrow U \not\perp\!\!\!\perp_P W \mid Z \cup Q \vee Z \not\perp\!\!\!\perp_P Q \mid U \cup W.$$

It implies that if learning about W yields knowledge about U, having beliefs concerning Z, and the same holds when we have beliefs about Q, then our knowledge about W is still relevant to our knowledge about U if we know both Z and Q, or our knowledge about both Z and Q makes Z and Q exchange knowledge. It is equivalent to

$$U \perp\!\!\!\perp_P W \mid Z \cup Q \wedge Z \perp\!\!\!\perp_P Q \mid U \cup W \Rightarrow U \perp\!\!\!\perp_P W \mid Z \vee U \perp\!\!\!\perp_P W \mid Q.$$

An example of chordality is depicted in Figure 2(d).

3 Graphical Representation of Independence

In this section, we discuss the representation of the independence relation by means of graphs, the rest of this paper will be devoted to this topic. In the previous section, the conditional independence relationship was defined in terms of a joint probability distribution P. In Section 3.1 closely related notions of graph separation are defined and informally linked to conditional independence. In Section 3.2, various special Markov properties are introduced and discussed, building upon the separation criteria from Section 3.1. Finally, in Section 3.3, possible relationships between conditional (in)dependences in joint probability distributions and the graph separation properties introduced earlier are established formally. This provides a semantic foundation for the various types of graphs in terms of the theory of statistical independence.

Let $G = (V, E)$ be an undirected graph and let V be a finite set and X_v be a *discrete* random variable corresponding to $v \in V$. Define $X_W = (X_v)_{v \in W}$ for any subset $W \subseteq V$; in particular $X = X_V = (X_v)_{v \in V}$. Let P denote a joint probability distribution of X.

3.1 Graph Separation and Conditional Independence

The independence relation defined earlier can be represented as a graphical model, where arcs and edges represent the dependences, and absence of arcs and edges represents the (conditional) independences. Arcs and edges represent roughly the same (in)dependence information; however, there are some differences between the meaning of arcs and edges. The actual interpretation is subtle, and is the topic of this and subsequent sections. In this section, we provide the foundation for representing conditional independence statements by graphs, and we cover the similarities between these principles for undirected, acyclic directed as well as for chain graphs.

In an undirected graph $G = (V, E)$ two vertices $u, v \in V$ are dependent if $u-v \in E$; if u and v are connected by a single path containing an intermediate vertex $z \in V$, $z \neq u, v$, then u and v are conditionally independent given z. This is the underlying idea of the following separation criterion (cf. [3]):

Definition 2. *(u-separation)* Let $G = (V, E)$ be an undirected graph, and $U, W, S \subseteq V$ be disjoint sets of vertices. Then if each path between a vertex in U and a vertex in W contains a vertex in S, then it is said that U and W are u-separated *by* S, denoted by $U \perp\!\!\!\perp_G W \mid S$. Otherwise, it is said that U and W are u-connected *by* S, denoted by $U \not\!\perp\!\!\!\perp_G W \mid S$.

The basic idea of u-separation can be illustrated by Figure 3; for example, p is u-separated from $\{r, t\}$ by $\{q, w\}$, i.e. $p \perp\!\!\!\perp_G \{r, t\} \mid \{q, w\}$, whereas p and $\{r, t\}$ are u-connected by vertex q, i.e. $p \not\!\perp\!\!\!\perp_G \{r, t\} \mid \{q\}$.

The independence relation represented by means of an ADG can be uncovered by means of one of the following two procedures:

- *d-separation*, as introduced by Pearl (cf. [10]);
- *moralisation*, as introduced by Lauritzen (cf. [7]).

First we discuss d-separation based on Ref. [9]. Let the distinct vertices $u, v, z \in V$ constitute an induced subgraph of the ADG $G = (V, A)$, with $(u \cdots z), (w \cdots z) \in A$ and u and v are non-adjacent. Because the direction of the arcs between u, z and w, z is unspecified, there are four possible induced subgraphs, which we call *connections*, illustrated in Figure 4.[2] These

---

[2] The terminology used in Figure 4 varies in different papers. Here the meaning of serial connection corresponds to *head-to-tail meeting*, divergent connection to *tail-to-tail* meeting and convergent connection to *head-to-head* meeting.

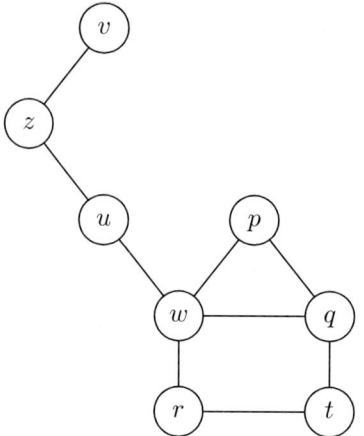

Fig. 3. Graphical illustration of u-separation. Vertex p and vertices $\{r,t\}$ are u-separated by vertices $\{q,w\}$, while vertex p and vertices $\{r,t\}$ are u-connected by q.

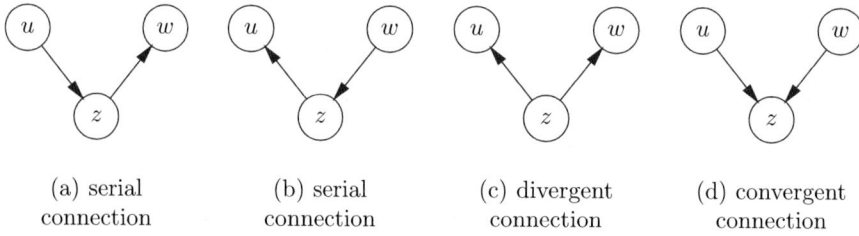

(a) serial connection   (b) serial connection   (c) divergent connection   (d) convergent connection

Fig. 4. The four possible connections for acyclic directed graph $G = (V, A)$ given vertices $u, w, z \in V$ with arcs $(u \cdots z), (w \cdots z) \in A$.

four possible connections offer the basis for the representation of conditional dependence and independence in ADGs. The two serial connections shown in Figure 4(a) and Figure 4(b) represent exactly the same independence information; this is also the case for the divergent connection represented in Figure 4(c). Figure 4(d) illustrates the situation where random variables X_u and X_v are initially independent, but become dependent once random variable X_z is instantiated.

Let $S \subseteq V$, and $u, w \in (V \setminus S)$ be distinct vertices, which are connected to each other by the trail τ. Then τ is said to be *blocked* by S if one of the following conditions is satisfied:

- $z \in S$ appears on the trail τ, and the arcs of τ meeting at z constitute a serial or divergent connection;
- $z \notin S$, $\delta(z) \cap S = \emptyset$ and the arcs meeting at z on τ constitute a convergent connection, i.e. if z appears on the trail τ then neither z nor any of its descendants occur in S.

The notion of d-separation exploits this notion of blocking, taking into account that vertices can be connected by more than one trail:

Definition 3. *(d-separation)* Let $G = (V, A)$ be an ADG, and let $U, W, S \subseteq V$ be disjoint sets of vertices. Then U and W are said to be d-separated by S, denoted by $U \perp\!\!\!\perp_G^d W \mid S$, if each trail τ in G between each $u \in U$ and each $w \in W$ is blocked by S; otherwise, U and W are said to be d-connected by S, denoted by $U \not\!\perp\!\!\!\perp_G^d W \mid S$.

As an example, consider the graph in Figure 5(a), where the vertices z and p are connected by the following three trails: $\tau_1 = z \to u \to w \leftarrow p$; $\tau_2 = z \to u \to w \to q \leftarrow p$ and $\tau_3 = z \to u \to w \to r \to t \leftarrow q \leftarrow p$. Then trail τ_1 is blocked by $S = \{u, v\}$; since v does not appear on this trail and the arcs on τ_1 meeting at u form a serial connection. As u blocks τ_2 and τ_3 following Definition 3, we conclude that S d-separates z and p. On the other hand, neither $S' = \{v, w\}$ nor $S'' = \{v, t\}$ block τ_1, because $u \to w \leftarrow p$ is a convergent connection, $w \in S'$; and t is a descendant of vertex w which occurs in S''; it also participates in a convergent connection with respect to τ_3. Thus not every trail between z and p in G is blocked by S' or S'', and z and p are d-connected by S' or S''.

Next, we discuss the procedure of moralisation. Recall that the procedure of moralisation of a graph G consists of two steps:

(i) non-adjacent parents of a common chain component become connected to each other by an edge, and
(ii) each arc becomes an edge by removing its direction, resulting in an undirected graph.

An example of moralisation is presented in Figure 5. Since acyclic directed graphs are chain graphs, moralisation can also be applied to ADGs, where each chain component contains exactly one vertex. Observe that during the first step of the moralisation procedure, there may be extra edges inserted into the graph. Since edges between vertices create a dependence between random variables, vertices which became connected in the first step have a dependence relation in the resulting undirected graph. For example, as u and p have a common child w, and r and q have a common child t, the graph in Figure 5(a) is extended by two extra edges $u - p$ and $r - q$. The resulting graph after the first step of moralisation is depicted in Figure 5(b). The moral graph, obtained by replacing arcs by edges, is shown in Figure 5(c). Observe that moralisation transforms the independences and dependences represented by d-separation (d-connection) into u-separation (u-connection). In the resulting moral graph, the vertices u and p and the vertices r and q have become dependent of one another, and thus, some independence information is now lost. This independence information, however, can still be represented in the underlying joint probability distribution such that it still holds that $z \perp\!\!\!\perp_P p \mid v$. However, it is also possible to parametrise the moralisation procedure on the vertices which potentially gives rise to extra dependences. This possibility is a

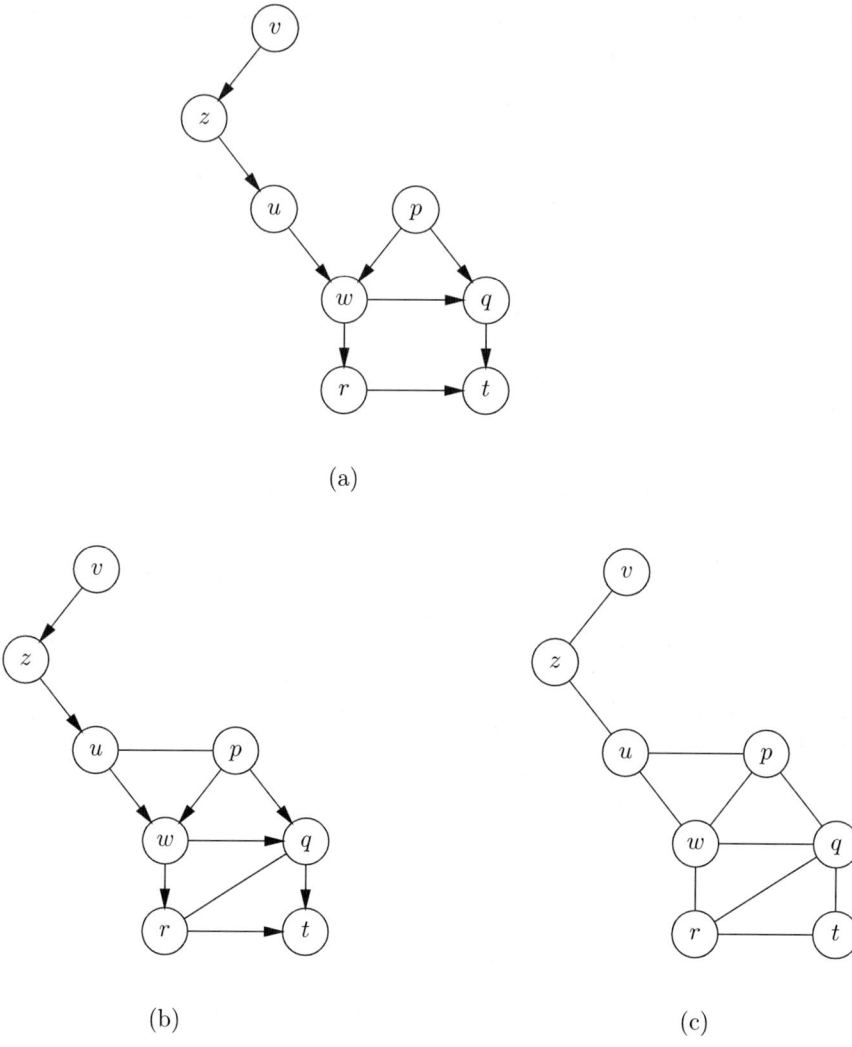

Fig. 5. An example of the moralisation procedure as applied to the graph shown in Figure (a). Graph (b) depicts the resulting graph after application of the first step of moralisation. Note that the vertices u and p and the vertices r and q are non-adjacent parents of the same child, therefore they became connected by an edge. Graph (c) results after changing the arcs in graph (b) into edges. Applying the definition of d-separation, it holds that $z \perp\!\!\!\perp_G^d p \mid v$ in graph (a); however for the moral graph in (c) we have that $z \not\perp\!\!\!\perp_{G^m} p \mid v$.

consequence of the meaning of a convergent connection $u \rightarrow z \leftarrow v$, because u and v are independent if z is not instantiated, and only become dependent if we know z. If this is the case, and if we also assume that we know u (or v), the dynamically created dependence between u and v gives rise to a type of reasoning known as *explaining away* [10]: v (u) becomes less or more likely if we know for certain that u (v) is the cause of z.

The moralisation procedure takes the presence of created dependences into account by means of the ancestral set, introduced in Section 2.1. Hence, this form of moralisation preserves all relevant (independence) information represented in the original ADG. The correspondence between d-separation and u-separation after moralisation is established in the following proposition:

Proposition 1. *Let $G = (V, A)$ be an acyclic directed graph with disjoint sets of vertices $U, W, S \subseteq V$. Then U and W are d-separated by S iff U and W are u-separated in the moral graph $G^m_{an(U \cup W \cup S)}$, where $an(U \cup W \cup S)$ is the smallest ancestral set of $U \cup W \cup S$.*

Proof: See Ref. [5], page 72. □

Figure 6 illustrates Proposition 1 by means of the conditional independence $z \perp\!\!\!\perp^d_G p \mid \{u, v\}$ and the conditional dependence $z \not\perp\!\!\!\perp^d_G p \mid \{y, w\}$ represented in the graph shown in Figure 5(a). We start by investigating the conditional independence $z \perp\!\!\!\perp^d_P p \mid \{u, v\}$. The smallest ancestral set of $\{z\} \cup \{p\} \cup \{u, v\}$ is $an(\{z, p, u, v\}) = \{z, p, u, v\}$; the graph depicted in Figure 6(a) contains all vertices of $an(\{z, p, u, v\})$ for the graph shown in Figure 5(a). We see that vertex p is disconnected from the subgraph $v \to z \to u$. The moral graph of this smallest ancestral set is shown in Figure 6(b). Observe that in graph (b) the vertices z and p are (unconditionally) independent, as there is no path between them. Therefore, $z \perp\!\!\!\perp^d_G p \mid \{u, v\}$ still holds, although now as $z \perp\!\!\!\perp_{G^m} p \mid \{u, v\}$, as in the original graph in Figure 5(a). The situation where we wish to keep the conditional dependence $z \not\perp\!\!\!\perp^d_P p \mid \{v, w\}$ is illustrated by Figures 6(c) and 6(d). In Figure 6(c) the subgraph associated with the smallest ancestral set $an(z \cup p \cup \{v, w\}) = \{z, p, u, v, w\}$ is shown, and Figure 6(d) gives the resulting moral graph of Figure 6(c). In the graph (d) we can see that vertices z and p are connected by a path, therefore, the created dependence between u and p is now represented in the moral graph of G.

Moralisation can also be applied to chain graphs; however, there is also another read-off procedure, called *c-separation*, introduced by Studený and Bouckhaert [8]. The concept of c-separation generalises both u-separation and d-separation.

The concept of c-separation takes into account the chain graph property that vertices may be connected by either edges or arcs. Let $G = (V, E)$ be a chain graph and let σ denote a section of the trail τ in G. Then σ is *blocked* by $S \subseteq V$, if one of the following conditions holds:

- $z \in S$ appears on the section σ, where σ has one head and one tail terminal and every slide of the tail terminal is mediated by z, or σ has two tail terminals and every slide of at least one of the two tail terminals is mediated by z;
- $z \in S$ does *not* appear on the section σ, where σ has two head terminals and $z \notin \delta(\sigma)$.

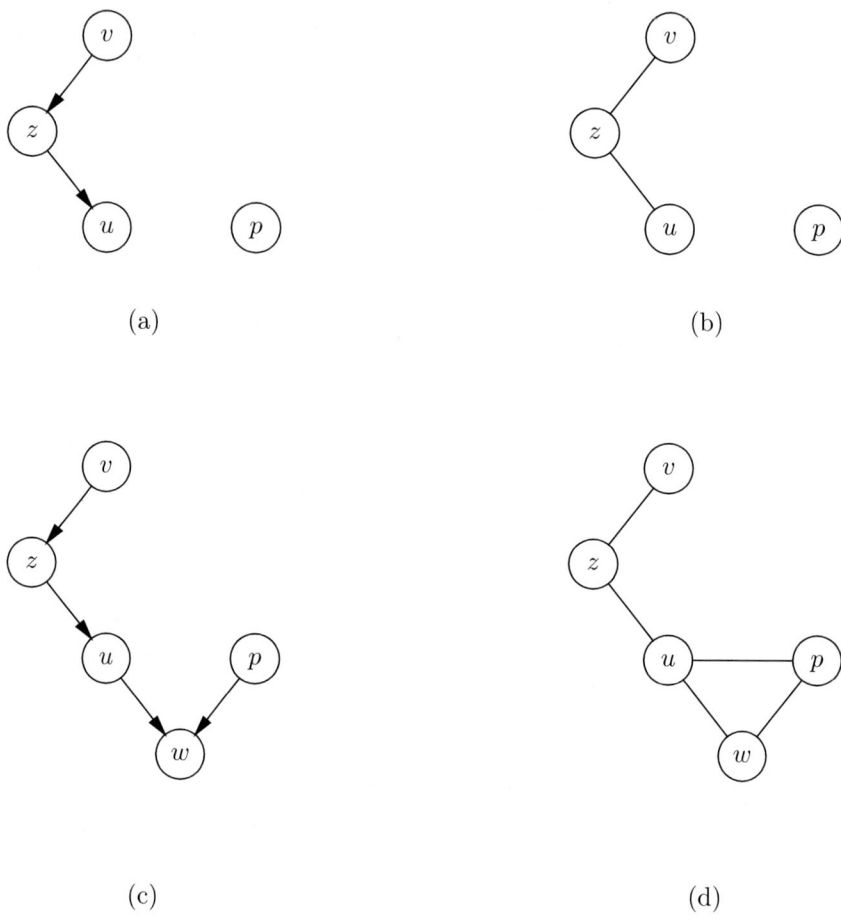

Fig. 6. Illustration of Proposition 1 with regard to the conditional independence $z \perp\!\!\!\perp_G^d p \mid \{u, v\}$ and the conditional dependence $z \not\!\perp\!\!\!\perp_G^d p \mid \{v, w\}$ which holds for the graph G shown in Figure 5(a). The induced subgraph H of this graph shown in Figure (a) above corresponds to $an(\{z, p, u, v\})$; in (b) its associated moral graph is represented. We see that the conditional independence $z \perp\!\!\!\perp_{H^m} p \mid \{u, v\}$ still holds. Figure (c) above shows the induced subgraph L of graph 5(a) corresponding to the smallest ancestral set of $\{z\} \cup \{p\} \cup \{v, w\}$. The graph L^m in (d) is the moral version of this graph. We see that the conditional dependence $z \not\!\perp\!\!\!\perp_{L^m} p \mid \{v, w\}$ holds for this moral graph. Hence, in both cases, all relevant (in)dependence information is preserved.

Based on these conditions we define the c-separation as follows.

Definition 4. *(c-separation) Let $G = (V, E)$ be a chain graph. Then two distinct vertex sets $U, W \in V$ are c-separated by $S \subseteq (V \setminus \{U \cup W\})$, if at least one of the sections of each trail τ between any vertices $u \in U$ and $v \in V$*

is blocked by S, written as $U \perp\!\!\!\perp_G^\kappa W \mid S$. Otherwise, U and W are c-connected by S and we write $U \not\perp\!\!\!\perp_G^\kappa W \mid S$.

As an example we use the chain graph presented in Figure 7(a). We examine whether $z \perp\!\!\!\perp_G^\kappa t \mid \{u, q, r\}$ (i.e. we have $S = \{u, q, r\}$). The following three trails between z and t will be investigated: $\tau_1 = z \to u - w \leftarrow p \to q \to t$ with sections $\sigma_{11} = u - w$ and $\sigma_{12} = q$; $\tau_2 = z \to u - w - q \to t$ with section $\sigma_{21} = u - w - q$ and $\tau_3 = z \to u - w \to r \to t$ with sections $\sigma_{31} = u - w$ and $\sigma_{32} = r$. In trail τ_1 section $u - w$ has two head-terminals and because $u \in S$ section σ_{11} does not block trail τ_1. In contrast to σ_{11}, σ_{12} has one head and one tail terminal (the terminals are both equal to vertex q) and slide $p \to q$ is mediated and therefore blocked by q. Since a trail is blocked if at least one of its sections is blocked by S, we conclude that trail τ_1 is blocked by $S = \{u, q, r\}$. Section $u - w - q$ in trail τ_2 has one head and one tail terminal, and satisfies the first blocking condition, because the slides $p \to q$ and $z \to u - w - q$ are both mediated by $q \in S$. Therefore, trail τ_2 is also blocked by S. This is also the case for trail τ_3 with section $u - w$, which has one head and one tail terminal and slides $z \to u - w \to r$ and $p \to q - w \to r$ are both mediated by $r \in S$, thus τ_3 is also blocked by S. There are also other trails between vertices z and t (e.g. $z \to u - w - q \leftarrow p \to q - w \to r \to t$), which are not mentioned here, because their sections are the same as in trails τ_1, τ_2 and τ_3. Therefore, these trails are also blocked by S. Thus, following Definition 4, the conditional independence relation contains $z \perp\!\!\!\perp_G^\kappa t \mid \{u, q, r\}$.

3.2 Markov Properties of Graphical Models

The dependence and independence relations determined by a joint probability distribution defined on the random variables corresponding to the vertices of a graph are graphically represented by the so-called *Markov properties*. We start by examining Markov properties for chain graphs, and next consider Markov properties for undirected and acyclic directed graphs, as these are special cases of chain graphs.

Each *chain Markov property* introduced below is illustrated by an example based on the chain graph shown in Figure 7(a). Vertices in the figures are presented using various shades depending on the role they play in visualising conditional independence properties.

A chain graph $G = (V, E)$ is said to obey:

- the *pairwise chain Markov property*, relative to G, if for any non-adjacent distinct pair $u, v \in V$ with $v \in \bar{\delta}(u)$:

$$u \perp\!\!\!\perp_G^\kappa v \mid \bar{\delta}(u) \setminus \{v\}. \tag{3}$$

Recall that $\bar{\delta}(u)$ is the set of non-descendants of u. For the corresponding example shown in Figure 7(b) it holds that $\bar{\delta}(u) = \{z, v, p, w\}$ and therefore this property expresses that $u \perp\!\!\!\perp_G^\kappa v \mid \{z, p, w\}$.

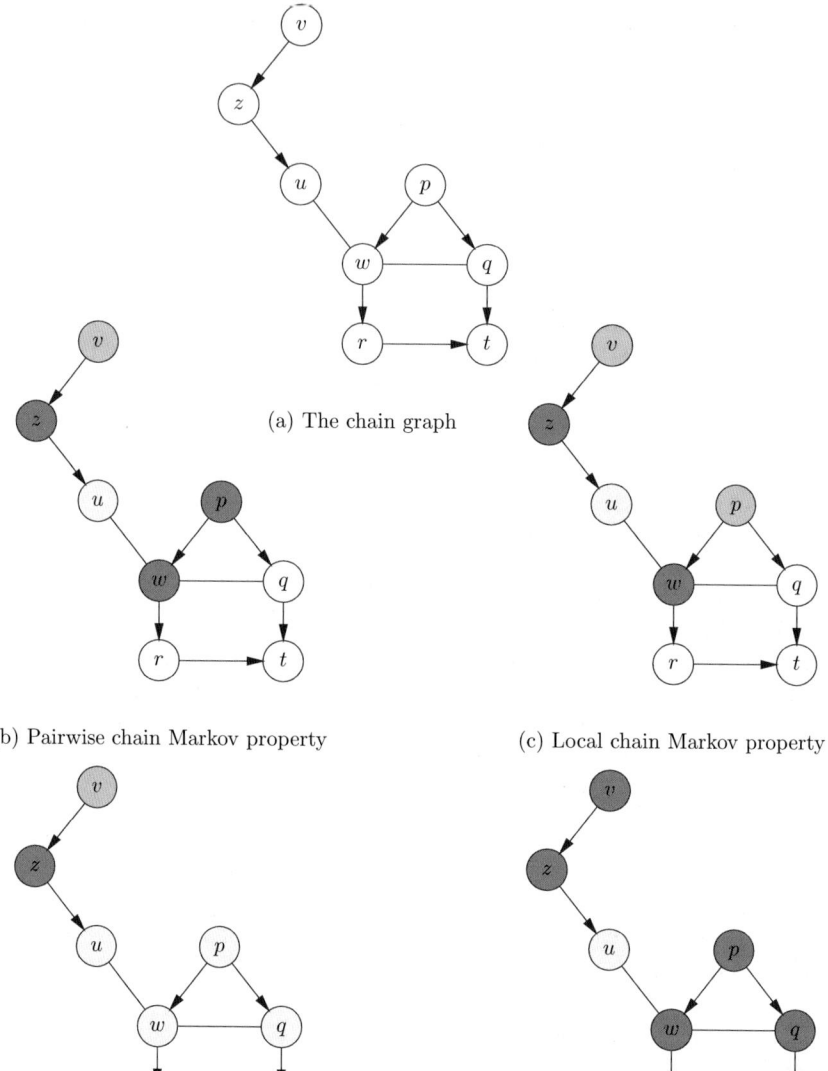

Fig. 7. Graphical illustration of the chain Markov properties, taking the chain graph in (a) as an example. Shown are (b): the pairwise chain Markov property $u \perp\!\!\!\perp_G^\kappa v \mid \{z, p, w\}$; (c): the local chain Markov property $u \perp\!\!\!\perp_G^\kappa \{v, p\} \mid \{z, w\}$; (d): the global chain Markov property $\{u, w, p, q, r, t\} \perp\!\!\!\perp_G^\kappa v \mid z$; (e): the block-recursive Markov property $u \perp\!\!\!\perp_G^\kappa r \mid \{v, z, w, p, q\}$.

- the *local chain Markov property*, relative to G, if for any vertex $v \in V$:

$$v \perp\!\!\!\perp_G^\kappa \bar{\delta}(v) \setminus bd(v) \mid bd(v). \tag{4}$$

Figure 7(c) illustrates this property by $u \perp\!\!\!\perp_G^\kappa \{v, p\} \mid \{z, w\}$.

- the *global chain Markov property*, relative to G, if for any triple of disjoint sets $U, V, Z \subseteq V$:

$$U \perp\!\!\!\perp_G^\kappa V \mid Z. \tag{5}$$

Figure 7(d) includes the following example of the global chain Markov property: $\{u, w, p, q, r, t\} \perp\!\!\!\perp_G^\kappa v \mid z$.

- the *block-recursive chain Markov property*, relative to G, if for any non-adjacent distinct pair $u, v \in V$:

$$u \perp\!\!\!\perp_G^\kappa v \mid C_{l^*} \setminus \{u, v\}, \tag{6}$$

where C_l is the set of concurrent variables of V_l and l^* is the smallest l with $u, v \in C_l$. This is shown in Figure 7(e). The well-ordered partitioning of a chain graph is not unique. In our example we take the following order of the partitioning: $\{v\} < \{z\} < \{p\} < \{u, w, q\} < \{r\} < \{t\}$, where corresponding to Section 2.1 $V_1 = \{v\}$, $V_2 = \{z\}, \ldots, V_6 = \{t\}$ (hence, by this ordering $l = 6$). Then based on this partitioning, the block-recursive chain Markov property states for example that it holds that $u \perp\!\!\!\perp_G^\kappa r \mid \{v, z, w, p, q\}$ with $l^* = 5$.

Based on the Markov properties for chain graphs, we will derive the related properties for undirected graphs, followed by acyclic directed graphs. As before, the undirected Markov properties are illustrated by means of figures. Here the graph in Figure 8(a) is taken as the example undirected graph.

Let $G = (V, E)$ be an undirected graph. Due to the fact that undirected graphs do not include arcs we cannot distinguish between ancestors and descendants of vertices; non-descendants $\bar{\delta}(v)$ in the chain Markov properties have to be replaced by the entire vertex set $V \setminus \{v\}$. In addition, the block-recursive chain Markov property makes no sense for the undirected graphs, because they do not have directionality. The undirected graph G is said to obey:

- the *pairwise undirected Markov property*, relative to G, if for any non-adjacent vertices $u, v \in V$:

$$u \perp\!\!\!\perp_G v \mid V \setminus \{u, v\}. \tag{7}$$

In this case the set of non-descendants $\bar{\delta}(u)$ from the chain property case is replaced by $V \setminus \{u\}$. The example in Figure 8(b) shows that $u \perp\!\!\!\perp_G v \mid \{Z, W, P, Q, R, T\}$.

22   Ildikó Flesch and Peter J.F. Lucas

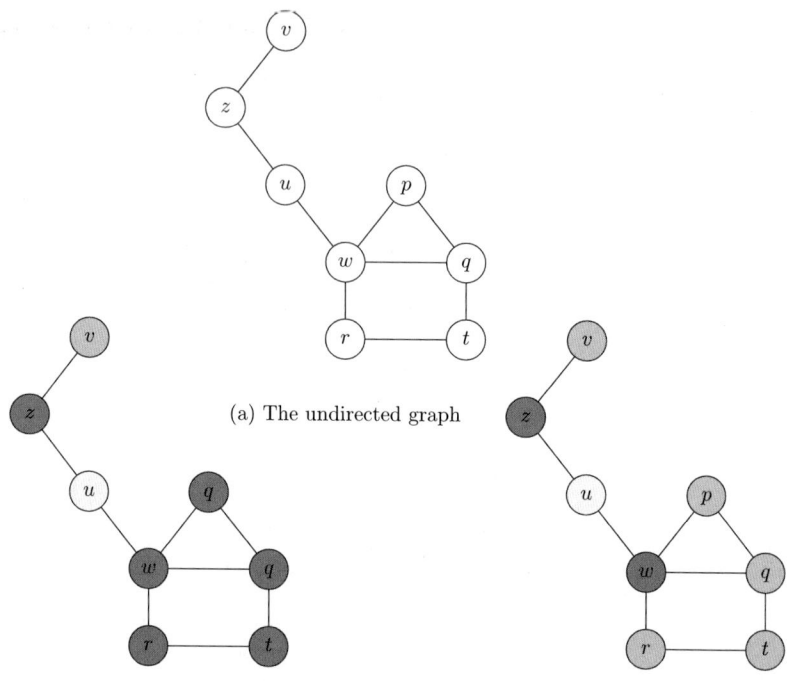

(a) The undirected graph

(b) Pairwise undirected Markov property

(c) Local undirected Markov property

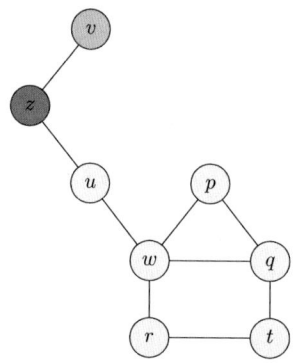

(d) Global undirected Markov property

Fig. 8. Graphical illustration of the undirected Markov properties, taking the UG from (a) as an example. Shown are: (b): the pairwise undirected Markov property $u \perp\!\!\!\perp_G v \mid \{z, w, p, q, r, t\}$; (c): the local undirected Markov property as $u \perp\!\!\!\perp_G \{v, p, q, r, t\} \mid \{z, w\}$; (d): the global chain Markov property $\{u, w, p, q, r, t\} \perp\!\!\!\perp_G v \mid z$.

- the *local undirected Markov property* relative to G, if for any $u \in V$:

$$u \perp\!\!\!\perp_G V \setminus cl(u) \mid bd(u), \qquad (8)$$

where $bd(u)$ is the boundary or *undirected Markov blanket* (the minimal boundary) of u and $cl(u)$ is the closure of u defined in Section 2.1. As was mentioned above $\bar{\delta}(u)$ of the chain property is replaced by V. Observe that in the local chain Markov property the expression $\bar{\delta}(u) \setminus bd(u)$ does not contain the random variable u, which would not be the case for $V \setminus bd(u)$. Therefore, the boundary set $bd(u)$ is replaced by $cl(u)$. Graph (c) in Figure 8 depicts $u \perp\!\!\!\perp_G \{v, p, q, r, t\} \mid \{z, w\}$, i.e. $bd(u) = \{z, w\}$.
- the *global undirected Markov property*, relative to G, if for any triple of disjoint sets $U, W, Z \subseteq V$:

$$U \perp\!\!\!\perp_G W \mid Z. \qquad (9)$$

For this property no changes need to be made with regard to the corresponding chain Markov property. An example for this property is given in Figure 8(d); here: $\{u, w, p, q, r, t\} \perp\!\!\!\perp_G v \mid z$.

Finally we consider Markov properties for ADGs, i.e. *directed Markov properties*. These are visualised using the ADG shown in Figure 9(a) as a basis. For the acyclic directed graph $G = (V, A)$ the local and global directed Markov properties are derived from the local, respectively global chain Markov properties, replacing the boundary by the parents of a vertex. Furthermore, the local chain Markov property generalises the blanket directed Markov property, and the ordered directed Markov property is derived from the block-recursive chain Markov property. The ADG G is said to obey:

- the *local directed Markov property*, relative to G, if for any $v \in V$:

$$v \perp\!\!\!\perp_G^d (\bar{\delta}(v) \setminus \pi(v)) \mid \pi(v). \qquad (10)$$

Note that the set $bd(v)$ from the chain property is replaced by $\pi(v)$ and, in addition, the expression $\bar{\delta}(v) \setminus bd(v)$ in the local chain Markov property is simplified to $\bar{\delta}(v) \setminus \pi(v)$. This property is illustrated in Figure 9(b); it expressed the conditional independence $u \perp\!\!\!\perp_G^d \{v, p\} \mid z$.
- the *blanket directed Markov property*, relative to G, which is derived from the local Markov property for chain graphs if we assume that for any $v \in V$:

$$v \perp\!\!\!\perp_G^d V \setminus (\beta(v) \cup v) \mid \beta(v), \qquad (11)$$

where $\beta(v)$ is the *directed Markov blanket*, defined as follows:

$$\beta(v) = \pi(v) \cup ch(v) \cup \{u : ch(u) \cap ch(u) \neq \varnothing; u \in V\}. \qquad (12)$$

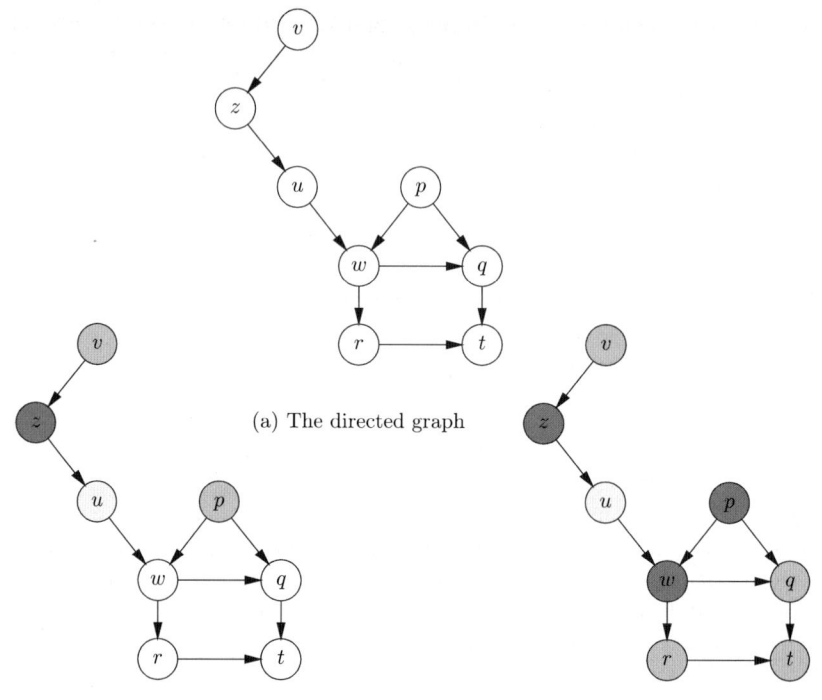

(a) The directed graph

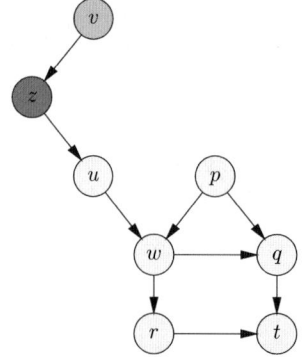

(b) Local directed Markov property

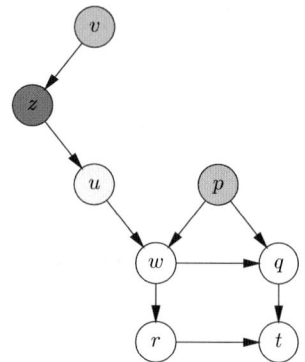

(c) Blanket directed Markov property

(d) Global directed Markov property

(e) Ordered directed Markov property

Fig. 9. Graphical illustration of the acyclic directed Markov properties, taking the ADG shown in (a) as an example. Shown are (b): the local directed Markov property $u \perp\!\!\!\perp_G^d \{v,p\} \mid z$; (c): the blanket directed Markov property $X \perp\!\!\!\perp_G^d \{v,q,r,t\} \mid \{z,w,p\}$; (d): the global directed Markov property $\{u,w,p,q,r,t\} \perp\!\!\!\perp_G^d v \mid z$; (e): the ordered directed Markov property $u \perp\!\!\!\perp_G^d \{v,p\} \mid z$.

This property can be derived from the blanket undirected Markov property easily, as v's children, parents and children's parents constitute the directed Markov blanket. An example is given in Figure 9(c); here we have for example $u \perp\!\!\!\perp_G^d \{v,q,r,t\} \mid \{z,w,p\}$.
- the *global directed Markov property*, relative to G, if for any triple of disjoint sets $U, W, Z \subseteq V$:

$$U \perp\!\!\!\perp_G^d W \mid Z. \tag{13}$$

This property need not be changed. Graph (d) in Figure 9 illustrates this property; for example, we have: $\{u,w,p,q,r,t\} \perp\!\!\!\perp_G^d v \mid z$.
- the *ordered directed Markov property*, relative to G, if for any $v \in V$:

$$v \perp\!\!\!\perp_G^d (pr(v) \setminus \pi(v)) \mid \pi(v), \tag{14}$$

where $pr(v)$ denotes the predecessor set of v. This property can be derived from the block-recursive chain property by the following idea: the acyclicity of graph G provides a well-ordering of its vertex set, in which each vertex can be seen as a chain component containing exactly one element. Figure 9(e) gives an example; based on the well-ordering $v < z < p < u < w < q < r < t$ it holds that $u \perp\!\!\!\perp_G^d \{v,p\} \mid z$.

3.3 D-map, I-map and P-map

In a graphical model it is not always the case that all independence information is represented, and it may also not be the case that all dependence information is represented. In this section the relationship between the representation of conditional dependence and independence by joint probability distributions and graphs is explored.

Let $\perp\!\!\!\perp_P$ be an independence relation defined on X_V for joint probability distribution P, then for each $U, W, Z \subseteq V$, where U, W and Z are disjoint:
- G is called an undirected *dependence map*, D-map for short, if

$$U \perp\!\!\!\perp_P W \mid Z \Rightarrow U \perp\!\!\!\perp_G W \mid Z,$$

- G is called an undirected *independence map*, I-map for short, if

$$U \perp\!\!\!\perp_G W \mid Z \Rightarrow U \perp\!\!\!\perp_P W \mid Z.$$

- G is called an undirected *perfect map*, or P-map for short, if G is both a D-map and an I-map, or, equivalently

$$U \perp\!\!\!\perp_P W \mid Z \iff U \perp\!\!\!\perp_G W \mid Z.$$

Observe that in a D-map each independence encoded in the joint probability distribution P has to be represented in the graph G. Using the equivalence

$a \rightarrow b \equiv \neg b \Rightarrow \neg a$, it holds for D-maps that each dependence encoded by the graph G has to be represented in the joint probability distribution P. This does not mean that each dependence represented in the joint probability distribution P is also discerned in the D-map. In contrast to D-maps, in I-maps each independence relationship modelled in the graph G has to be consistent with the joint probability distribution P and each dependence relationship represented in the joint probability distribution P has to be present in the graph representation G. Clearly, a perfect map is just a combination of a D-map and an I-map.

The notions of D-map, I-map and P-map can easily be adapted to similar notions for ADGs and chain graphs, and thus we will not include the definitions here. Consider the following example, illustrated by Figure 10. Let $V = \{u, v, z, w\}$ be the set of random variables with joint probability distribution: $P(X_u, X_v, X_z, X_w) = P(X_u \mid X_z)P(X_v \mid X_z)P(X_w \mid X_z)P(X_z)$. The associated conditional independence set consists of three members: $X_u \perp\!\!\!\perp_P \{X_v, X_w\} \mid X_z$, $X_v \perp\!\!\!\perp_P \{X_u, X_w\} \mid X_z$ and $X_w \perp\!\!\!\perp_P \{X_u, X_v\} \mid X_z$. Then,

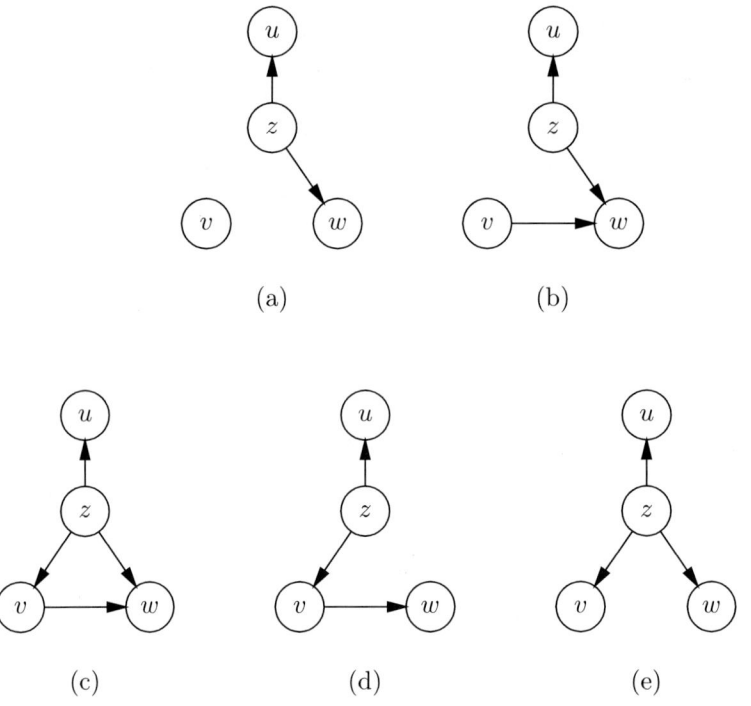

Fig. 10. Given the joint probability distribution $P(X_u, X_v, X_z, X_w) = P(X_u \mid X_z) \; P(X_v \mid X_z)P(X_w \mid X_z)P(X_z)$ with conditional independence set: $X_u \perp\!\!\!\perp_P \{X_v, X_w\} \mid X_z$, $X_v \perp\!\!\!\perp_P \{X_u, X_w\} \mid X_z$ and $X_w \perp\!\!\!\perp_P \{X_u, X_v\} \mid X_z$, graph (a) is a D-map, graph (b) is neither a D-map nor an I-map, graph (c) is an I-map, graph (d) is neither a D-map nor an I-map and graph (e) is a perfect map.

the graph in Figure 10(a) is a D-map of P whereas graph (b) is not a D-map of P, since it describes a dependence $v \to w$, which is not in P. Graph (b) is also not an I-map, since it does not include the arc $z \to v$. Graph (c) is an I-map of P but not a perfect map, because it includes the dependence $v \to w$, which is not part of P. Graph (d) is not an I-map by the fact that it does not represent the dependence between vertices z and w (i.e. it does not contain arc $z \to w$) and it is also not a D-map. Graph (e) is a perfect map of the joint probability distribution P. In the remainder of this section we investigate the correspondence between the above-mentioned properties of conditional independence and undirected, respectively, directed perfect maps. The following theorem establishes conditions for the existence of an undirected perfect map for any joint probability distribution.

Theorem 1. *The conditional independence relations associated with a joint probability distribution P need to satisfy the necessary and sufficient conditions of (i) symmetry, (ii) decomposition, (iii) intersection, (iv) strong union, and (v) strong transitivity, to allow their representation as an undirected perfect map.*

As mentioned above, any joint probability distribution obeys the semi-graphoid properties (symmetry, decomposition, weak union and contraction). According to Theorem 1 in addition to the properties of symmetry and decomposition, the properties of intersection, strong union and strong transitivity should hold, which however, are not semi-graphoid properties. Thus, not every joint probability distribution will have a corresponding undirected graphical representation as a perfect map. Furthermore, for directed perfect maps we have a number of necessary conditions, but these are not always sufficient.

Theorem 2. *Necessary conditions for the conditional independence relations associated with a joint probability distribution P to allow representation as a directed perfect map are: (i) symmetry, (ii) contraction, (iii) decomposition, (iv) weak union, (v) intersection, (vi) weak transitivity, and (vii) chordality.*

Theorem 2 indicates that similar to the undirected case, the independence relations corresponding to a joint probability distribution need not always allow representation as a directed perfect map. In many practical situations, it will not be possible to find a perfect map of a joint probability distribution. Therefore we wish to focus on graphical representations that are as sparse as possible, and thus do not encode spurious dependences, which is something offered by *minimal* I-maps.

Definition 5. *(**minimal I-map**) A graph is called a* minimal I-map *of the set of independence relations of the joint probability distribution P, if it is an I-map and removing any arc of the graph will yield a graph which is no longer an I-map.*

Minimising the number of arcs in a graphical model is not only important for representation reasons, i.e. in order to keep the amount of probabilistic information that has to be specified to the minimum, but also for computational reasons. It has been shown that every joint probability distribution P for which the conditional independence relations satisfy the conditions of symmetry, decomposition, and intersection has a minimal undirected I-map, whereas any joint probability distribution P with associated conditional independence relations satisfying the conditions of symmetry, decomposition, weak union and contraction has a minimal directed I-map representation [3]. This implies that each graphoid has a corresponding minimal undirected I-map, as well as a minimal directed I-map, and each semi-graphoid has a minimal directed I-map as graphical representation. As for every joint probability distribution the semi-graphoid properties hold, we can conclude that each joint probability distribution has a directed minimal I-map.

4 Equivalence of Bayesian Networks

In this section we return to the question which acted as the main motivation for writing this paper: how can equivalence of Bayesian networks be characterised best? It appears that in particular the concept of essential graphs plays a pivotal role in this. Before discussing essential graphs, we start by reviewing the definition of a Bayesian network in Section 4.1. Subsequently, in Sections 4.2 and 4.3, the equivalence relation on Bayesian networks which forms the basis for the concept of essential graphs will be studied.

4.1 Bayesian Networks

As before, let V be a finite set and let X_v be a *discrete* random variable corresponding to $v \in V$. Define $X_W = (X_v)_{v \in W}$ for any subset $W \subseteq V$; in particular $X = X_V = (X_v)_{v \in V}$. Let P denote a joint probability distribution, or JPD for short, of X.

Formally, a *Bayesian network* is a pair $\mathcal{B} = (G, P)$, where $G = (V, A)$ is an acyclic directed graph and P is a joint probability distribution defined on a set of random variables X.

As mentioned above, the set of arcs A describes the dependence and independence relationships between groups of vertices in V corresponding to random variables. If a joint probability distribution P admits a recursive factorisation then P can be defined on the set of random variables X_V as follows:

$$P(X) = \prod_{v \in V} P(X_v \mid X_{\pi(v)}). \tag{15}$$

Equation (15) implies that a joint probability distribution over a set of random variables can be defined in terms of local (conditional) joint probability distributions $P(X_v \mid X_{\pi(v)})$. Considerable research efforts have been made to exploit the structure of such a joint probability distribution for achieving computational savings. A Bayesian network is by definition a directed I-map.

What is interesting about Bayesian networks, and which is a main difference between directed and undirected graphical models, is that by instantiating vertices in the directed structure independences may change to dependences, i.e. stochastic independence has specific *dynamic* properties. In Section 3.1 we have called the type of reasoning associated with this 'explaining away'. This dynamic property is illustrated by Figure 4(d), where random variables X_1 and X_2, $V = \{1, 2, 3\}$ are independent of one another if random variable X_3 is unknown, but as soon as X_w becomes instantiated, a dependence between X_1 and X_2 is created. However, similar to undirected graphs, part of the independence information represented in the graphical part of a Bayesian network is *static*. The structure of a Bayesian network allows reading off independence statements, essentially by using the notions of d-separation and moralisation treated in the previous section.

Our motivation to study the Markov properties associated with graphs arises from our wish to understand the various aspects regarding the representation of independence in Bayesian networks. The following proposition establishes a very significant relationship between Markov properties on the one hand, and joint probability distributions on the other hand; it is due to Lauritzen [5]:

Proposition 2. *If the joint probability distribution admits a recursive factorisation according to the acyclic directed graph $G = (V, A)$, it factorises according to the moral graph G^m and therefore obeys the global Markov property.*

Proof: *See Ref. [5], page* 70. □

Proposition 2 implies an important correspondence between a recursive factorisation according to graph G and the global Markov property. This proposition can be extended resulting in the following theorem, also by Lauritzen [5]:

Theorem 3. *Let $G = (V, A)$ be an acyclic directed graph. For the joint probability distribution P the following conditions are equivalent:*

- *P admits a recursive factorisation according to G;*
- *P obeys the global directed Markov property, relative to G;*
- *P obeys the local directed Markov property, relative to G;*
- *P obeys the ordered directed Markov property, relative to G.*

Proof: *See Ref. [5], page* 74. □

Theorem 3 establishes the relation between a recursive factorisation of joint probability distribution P and the directed Markov properties introduced in Section 3.2, and therefore explains why the Markov properties and their relations are relevant in the context of Bayesian networks and thus to structure learning.

4.2 The Equivalence Relation on Acyclic Directed Graphs

In this section we introduce some notions required to study equivalence among Bayesian networks. We start by the definition of Markov constraints [14].

Definition 6. *(Markov constraints)* Let $G = (V, A)$ be an ADG. Then the Markov independence constraints, Markov constraints *for short, are the set of independence relations defined by the global directed Markov property.*

The Markov independence constraints allow us to define an equivalence relation on ADGs, as follows:

Definition 7. *(Markov equivalent)* Two ADGs are Markov equivalent *if they have the same set of Markov constraints.*

However, this definition is far removed from a procedural recipe: it is difficult to imagine how we can actually determine whether two ADGs are equivalent without enumerating all triples in the independence relations defined using these graphs. However, the following two definitions allow us to look at the problem from a different, and practically more useful, angle.

Definition 8. *(skeleton)* Let G be an ADG. The undirected version of G is called the skeleton *of G.*

For example, the graph in Figure 11(b) is the skeleton of graph (a).

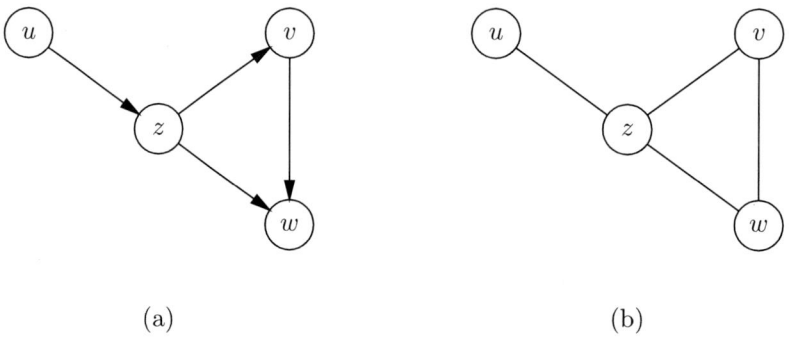

Fig. 11. An acyclic directed graph (a) and its skeleton (b).

Definition 9. *(immorality)* An induced subgraph in an ADG G with $u, v, z \in V$ is called an immorality, *if the graph contains the arcs $u \to z$ and $v \to z$, and vertices u and v are non-adjacent.*

Definition 9 implies that the concept of immorality is equivalent to that of convergence connection (cf. Figure 4(d)); both describe conditional dependence between random variables. (Immorality is synonymous with *v-structure* introduced by Verma and Pearl (cf. [14]).) However, immoralities are also the smallest induced subgraphs for the representation of conditional dependence. Observe that if the direction of one or both arcs of an immorality is reversed, the conditional dependence would turn into conditional independence, and thus would destroy the original meaning of the graph. Therefore to keep the independence relation defined on the random variables unchanged, it is not allowed to reverse the direction of these arcs.

Definition 10. *(essential arcs) Arcs that cannot be reversed without changing the conditional dependence and independence relations are called* essential arcs.

By Definition 10 both arcs of an immorality are essential arcs.

Applying Definition 8 and Definition 9, Markov equivalence is redefined in terms of the concepts of skeleton and immoralities by the following theorem, originally introduced by Verma and Pearl, which establishes the connection between these notions (cf. [14]):

Theorem 4. *Two ADGs are* Markov equivalent *with each other if and only if they have the same skeleton and they consist of the same set of immoralities.*

An example of Markov equivalence is given in Figure 12. Graph (a), (b) and (c) are equivalent by Theorem 4, but graph (d) is not equivalent to graphs (a), (b) and (c) since it contains, in contrast to the other graphs, the immorality $u \rightarrow z \leftarrow w$.

Let us try to explain why Theorem 4 plays a significant role in the field of structure learning. Recall that an immorality describes an independence relationship between random variables and it is also the smallest induced subgraph reflecting conditional dependence. The purpose of structure learning is to find the relations between the random variables of the problem domain based on the data. Thus, if we have the entire set of independence relationships (the Markov constraints) or the entire set of dependence relationships over the random variables our aim has been achieved. In the graphical representation of dependence there are two kinds of dependences that can be distinguished:

(i) static, and
(ii) dynamic dependence.

By static dependences we mean the existence of a direct connection (i.e. an arc or edge) between vertices. Since the joint probability distribution on two static dependent random variables $u \rightarrow v$ is the same as $v \rightarrow u$, according to Bayes' theorem, this dependence can be represented by an arc. In contrast to static dependences, dynamic dependences are conditionally dependent on the instantiation of random variables associated with vertices with convergent connections. Therefore, these arcs have to preserve their direction. This is exactly what is said by Theorem 4.

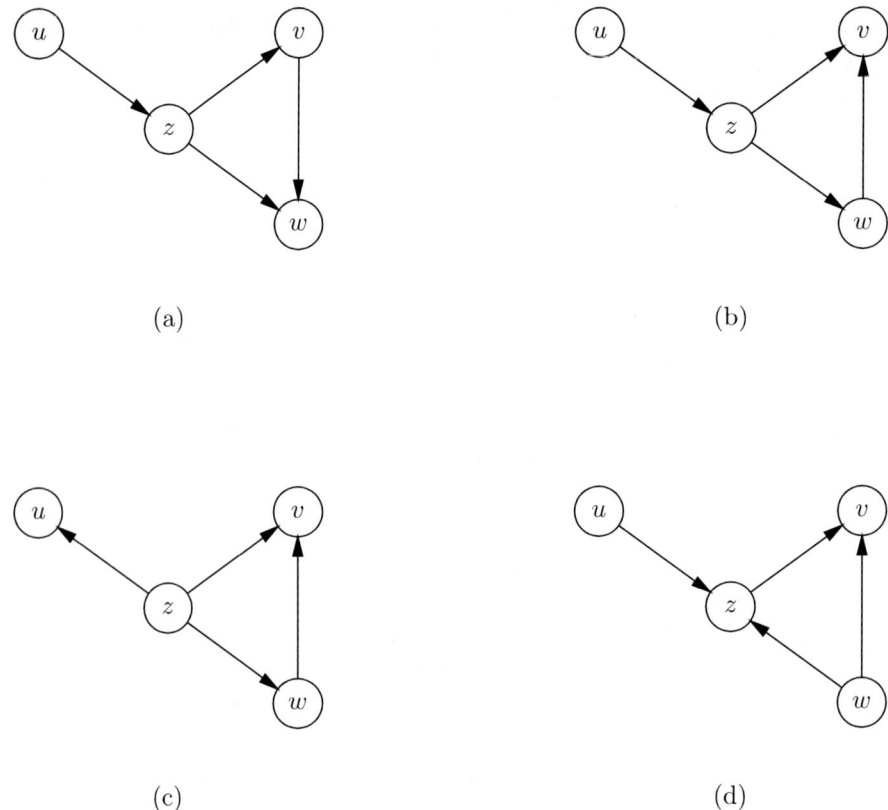

Fig. 12. An example of Markov equivalence. Graph (a), (b) and (c) are equivalent since they have the same skeleton and the same set of immoralities. Graph (d) has also the same skeleton as graph (a), (b) and (c), but graph (d) also contains an immorality $u \to z \leftarrow w$ which does not occur in the other graphs. Therefore graph (d) is not equivalent to graph (a), (b) and (c).

4.3 Essential Graphs

Taking Theorem 4 as a foundation, in this section we will study the important problem of equivalence of ADGs. Recall that equivalent ADGs have the same immoralities, and these immoralities consist of essential arcs, which in each equivalent ADG have the same direction. In contrast, if one wishes to build an ADG from a skeleton and a collection of immoralities, there are normally different choices possible for edges which do not participate in an immorality, to the extent that choices that give rise to a directed cycle or to a new immorality are not allowed. We can therefore conclude that the difference between equivalent ADGs is entirely based on the difference in the direction of their non-essential arcs.

It now appears that classes of Markov equivalent ADGs can be uniquely described by means of chain graphs, called essential graphs, which thus act as class representatives [1]; they are defined as follows:

Definition 11. (*essential graph*) *Let E denote the equivalence class of ADGs that are Markov equivalent. The* essential graph G_E^* *is then the smallest graph larger than any of the ADGs G in the equivalence class E; formally:*

$$G_E^* := \bigcup \{G \mid G \in E\}. \tag{16}$$

This definition implies that any of the non-essential arcs in any of the ADGs $G \in E$ is replaced by an edge (which means that to the arc $(u, v) \in A$ an arc (v, u) is added), and this explains why an essential graph is as large or larger than any of the members of the equivalence class E which it represents. Of course, as an essential graph is a chain graph, it may not be (and usually is not) an ADG, and therefore usually not a member of the equivalence class it represents.

It has been established that if an arc is part of a particular subgraph with a specific structure, then we know that the arc must be essential. There are four different (sub)graphs where $u \to v$ will always be an essential arc; these are shown in Figure 13. As mentioned above, a serial or divergent connection mirrors conditional independence, while a convergent connection reflects a potential dependence relationship between random variables (see Figure 4). Clearly, it is not allowed to express a dependence represented in the ADGs of an equivalence class as an independence in the associated essential graph, and vice versa. This is illustrated by the subgraphs (a) and (b) in Figure 13. Case (a) means that we have a serial connection, which would be turned into convergent connection if the direction of $u \to v$ is reversed. Therefore $u \to v$ is an essential arc. In contrast, changing the direction of $u \to v$ in case (b) would destroy an immorality, as a convergent connection would be changed into a serial connection. Even though any graph $G \in E$ is acyclic, reversing an arc might create a directed cycle. Clearly, reversing the direction of such arcs is not allowed, i.e. it is also an essential arc. This is shown in Figure 13(c).

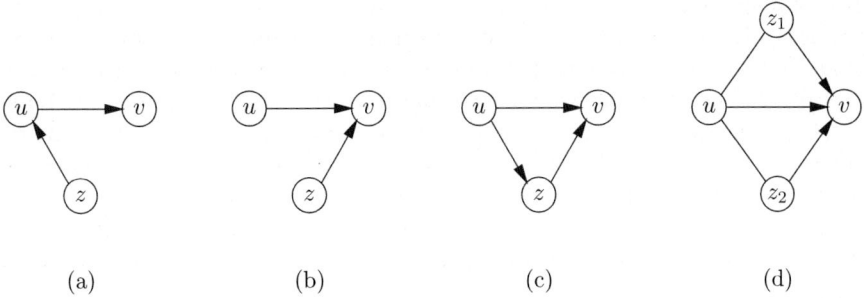

Fig. 13. The four possible induced subgraphs, where arc $u \to v$ is strongly protected.

Finally, in case (d) $u \to v$ is an essential arc and the two other essential arcs $z_1 \to v$ and $z_2 \to v$ are participating in the immorality $z_1 \to v \leftarrow z_2$ (i.e. they are irreversible), the direction of the arc $u \to v$ cannot be reversed to ensure that vertices z_1 and z_2 will not become dependent when conditioning on u.

Definition 12. *(strongly protected) An arc $u \to v$ is called* strongly protected *if it is part of one of the four induced subgraphs shown in Figure 13.*

In the next part of this section we turn our attention to the characterisation of essential graphs. First of all we consider two of the most significant properties of essential graphs.

Lemma 1. *The essential graph G_E^* representing the equivalence class E is a chain graph, i.e. G_E^* comprises no directed cycles.*

Proof: *In this proof we suppose that the essential graph G_E^* has a directed cycle and then show that this assumption results in a contradiction.*

Suppose that G_E^* has a directed cycle $u, v, z_1, \ldots, z_k \equiv u$ with $k \geq 2$. Observe that this directed cycle has at least: (i) one arc, which follows from the definition of directed cycle and (ii) one edge, otherwise some $G \in E$ would have a directed cycle which does not fit their property of acyclicity. Thus the cycle can be written as $u \to v, z_1, \ldots, z_k \equiv u$, with $k \geq 2$, as shown in Figure 14(a). Because there is at least one edge inside this cycle, assume that this is the edge $z_{i-1} - z_i$ in G_E^* with $i \leq k$. Due to the fact that z_{i-2} and z_{i-1} can be connected by an edge (i.e. $z_{i-2} - z_{i-1}$) or by an arc directioned into z_{i-1} (i.e. $z_{i-2} \to z_{i-1}$) there must exist at least one $G \in E$ with substructure $z_{i-2} \to z_{i-1} \leftarrow z_i$ (deduced from $z_{i-2} \to z_{i-1} - z_i$). As this cannot be an immorality there has to be a connection $z_{i-2} \cdots z_i$ (hence, the open possibility for either an edge or an arc directed to z_i is denoted by $\cdots$). But this means that there is a smaller cycle such that $u \to v, z_1, \ldots, z_{i-2}, z_i, \ldots, z_k \equiv u$, with $k \geq 2$. If we continue to reduce this directed cycle using the same idea, we observe that $u \to v \cdots z_1 \cdots z_2 \equiv u$ which is equivalent to $u \to v \cdots z \cdots u$ with $z_1 = z$. Figure 14(b) depicts the reduced variant of the original directed cycle in Figure 14(a).

Next we show that $u \to v \cdots z \cdots u$ cannot be an induced subgraph of the essential graph G_E^*. Our assumption says that G_E^* contains a directed cycle. Then there exist four possible structures in G_E^* deduced from $u \to v \cdots z \cdots u$, shown in Figure 15. For each of these cases there exists at least one $G \in E$ equivalent to G_E^* containing a directed cycle, thus contradicting the acyclicity property: case (a) $\exists G \in E$ containing arc $v \to z$; case (b) $\exists G \in E$ containing arc $z \to u$; case (c) $\exists G \in E$ containing arcs $v \to z$ and $z \to u$; case (d) is already a directed cycle.

Due to the fact that each $G \in E$ should be an ADG, G_E^* cannot contain any of the substructures from Figure 15; therefore, the essential graph is an chain graph, which completes our proof. □

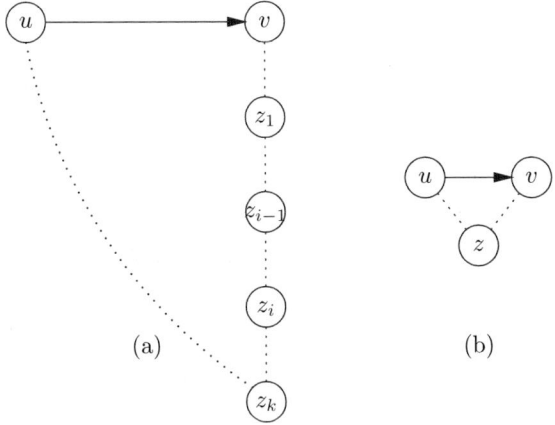

Fig. 14. The directed cycle (a) and the reduced variant (b).

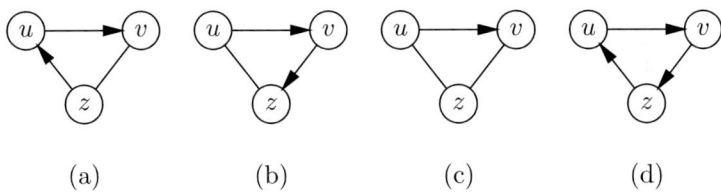

Fig. 15. The possible graphs obtained by replacing the symbol $\cdots$ in $u \to v \cdots z \cdots u$ by an edge or an arc, taking the requirement into account that the resulting graph should contain a directed cycle.

The essential graph has another very important property which is stated in the following lemma.

Lemma 2. *Let G_E^* be the essential graph which represents the equivalence class E and let $V(l)$ be a the chain component of G_E^*, then $V(l)$ is chordal.*

Proof: Suppose that there is an undirected cycle with $k \geq 4$ in chain component $V(l), l \leq L$. Then there should exists at least one ADG $G \in E$, which by its property of being acyclic should consist of an immorality. Since G_E^* has to represent each immorality of E, thus the assumption above results in a contradiction. Therefore the chain components of the graph G^* are chordal. □

Lemma 1 and Lemma 2 concern two fundamental properties of essential graphs. In addition, an essential graph is meant to preserve dependence information from the ADGs it represents. As mentioned above, immoralities are meant to represent conditional dependence information. To preserve these immoralities in an essential graph, the concept of strongly protected arcs have been introduced. In the following lemma, this notion is used to further characterise essential graphs.

Lemma 3. Let G_E^* be the essential graph corresponding to the equivalence class E. Then each arc in G_E^* is a strongly protected arc.

Proof: Suppose that $u \rightarrow v$ is not a strongly protected arc in G_E^*. This means that its direction is reversible. Thus, there exists a graph $G \in E$ with arc $u \leftarrow v$. But then G_E^* should comprise $u - v$, leading to a contradiction. $\square$

As was discussed above in relationship to Figure 13(a) and 13(b) it is not permitted that an immorality is changed into a divergent or serial connection, and vice versa.

Lemma 4. Let G_E^* be the essential graph corresponding to the equivalence class E. Then G_E^* cannot contain the structure $u \rightarrow v - z$ as an induced subgraph.

Proof: Suppose $u \rightarrow v - z$ is an induced subgraph in G_E^*. Then, there exists a $G \in E$ such that $u \rightarrow v \leftarrow z$. But this is an immorality which should be included in G_E^*, leading to a contradiction; this completes the proof. $\square$

Combining the lemmas (1), (2), (3) and (4) leads to a full characterisation of essential graphs in the next theorem.

Theorem 5. Let G_E^* be the essential graph corresponding to the equivalence class E. Then G_E^* satisfies the following four conditions:

- G_E^* is a chain graph;
- each chain component of G_E^* is chordal;
- each arc in G_E^* is strongly protected;
- there exists no induced subgraph $u \rightarrow v - z$ in G_E^*.

Proof: The lemmas 1, 2, 3 and 4 are used subsequently to prove the statements mentioned above, exactly in this order.

5 Conclusions

This paper is meant as a guide to the field of probabilistic graphical models, where in particular we have tried to offer a balanced view of the various issues involved in the study of stochastic dependence and independence, and its role in Bayesian-network structure learning. As there are many different ways in which (in)dependence information can be represented, e.g. as a joint probability distribution, as logical statements, or in the form of different types of graphs, we have focused on the relationships between these different representations.

There were a number of key results given attention to in the paper that are worth recalling. The independence relation may be looked upon as a logical relation, where special properties of the relation can be defined axiomatically. Unfortunately, the Independence relation does not permit finite axiomatisation. Nevertheless, there are a number of axioms that are worth knowing, as they support our understanding of the nature of independence; the most familiar axioms were covered in the paper.

The subtle differences between representing stochastic independence using undirected, acyclic directed and chain graphs was another related topic also studied in this paper. The process of moralisation transforms acyclic directed graphs and chain graphs into undirected graphs, which allows us to determine the semantic relationships between these different graphical ways to represent stochastic independence. Linked to this topic, a number of reading-off methods specific for particular types of graph were discussed, which supported reasoning about the independence information represented in a graph solely in terms of the graph structure.

Ways to identify and represent Markov equivalence in Bayesian networks were the last topics studied. In particular, the concept of the essential graph yields a significant insight into this matter, as an essential graph summarises a class of Markov equivalent networks, and thus renders it possible to determine which arcs in a Bayesian network are really significant. Bayesian networks contain static and dynamic dependences. For the case of static dependences changing directionality of arcs has no effect on the dependences in the entire network, as long as it does not give rise to the creation of immoralities. On the other hand, dynamic dependences are captured by the structure of the immoralities and as these cannot be changed without changing the meaning of a probabilistic graphical model, we have to maintain the direction of arcs in this case. Therefore, the equivalent relation on Bayesian networks is defined in terms of the structure of the skeleton and the associated set of immoralities contained in the graphs. The concept the of essential graph has given rise to much research activity, in particular in areas devoted to the development of algorithms for searching the equivalence space of Bayesian networks (instead of the entire space of Bayesian networks) to determine the Bayesian network that best fits the data from a given domain. What is clear is that probabilistic graphical models offer a rich and complicated landscape of probabilistic representations, which will remain a topic of research in the future.

References

[1] S.A. Andersson, D. Madigan and M.D. Perlman. A Characterization of Markov Equivalence Classes for Acyclic Digraphs. *Annals of Statistic*, 25:505–541, 1997.

[2] A.P. Dawid. Conditional Independence in statistical theory. *Journal of the Royal Statistical Society*, 41:1–31, 1979.

[3] E. Castillo, J.M. Gutiérrez and A.S. Hadi. *Expert Systems and Probabilistic Network Models*. Springer-Verlag New York, 1997.

[4] D.M. Chickering. Learning Equivalence Classes of Bayesian Network Structures. *Journal of Machine Learning Research*, 2:445–498, 2002.

[5] R.G. Cowell, A. Philip Dawid, S.L. Lauritzen and D.J. Spiegelhalter. *Probabilistic Networks and Expert Systems*. Springer-Verlag New York, 1999.

[6] E. Kreyszig. *Introductory Mathematical Statistics*. Wiley, New York, 1970.

[7] S.L. Lauritzen. *Graphical models*. Clarendon Press, Oxford, 1996.

[8] M. Studený and R.R. Bouckaert. On chain graph models for description of conditional independence structures. *Annals of Statistics*, 26(4):1434–1495, 1998.

[9] R.E. Neapolitan. *Learning Bayesian Networks*. Prentice Hall, New Jersey, 2003.

[10] J. Pearl. *Probabilistic Reasoning in Intelligent Systems:Networks of Plausible Inference*. Morgan Kauffman, San Francisco, CA, 1988.

[11] R.W. Robinson. Counting unlabeled acyclic graphs. In *LNM 622*, pages 220–227. Springer, NY, 1977.

[12] M. Studený. Multiinformation and the Problem of Characterization of Independence Relations. *Problems of Control and Information Theory*, 3:3–16, 1989.

[13] M. Studený. Conditional Independence Relations Have No Finite Complete Characterization. In S. Kubíck and J.A. Vísek, editors, *Information Theory, Statistical Decision Functions and Random Processes:Transactions of 11th Prague Conference*, pages 377–396. Kluwer, Dordrecht, 1992.

[14] T. Verma and J. Pearl. Equivalence and synthesis of causal models. In M. Henrion, R. Shachter, L. Kanal, and J. Lemmer, editors, *Uncertainty in Artifical Intelligence Proceedings of the Sixth Conference*, pages 220–227, San Francisco, CA, 1990. Morgan Kaufmann.

A Causal Algebra for Dynamic Flow Networks

James Q. Smith[1] and Liliana J. Figueroa[2]

[1] University of Warwick, Coventry CV4 7AL, UK
[2] Banco Santander Serfin, Mexico City 01219, Mexico

Summary. Over the past decades there has been considerable research to develop various dynamic forms of Bayesian Networks (BNs) and a parallel development in the field of causal BNs. However, linking these two fields is subtle. In this paper we demonstrate that, for classes of models exhibiting mass balance, it is necessary to first redefine the stochastic variables in a process using a decomposition which gives rise to a class of particular Dynamic Linear Models. These models on the transformed space can be interpreted as a dynamic form of a causal BN. A manipulation algebra is then defined to enable the prediction of effects of interventions that would not have been obtainable using the current Causal algebras. The necessary deconstruction of the processes and the algorithms to determine the effects of manipulations on the original process are demonstrated using a simple example of a supply chain in a hypothetical product market.

1 Introduction

There are now many classes of dynamic Bayesian Networks developed for a variety of uses (see for example [14]). These can be used purely to describe the dynamics of a multivariate time series. However, if we intend to use these models for on-line decision making and forecasting, then, ideally, we would like such models to support forecasts when the system is externally manipulated. For example, if a model described the trade of a commodity around the world and there was a conflict such as an embargo or a war which disrupted this trade, it would be very useful if the model was formulated in such a way that its forecasts could be simply adapted to accommodate this external information.

In a non-dynamic setting these issues have been addressed through the construction of Causal Bayesian Networks (CBN) ([15] and [13]). Explicitly, let $X_1, X_2, .., X_n$ be random variables whose density $p(\mathbf{x})$ is defined through a BN Γ on these variables so that

$$p(\mathbf{x}) = \prod_{i=1}^{n} p(x_i | pa_i) ,$$

where pa_i denote the parents of X_i in Γ, $1 \leq i \leq n$. If Γ is a CBN and $\mathbf{x} = (\mathbf{x}_{\hat{j}}, x_j)$, then the density $p(\mathbf{x}_{\hat{j}}||x_j)$ of the remaining variables $\mathbf{x}_{\hat{j}}$ given you manipulate the variable X_j to the value x_j is given by

$$p(\mathbf{x}_{\hat{j}}||x_j) = \prod_{i=1, i \neq j}^{n} p(x_i|pa_i(x_j))$$

for $1 \leq j \leq n$ where $pa_i(x_j)$ denotes the parents of X_i with the value of X_j set to x_j. This formula is the one you would naturally adopt if you believed that the variables after X_j in the partial order induced by Γ, i.e. those that might be affected by the "cause" X_j, were changed in the same way as if you had observed x_j, whilst other variables – not downstream of the cause – would be unaffected by this manipulation. A stochastic analogue of this natural rule, i.e. the study of the effects of manipulations of variables in time series, has been studied for a long time in the context of the Dynamic Linear Model –see e.g. the work of Harrison and Stevens [9], and West and Harrison [18; 19; 20].

In this paper we demonstrate a new technology which draws together these two areas for multivariate systems. We illustrate that it is very often necessary to preprocess a multivariate time series, transforming the processes appropriately so that the model can be used for causal as well as observational inferences. Here we illustrate this technique with a hierarchical supply chain time series model, utilising formal methodologies developed in [4] and [6] for assessing the effects of a manipulation in such time series. A particular feature of series in the more general flow network model, previously studied in relation to unmanipulated transport problems in [17; 21] and [22], is mass balance: the quantity of mass leaving an outlet must equal the mass going in [10]. Mass balance complicates the dependence structure across the component time series and makes the naive use of dynamic BNs inappropriate in this context.

The transformation we propose defines states in terms of the paths of a graph called a Flow Graph. The associated decomposition leads to a conventional dynamic BN [11]. Although its associated recurrences are non-standard, fast closed-form solutions for the prediction of Gaussian processes of this type are now available [6]. With some important caveats, it is also possible to build a causal algebra for these processes which is analogous, but not the same as, the causal algebras developed in [13] and [15] using this transformed space. We demonstrate below how very different predictions of the effects of manipulation are from their naive analogues associated with untransformed processes, and how they often need additional external information before they can be successfully accommodated.

2 Flow Graphs of Supply Chains

The types of supply chain we study below have been used in traffic, education, environmental and medical models as well as for commodities. Such chains regularly experience traumatic disruption, provoked by events external to the definition of the chain, changing the flows between elements of the chain. Here our running example will be of a hypothetical product market. Monthly output of the product is sold on by a producer to a consumer through an intermediary – a trading company. This results in a supply chain with a strict hierarchy: a purchaser at one level of this hierarchy buying from a supplier at the previous level, i.e. the market is *a hierarchical flow network (HFN)*.

Thus suppose such a product supply chain has 3 producers $z(1,i)$, $1 \leq i \leq 3$, (at level 1); 2 traders $z(2,i)$, $1 \leq i \leq 2$, (at level 2), and 2 consumers $z(3,i)$, $1 \leq i \leq 2$, (at level 3). Although not necessary for the development of this general architecture, for simplicity we assume in this paper that the movement of the product from a seller to a purchaser takes one unit of time.

In general, not all purchasers in the HFN will be able to buy from all sellers in the level below. This gives an underlying structure to a market in a HFN which can be conveniently represented by a Flow Graph. Thus in our example, if, under current circumstances, trader $z(2,1)$ buys only from producers $z(1,1)$ and $z(1,2)$, selling to both market consumers $z(3,1)$ and $z(3,2)$, whilst trader $z(2,2)$ buys only from $z(1,2)$ and $z(1,3)$ and sells only to $z(3,2)$, then this would be represented by the Flow Graph G_f of Fig. (1).

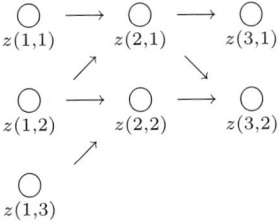

Fig. 1. Flow Graph G_f

Unanticipated external events such as wars, embargoes, unusual weather conditions, can disrupt trading of the product, provoking knock-on effects between the market players and their associated purchases and sales. Such external manipulations of the chain cannot *systematically* feature in a time series model but may well need to be accommodated when they are imminent or immediately after they have happened.

The first question to ask is whether the Flow Graph, itself a directed acyclic graph, is in fact a valid dynamic analogue of a BN and if this is so, whether it is also causal. This would then allow us to deduce how to adapt the distributions of flows in the light of a disrupting event and be of significant value to an observer of the market. Unfortunately, this conjecture falls at the first hurdle. Although it has been suggested in the past that a flow graph like the one above is a BN [21], this is not so in general for standard dynamic analogues of BNs. To appreciate this, suppose the total amount of the product sold by the three producers in our example were known to you at time $t-1$. Then you would know for sure the total amount of product all traders owned at time t: a law of mass conservation. So, on being told how much $z(2,1)$ held at time t would then give you perfect information about the amount held by $z(2,2)$. Therefore, for the Flow Graph to be a BN, you would certainly appear to need an edge between $z(2,1)$ and $z(2,2)$. Although it is important to show care with definitions here, since we are dealing with processes and not random variables, there appears no way in which the flow graph above can be simply interpreted as a BN (see [6] for a formal treatment) even when there is no intervention. Therefore, some further work needs to be done before the machinery of BNs can be used successfully.

3 A Path Decomposition of G_f

Let $\phi'_t[l] = (\phi_t(l,1), \phi_t(l,2), ..., \phi_t(l,n_l))$ be the *node states* vector for level l, $1 \leq l \leq m$, where $\phi_t(l, j_l)$ represents the mass owned by player $z(l, j_l)$ during time t in G_f, $1 \leq j_l \leq n_l$, and $\phi'_t = (\phi'_t[1], \phi'_t[2], \ldots, \phi'_t[m])$. In the example above, $n_1 = 3, n_2 = 2, n_3 = 2$, and $m = 3$.

Suppose we have available time series of a vector $\mathbf{X}'_t = (\mathbf{X}'_t[1], \mathbf{X}'_t[2], \mathbf{X}'_t[3])$ of noisy observations of the vector of actual product amounts ϕ_t at time t owned by each of the players, so that

$$\mathbf{X}_t[l] = \phi_t[l] + \mathbf{v}_t[l]; \ \mathbf{v}_t[l] \sim N(\mathbf{0}, \mathbf{V}_t[l]) \ . \tag{1}$$

Our problem here is the mass conservation in the system: it induces severe dependencies between the component processes of ϕ_t. An important insight is that a unit of the product can only pass from one producer to a consumer along one path. Thus, to obtain functional independence, we simply need to represent the processes above as aggregates of the product amounts taking the different root-to-sink paths $\pi(j)$ of G_f. In the example, these trading routes are:

$$\pi(1) = \{z(1,1), z(2,1), z(3,1)\}$$
$$\pi(2) = \{z(1,1), z(2,1), z(3,2)\}$$
$$\pi(3) = \{z(1,2), z(2,1), z(3,1)\}$$
$$\pi(4) = \{z(1,2), z(2,1), z(3,2)\}$$
$$\pi(5) = \{z(1,2), z(2,2), z(3,2)\}$$
$$\pi(6) = \{z(1,3), z(2,2), z(3,2)\} \ .$$

This motivates the following definition.

Definition 1. *The outlet-path incidence matrix $\mathbf{S}(G_f)$ of Flow Graph G_f, called the superposition matrix, is an $(n \times \rho)$ matrix whose rows correspond to the n nodes of G_f and whose columns correspond to its ρ root-to-sink paths (in the order specified above), and in which $\{\mathbf{S}(G_f)\}_{(i,j)} = 1$ iff the ith listed node lies on the jth path, and zero otherwise.*

The superposition matrix for the flow graph in Fig. (1) is given by

$$\mathbf{S}(G_f) = \begin{bmatrix} 1 & 1 & 0 & 0 & 0 & 0 \\ 0 & 0 & 1 & 1 & 1 & 0 \\ 0 & 0 & 0 & 0 & 0 & 1 \\ 1 & 1 & 1 & 1 & 0 & 0 \\ 0 & 0 & 0 & 0 & 1 & 1 \\ 1 & 0 & 1 & 0 & 0 & 0 \\ 0 & 1 & 0 & 1 & 1 & 1 \end{bmatrix}.$$

Because of the hierarchical structure of the system, $\mathbf{S}(G_f)$ can be decomposed into full-rank matrices $\mathbf{S}[1], ..., \mathbf{S}[m]$, where $\mathbf{S}[l]$ ($1 \leq l \leq m$) is of dimension $(n_l \times \rho)$ and has columns with exactly one non-zero entry. These submatrices represent the outlet-path incidence over level l of G_f ($1 \leq l \leq m$). In the example above, $\mathbf{S}[1], \mathbf{S}[2]$ and $\mathbf{S}[3]$ are the first three, the fourth and fifth, and the last two rows, respectively, of $\mathbf{S}(G_f)$.

If to each of the paths $\boldsymbol{\pi}(j)$ of G_f we assign a time series of *path-flow state* components $\varphi_t(j)$ of the mass emerging from a root node over the time period t and sold along the path $\boldsymbol{\pi}(j)$, then we can write

$$\boldsymbol{\phi}_t[l] = \mathbf{S}[l]\boldsymbol{\varphi}_{t-l+1}; \quad t \geq 1, \quad l = 1, 2, ...m, \qquad (2)$$

where $\boldsymbol{\varphi}'_t = (\varphi_t(1), \varphi_t(2), \ldots, \varphi_t(\rho))$.

By re-expressing $\boldsymbol{\phi}_t$ in terms of the functionally independent vector $\boldsymbol{\varphi}_t$ we can write this process as a Gaussian multivariate multilevel Dynamic Linear Model [7] with observation equation as in (1). Path flows are expressed as noisy linear functions of a second vector of descriptive states $\{\boldsymbol{\theta}_t\}_{t\geq 1}$, called *core states*, representing levels, seasonal factors and so on, which are assumed to evolve through time as a Markov chain. Thus, for each time $t \geq 1$,

$$\begin{aligned} \boldsymbol{\varphi}_t &= \mathbf{F}'_t\boldsymbol{\theta}_t + \mathbf{u}_t; \quad \mathbf{u}_t \sim N(\mathbf{0}, \mathbf{U}_t) \\ \boldsymbol{\theta}_t &= \mathbf{G}_t\boldsymbol{\theta}_{t-1} + \mathbf{w}_t; \quad \mathbf{w}_t \sim N(\mathbf{0}, \mathbf{W}_t) , \end{aligned} \qquad (3)$$

with prior

$$(\boldsymbol{\theta}_0|D_0) \sim N(\mathbf{a}_0, \mathbf{R}_0) \qquad (4)$$

and error sequences $\{\mathbf{u}_t\}_{t\geq 1}, \{\mathbf{v}_t\}_{t\geq 1}, \{\mathbf{w}_t\}_{t\geq 1}$ mutually and internally independent. Note that in the above, $\mathbf{F}_t$ is a known $(r \times \rho)$ design matrix, specifying the relationships between path flows and the underlying core states; $\mathbf{G}_t$ is a known $(r \times r)$ block matrix, governing the dynamic evolution of the core states.

The model defined by (1)-(4) is analogous to the Hierarchical Dynamic Model with aggregation [7], the difference being that in our model some of the information is delayed. For instance, in our illustrative market, we only learn about the amount of material sold by $z(1,1)$ to $z(3,1)$ (via traders) once $z(3,1)$ receives it, i.e. two time steps after the material left its first source. This fact makes this new model different to the standard Dynamic Linear Model [9] in that the state $\boldsymbol{\theta}_t$ no longer summarises the information relevant for forecasting at time t. Nonetheless, new forecasting and updating recurrences for an external observer can be obtained based on a reparameterisation of the core states [6]. Note also that in practice the external observer will usually only have a proper subset of the information available to all producers, traders and consumers [4].

4 Dynamic Bayesian Networks of Flow Systems

The new state space representation can now be used to define a slightly adapted 2-time-slice BN (2-TBN) ([2; 11]) – one of the more useful classes of dynamic BN for multiple time series.

Let the *real state vector* $\boldsymbol{\alpha}_t$ of an HFN be the vector of states which retains the usual Markov property

$$\{\mathbf{X}_s\}_{s \geq t} \perp\!\!\!\perp \{\mathbf{X}_s\}_{1 \leq s \leq t-1} \mid \boldsymbol{\alpha}_t \tag{5}$$

of the Dynamic Linear Model, where $A \perp\!\!\!\perp B \mid C$ reads "A is independent of B given C". Note from the defining equations above that, because of the delays in learning about aggregated flows, a real state vector $\boldsymbol{\alpha}_t$ for an m-level HFN is

$$\boldsymbol{\alpha}_t = (\boldsymbol{\theta}_t, \ldots, \boldsymbol{\theta}_{t-m+1}, \boldsymbol{\varphi}_t, \ldots, \boldsymbol{\varphi}_{t-m+1}). \tag{6}$$

So although G_f is not a BN, we are able to define a (slightly adapted) 2-TBN for our HFN using the real state vector components as arguments. Like the Flow Graph G_f for an HFN at untraumatic times, this 2-TBN is time homogeneous and gives sufficient information to reconstruct the original infinite BN, since it gives the BN of state relationships from time $t-1$ to time t and the relationships between states and observations at time t.

The relevant 2-TBN for the three level HFN in Fig. (1) is given in Fig. (2) below.

The tree structure of this BN means that dynamic propagation algorithms can be designed to be extremely quick [12]. Also, in contrast to the 2-TBNs studied in [11], because of the overlap of components in the real state vector, the graphical structure of these HFNs does not degenerate over time. The BN architecture also allows us to adapt these algorithms painlessly when observation on some of the components of the observation vector are systematically missing (for details, see [5]).

$$\boldsymbol{\theta}_{t-3} \to \boldsymbol{\theta}_{t-2} \to \boldsymbol{\varphi}_{t-2} \to \mathbf{X}_t[3]$$
$$\downarrow$$
$$\boldsymbol{\theta}_{t-1} \to \boldsymbol{\varphi}_{t-1} \to \mathbf{X}_t[2]$$
$$\downarrow$$
$$\boldsymbol{\theta}_t \dashrightarrow \boldsymbol{\varphi}_t \to \mathbf{X}_t[1]$$

Fig. 2. 2-TBN for 3-level HFN

5 Causality in Supply Chains

In a forecasting system for a Flow System, we will be interested in a variety of interventions along the different stages of the model hierarchy. For example, we will be concerned with interventions at time t on the subaggregate $\phi_T(l, j_l)$ that might arise due to an event that modifies the supply of material. A different situation occurs when players in a market decide to stop trading through a specific supply chain route (an embargo is a typical example of this type of intervention). In this case, a path flow $\varphi_T(j)$ is to be manipulated. Another type of intervention on unobservable processes to be considered will be on the core states. This is the usual intervention described in [20] and is that concerned with changes in level, seasonal patterns, growth, etc.

In this manipulation context, we maintain the necessity of transforming the series through the path decomposition of Sect. 3. Take our running example: if $z(1,2)$ were prevented from trading, this situation could be easily incorporated into a model in which only relationships between states ϕ were used. Notice that this could be represented by erasing the node $z(1,2)$ and its two connecting edges from G_f in Fig. (1) to give new flow graph G_f^+, represented in Fig. (3) below. However, there is no obvious and simple way to represent an embargo by $z(3,1)$ on trade from $z(1,2)$ solely through relationships between node states. The path decomposition becomes paramount in this setting.

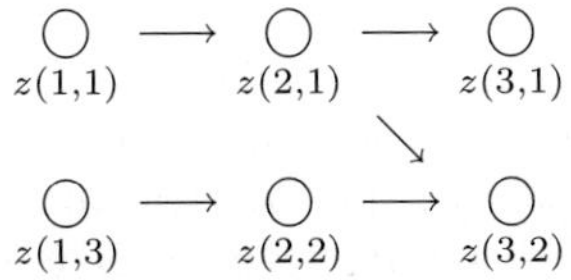

$$\begin{array}{ccccc}
z(1,1) & \to & z(2,1) & \to & z(3,1) \\
 & & & \searrow & \\
z(1,3) & \to & z(2,2) & & z(3,2)
\end{array}$$

Fig. 3. Flow Graph G_f^+

The 2-TBN of Fig. (2) is certainly "causal" on its states in the coarse sense that, for obvious physical reasons, any manipulation at a time t on trading cannot change the amount of oil currently in the system or indeed the amounts that have flowed along its paths to that time. So the distribution of such quantities, typically concerning states not in $\{\boldsymbol{\alpha}_s\}_{s \geq t}$, should not be changed by and at the time of the intervention. On the other hand, to reflect how manipulations affect this environment, the way in which the distribution of current and future states might change depends on the type of manipulation performed and may not be achieved by simple substitution of a particular value into some of the components of the states. Furthermore, the appropriate manipulation will typically depend on additional information not required when for forecasting the observed time series. However, it is possible to systematically use the modularity of the structure of the process, represented both in its flow graph and its 2-TBN with its distribution of the current states, as a framework for revising forecasts in the light of the unexpected trauma. We illustrate this process below and hence elaborate and extend a Spirtes/Pearl type causal algebra, pertinent to this particular domain. Explicitly, we need to specify the distribution of $(\boldsymbol{\theta}_t^+, \boldsymbol{\varphi}_t^+)$, i.e. the distribution of $(\boldsymbol{\theta}_t, \boldsymbol{\varphi}_t)$ after the intervention.

5.1 The Fast Response

A Motivating Example

Assume that under our running example the market structure is such that consumers will always be able to satisfy their demand. The simplest way of incorporating this into our model is to let the core states $\boldsymbol{\theta}_t = (\theta_t(1), \theta_t(2))'$ satisfy $\boldsymbol{\theta}_t = \phi_{t+2}[3]$ so that $\theta_t(i)$ represents the purchases two steps ahead for the level 3 players $z(3, i)$, $i = 1, 2$.

Now assume that at time $t - 1$, $z(3, 1)$ establishes an embargo on sales by $z(1, 2)$ for time t. The immediate effect of the embargo is to set to zero the flow along path $\pi(3)$, so we have set

$$p^+(\varphi_t(3) \mid D_{t-1}) = N(0, H), \qquad (7)$$

where H indicates our uncertainty about the perfect enactment of the cancelled sales.

From (2), we know that, in particular, after the intervention

$$\phi_{t+2}^+(3, 1) = \varphi_t^+(1) + \varphi_t^+(3). \qquad (8)$$

If the sole information available to us is that the purchases from $z(1, 2)$ are expected to be cancelled, then this would result in $\varphi_t^+(1) = \varphi_t(1)$ and $\phi_{t+2}^+(3, 1)$ having a distribution with the same mean as the distribution of $\varphi_t^+(1)$ – a lower purchase level, but with increased uncertainty. However, if we know that $z(3, 1)$ will seek to maintain its pre-intervention purchase level and expert knowledge indicates that $z(3, 1)$ will actually be able to obtain extra supplies from the sellers, then we can conclude that $\phi_{t+2}^+(3, 1) = \phi_{t+2}(3, 1)$. Therefore, if $z(3, 1)$ seeks no new trading relationships (i.e. purchasing from $z(1, 3)$ through $z(2, 2)$), it will require $z(1, 1)$ to increase her sales. In this example, we notice that, more generally, $\boldsymbol{\theta}_t^+ = \boldsymbol{\theta}_t$. We now update the remaining flows conditional on these events and have the result of the fast dynamic, in which at the time of the intervention, only node states for $z(1,1)$ and $z(1,2)$, and flows through paths $\pi(1)$ and $\pi(3)$ have different pre- and post-manipulation marginal distributions. Post-intervention forecasts are computed to obtain the distribution of $\mathbf{X}^+$. A fast elaboration of the type described above will modify the 2-TBN to:

$$\begin{array}{ccccc}
\boldsymbol{\theta}_{t-3} & \to & \boldsymbol{\theta}_{t-2} & \to & \varphi_{t-2} & \to & \mathbf{X}_t[3] \\
& & \downarrow & & & & \\
& & \boldsymbol{\theta}_{t-1} & \to & \varphi_{t-1} & \to & \mathbf{X}_t[2] \\
& & \downarrow & & & & \\
& & \boldsymbol{\theta}_t^+ & \to & \varphi_t^+ & \to & \mathbf{X}_t^+[1]
\end{array}$$

Fig. 4. 2-TBN for manipulated HFN at the time of intervention

Note that, because we have assumed the joint distribution of all the variables is Gaussian, this transformation is trivial to calculate.

Other considerations are the possible manipulations to data that can take place as a consequence of manipulations to the flows in the system. In a general embargo declared on a producer, some players might choose not to manipulate the flows themselves, i.e. enacting the embargo, but rather simply decide to provide data that will indicate they have done so. This type of manipulations and their effects, i.e. the resulting distribution of $\mathbf{X}^+$, have to be included in the model in order to obtain more precise forecasts.

We will show next how to compute in general the effects of manipulations on the processes φ, ϕ and $\mathbf{X}$, where the form of the intervention and its possible additional knock-in effects are guided by expert knowledge.

Manipulation of Path Flow and Observation Processes

First notice that if at time $t-1$ we solely manipulate the supply chain vector φ_T, i.e. those root to sink flows leaving the first source at time T, the forecast for $\mathbf{X}^T = (\mathbf{X}'_T[1], \mathbf{X}'_{T+1}[2], \ldots, \mathbf{X}'_{T+m-1}[m])'$ will consequently differ from the unmanipulated prediction. However, this intervention will not immediately modify our beliefs about other (future) supply chains – as is the case when intervention on $\boldsymbol{\theta}$ is performed (see Sect. 5.2). Outlying behaviour of a flow is accommodated into the model in this way in order not to modify (at the time of intervention) our beliefs about the state process $\{\boldsymbol{\theta}_k\}_{k \geq T}$. This is because we do not believe that other (future) path flows will also change.

Thus, assume first that based on new information obtained at time $t-1$ relevant to a change in our beliefs about the flow through path $\pi(j)$ for time T ($T > t-1$), we can represent our new beliefs by

$$p^+(\varphi_T \mid D_{t-1}) = N(\mathbf{a}_T^+, \mathbf{R}_T^+) \,. \tag{9}$$

The unmanipulated distribution obtained with the routine model was

$$p(\varphi_T \mid D_{t-1}) = N(\mathbf{a}_T, \mathbf{R}_T).$$

Thus we construct the post-intervention distribution of the error term $\mathbf{u}_T$ in (3) as

$$p^+(\mathbf{u}_T \mid D_{t-1}) = N(\mathbf{h}_T, \mathbf{U}_T + \mathbf{H}_T) \tag{10}$$

so that it is consistent with (9). We further assume that the sigma-algebra generated by $\mathbf{u}_T$ is sufficient for the pair $\{p, p^+\}$.

Thus, for instance,

$$p(\varphi_T \mid \mathbf{u}_T, D_{t-1}) = p^+(\varphi_T \mid \mathbf{u}_T, D_{t-1}) \,,$$

which in turn implies that

$$p^+(\varphi_T \mid D_{t-1}) = \int p(\varphi_T \mid \mathbf{u}_T, D_{t-1}) p^+(\mathbf{u}_T \mid D_{t-1}) d\mathbf{u}_T \,.$$

We obtain the post-manipulation predictive distributions in the same way and notice that, as desired, the distribution of the core states for time T are not affected at time $t-1$ by this manipulation since $\boldsymbol{\theta}_T \perp\!\!\!\perp_p \mathbf{u}_T \mid D_{t-1}$. In general, the conditional independence conditions represented by the Dynamic Bayesian Network defined by the 2-TBN will determine the subset of processes that will be affected by the intervention at time $t-1$.

Notice also that we will want to discriminate between the decision of an agent to change the true level of sales (say) and to change only the observed data – the two interventions being different in nature. As mentioned above, including in a model the belief that data is being manipulated is relevant since the arrival of more reliable oncoming data related to the same path flow or state process will help to derive more accurate forecasts and revised beliefs about the latent processes. If it is not known how data is being manipulated, to just indicate greater uncertainty than that assumed in routine is still useful.

Following the same argument used for the manipulation of the path flow vector φ, we can obtain the post-intervention distributions when manipulating the observation $\mathbf{X}_T[l]$ at time $t-1$. This time the intervention is triggered by the new distribution of the error $\mathbf{v}_T[l]$ in (1) constructed in a similar way to (10). Thus, expert knowledge postulates that our post-intervention beliefs can be expressed by

$$p^+(\mathbf{v}_T \mid D_{t-1}) = N(\mathbf{h}_T, \mathbf{V}_T + \mathbf{H}_T)$$

and assume that the sigma-algebra generated by $\mathbf{v}_T$ is sufficient for the pair $\{p, p^+\}$.

Manipulation of Node Flow Processes

When considering a broader set of interventions for a Flow System, it is important to also concentrate directly on the node process $\{\phi\}_{t\geq 1}$ since decisions can be directly concerned with it and, as a consequence, affect the path flow process. For example, a player in a market might choose to decrease his supply of material and will, therefore, modify his sales through certain supply chains. If the expert had some knowledge on how the specific path flows will be modified, the intervention could be accommodated by the methods explained above. However, if this is not available, he needs a methodology to manipulate directly the node processes while assuming the same path-node structure prevailing before the player's manipulation.

Our purpose is, therefore, to define an intervention calculus that can express the direct manipulation of the state random vectors of the true node process. When explicitly considering the node process $\{\phi_t\}_{t\geq 1}$, however, we note that the joint probability distribution function $p(\boldsymbol{\theta}_t, \boldsymbol{\varphi}_t, \boldsymbol{\phi}^t, \mathbf{X}^t)$ is singular and the Markov condition alone will not entail the conditional independence properties of the processes in the model. Thus the causal calculus of Pearl and Spirtes et al. are inapplicable for the computation of intervention effects.

We also note that in flow networks at certain times the path flows will be determined first and will, therefore, perfectly determine the node flows. On the other hand, at other times the value of a subaggregate $\phi(l, j_l)$ will be set first and will determine the value (or a set of possible values) for the different path flows that compose it. We cannot, as a consequence, specify a single immutable causal direction between ϕ and φ because this direction is determined by the nature of the manipulation we employ. Obviously such reversible causal systems, also addressed in [16], cannot be represented in the single cross sectional DAG of [15]. The semantics we have described above, which elaborate the description with the underlying flow graph, are however sufficient for this purpose.

Thus, if $\boldsymbol{\phi}_T[l]$ determines $\boldsymbol{\varphi}_{T-l+1}$, then

$$\boldsymbol{\varphi}_{T-l+1} \perp\!\!\!\perp \boldsymbol{\theta}_{T-l+1} \mid \boldsymbol{\phi}_T[l], \qquad (11)$$

while if $\boldsymbol{\varphi}_{T-l+1}$ determines $\boldsymbol{\phi}_{T-l+i}[i]$ $(1 \leq i \leq m, i \neq l)$ then

$$\boldsymbol{\phi}_{T-l+i}[i] \perp\!\!\!\perp \boldsymbol{\theta}_{T-l+1} \mid \boldsymbol{\varphi}_{T-l+1}. \qquad (12)$$

When computing the changes motivated by a change in our beliefs about a specific subaggregate $\phi_T(l, j_l)$ we first note that our beliefs about flows in the subset

$$\mathcal{E}^{(l,j_l)} = \{\varphi_{T-l+1}(i); 1 \leq i \leq \rho, (\mathbf{S}[l])_{j_l, i} = 1\} \qquad (13)$$

(determined by the topology of G_f) will change, and the path flows may become independent of $\boldsymbol{\theta}_{T-l+1}$ because our beliefs about the subaggregate could completely determine our new beliefs about the relevant path flows. We can, therefore, feed this new behaviour into paths in $\mathcal{E}^{(l,j_l)}$, transform the intervention to a manipulation of the error $\mathbf{u}_{T-l+1}$ and make use of the methodology demostrated above. Afterwards we feed the resulting information on paths bottom-up to other subaggregates in the flow network. Similar techniques, albeit in a non-dynamic setting, are shown in [20] for the combination of forecasts with the purpose of having a consistent system across the hierarchy. Our approach, however, does not combine independent forecasts, but rather assumes that our beliefs about the system flows have changed and some forecasts will not be valid anymore. Moreover, our methodology differs from theirs in that we transform the intervention on the subaggregate into a corresponding intervention of the error process $\{\mathbf{u}_t\}_{t \geq 1}$ with the purpose of maintaining the use of routine algorithms for deriving prior, predictive and posterior distributions.

This is achieved by assuming simply that

$$p^+(\varphi_{T-l+1}^{(l,j_l)} \mid \phi_T(l,j_l), D_{t-1}) = p(\varphi_{T-l+1}^{(l,j_l)} \mid \phi_T(l,j_l), D_{t-1}), \qquad (14)$$

where $\varphi_{T-l+1}^{(l,j_l)}$ denotes the vector of path flows in $\mathcal{E}^{(l,j_l)}$.

Because both pre- and post-intervention distributions share the same null sets (namely, when $\phi_T[l] \neq \mathbf{S}[l]\varphi_{T-l+1}$, $1 \leq l \leq m$), we can compute the post-intervention distribution $p^+(\varphi_{T-l+1}^{(l,j_l)} \mid D_{t-1})$ as

$$\int p(\varphi_{T-l+1}^{(l,j_l)} \mid \phi_T(l,j_l), D_{t-1}) p^+(\phi_T(l,j_l) \mid D_{t-1}) d\phi_T(l,j_l), \qquad (15)$$

with the marginal distributions for the path flows not in $\mathcal{E}^{(l,j_l)}$ not affected. This is because the covariances between paths that belong to $\mathcal{E}^{(l,j_l)}$ and those that do not can be set to zero, according to the modeller's expectations under his new beliefs.

Affected subaggregates at level i are signalled up by the elements of $\mathbf{S}[l]\mathbf{S}'[i]$. That is, if element $(\mathbf{S}[l]\mathbf{S}'[i])_{j_l,j_i} = 1$, then our beliefs about subaggregate $\phi_{T-l+i}(i,j_i)$ will also change immediately. This is because some path flows affected by the intervention on $\phi_T(l,j_l)$ also pass through outlet $z(i,j_i)$. Notice that $\mathbf{S}[l]\mathbf{S}'[l]$ is a diagonal matrix indicating that intervention on $\phi_T(l,j_l)$ alone will not affect other contemporary subaggregates in the same level.

Joint Interventions

One or more players in a flow network might decide to modify at the same time the behaviour of the different processes subject to their control, possibly as a consequence of another player's manipulation. This situation with the fast response of other players was exemplified above. To accommodate these manipulations into a model we make use of joint interventions.

Thus, for instance, if at time $t-1$, $\phi_T(l, j_l)$ and $\varphi_{T-l+1}(k)$ ($1 \le k \le \rho$, $t-1 < T-l+1$) are intervened to have post-intervention distribution $p^+(\phi_T(l, j_l), \varphi_{T-l+1}(k) \mid D_{t-1})$, then

1. If $(\mathbf{S}[l])_{j_l, i} = 1$ ($1 \le i \le \rho, i \ne k$) and $(\mathbf{S}[l])_{j_l, k} = 0$, then we simply compute the effects following the intervention methodology shown before. Note that for all j such that $(\mathbf{S}[l])_{j_l, j} = 0$ ($j \ne k$), then $p^+(\varphi_{T-l+1}(j) \mid D_{t-1}) = p(\varphi_{T-l+1}(j) \mid D_{t-1})$.
2. If $(\mathbf{S}[l])_{j_l, i} = 1$ ($1 \le i \le \rho, i \ne k$) and $(\mathbf{S}[l])_{j_l, k} = 1$, and we call φ_{T-l+1}^{-k} the vector of path flows in $\mathcal{E}^{(l, j_l)}$ not including $\varphi_{T-l+1}(k)$, then $p^+(\varphi_{T-l+1}^{-k} \mid D_{t-1})$ is obtained from

$$\int p(\varphi_{T-l+1}^{-k} \mid \phi_T(l, j_l), \varphi_{T-l+1}(k), D_{t-1}) \\ p^+(\phi_T(l, j_l), \varphi_{T-l+1}(k) \mid D_{t-1}) d\phi_T(l, j_l) d\varphi_{T-l+1}(k) \,.$$

With this new post-intervention distribution, we simply follow the procedures to transform this intervention into an error intervention and compute, accordingly, the effects of this intervention at $t-1$ and other times on the processes of the Flow Network.

Knowledge of the system is important here if we want to choose an appropriate transformation of the process after intervention. In the example above, we needed to know that $z(3, 1)$ would seek extra supplies from its usual supplier. However, he could have decided to purchase from $z(2, 2)$ under such circumstances, thus creating a new path in G_f. This situation is easily accommodated by extending the superposition matrix and, as a consequence, the path flow vector φ, which in turn may require new core states to represent its behaviour – see [20] for further details on the extension of the dimension of $\boldsymbol{\theta}$.

5.2 The slow dynamic

The next stage, the adjustment of the system equation, is analogous to the intervention methodology established in [20]. At this stage we need to model further phenomena such as the possible depressing effect on demand of an increase in price. This would result in a necessary change in the system equations: a change in conditional distributions of certain states given the past. This can be achieved by changes to the system matrix $\mathbf{G}_{t+1}$ to $\mathbf{G}_{t+1}^+$ and/or to the distribution of $\mathbf{w}_{t+1}$. In this way,

$$(\boldsymbol{\theta}_{t+1}^+ \mid \boldsymbol{\theta}_t^+) \sim N(\mathbf{G}_{t+1}^+ \boldsymbol{\theta}_t^+ + \mathbf{h}_t, \mathbf{W}_{t+1}^+) \,,$$

instead of the pre-intervention

$$(\boldsymbol{\theta}_{t+1}^+ \mid \boldsymbol{\theta}_t^+) \sim N(\mathbf{G}_{t+1} \boldsymbol{\theta}_t^+, \mathbf{W}_{t+1})$$

The effects of such interventions are computed as in routine (i.e. when no intervention has happened) through the usual forecasting and updating recurrences of the DLM.

The total effects on the original series of this embargo are now possible to calculate. The effects are clearly non-atomic in Pearl's sense of the term and furthermore act on states rather than observations. The fast effects on the paths using this algebra are often causal in the Spirtes/Pearl sense, although the appropriate manipulation will depend on extraneous information: is the market currently demand or supply driven, do the players have a storage capacity, and if so how big is this? These and other considerations determine the appropriate transformation to use. And certainly in this type of market there is nearly always a slow dynamic, as players adapt their purchasing strategies in the light of the new environment. Moreover, as discussed above, new processes might be born due to the manipulation of existing processes. In the market described above, although a systematic intervention algorithm has been built, the distribution of no observation is set to a pre-assigned value, nor indeed any linear function of observables.

Furthermore, the effects of the manipulation of a single player can affect the series of transactions of *every* other player, whether or not they are parents of the original enactor in G_f, in a way that is highly dependent on prior information about the trauma scenario, and in a way that is not purely deducible from the idle network.

A second important point to note is that the nature of the multivariate process describing the supply system could be completely transformed after the manipulation. This is not only true of the distributions but often the conditional independences across the processes. For example, in the case of a general embargo on $z(1,2)$ and under the situation when $z(2,1)$ chooses not to sell to $z(3,2)$, although before the intervention the multivariate series were all dependent on each other, after intervention the market has segmented. The transactions of $z(1,1)$, $z(2,1)$ and $z(3,1)$ are all now independent of the transactions of $z(1,3)$, $z(2,2)$ and $z(3,2)$. Thus, if we write $\mathbf{X}_s^{(a)} = (X_s(1,1), X_{s+1}(2,1), X_{s+2}(3,1))$ and $\mathbf{X}_s^{(b)} = (X_s(1,3), X_{s+1}(2,2), X_{s+2}(3,2))$, then

$$\{\mathbf{X}_s^{(a)}\}_{s \geq t+3} \perp\!\!\!\perp \{\mathbf{X}_s^{(b)}\}_{s \geq t+3} \mid \boldsymbol{\theta}_t \ .$$

Notice also that, unlike the causal algebra developed by Pearl, we do not destroy the dependence structure of the processes in the Flow System, but rather manipulate the distributions of the error processes. We are, therefore, able to compute the post-intervention retrospective distributions of the system, i.e. distributions of the form $p^+(\mathbf{z}_s \mid D_T)$, $(s \leq T)$ of any flow process $\{\mathbf{z}_s\}_{s \leq T}$ after we have intervened the system at time $t-1$ and observed data up to time T. These distributions will differ from the unmanipulated $p(\mathbf{z}_s \mid D_T)$ for any manipulation (either on the core, path flow or node states, as will be the case for the observations), when the relevant intervention was performed for the state or observation for time T. Thus, we agree with the fact that an intervention will not change the past, either of the process or what we observed, but it can certainly modify the way we view the past.

6 Discussion

In this paper we have demonstrated that dynamic causal structures are complex and need extra machinery to help guide the appropriate adaptation of the process before the trauma took place. In the simple supply chain model above, for example, we saw that the appropriate transformation of the multivariate time series of the observed market, was certainly quite different from one which simply plugged in certain values to the observed "causal" observations. This is because the effects of such manipulations were not considered by the routine model – adequate for reflecting non-embargo situations.

To predict the effects of an intervention, we have used more information than that provided by previous observations of idle series. We believe that this suggests that routine methods of model selection (like [1] and [3]), whilst providing a helpful role in exploring complex multivariate time series, have rather limited scope for the analysis of market which regularly experience trauma. These not only focus on the relationships between observations rather than states but also because these, perforce, through their relation to Granger-causality models [8], blur the distinction between intrinsic perturbations and external manipulation.

However, the use of qualitative information, like that in a flow graph, can guide us towards tailored classes of filtered series which have high predictive power under control. Valuable external information associated with the predicted effects of an intervention can be incorporated into the moments of the states of the system in a straightforward way and realistic predictions obtained. Such methods are obviously context-specific, but provide useful and intellectually fascinating new extensions to studies of causality within models which have an underlying dynamic BN structure. A development of such a systematic methodology customized to a general class of supply chain with an application to the oil market is presented in [6].

References

[1] Dalhaus, R., Eichler, M. (2003). Causality and graphical models for time series. In *Highly structured stochastic systems.* P. Green, N. Hjort, and S. Richardson (Eds.) University Press, Oxford.
[2] Dean,T., Kanazawa, K. (1988). Probabilistic Temporal Reasoning. Proc. AAAI-88, *AAAI*, 524–528.
[3] Eichler, M. (2002). Granger-causality and path diagrams for multivariate time series. To appear in *Journal of Econometrics*.

[4] Figueroa, L. J., Smith, J. Q. (2002). Bayesian manipulation and forecasting of hierarchical flow networks. *Research Report* 404, Department of Statistics, University of Warwick.
[5] Figueroa, L. J., Smith, J. Q. (2003). Bayesian forecasting of hierarchical flows in supply chains networks. *Research Report* 407, Department of Statistics, University of Warwick.
[6] Figueroa, L. J. (2004). *Bayesian Forecasting and Intervention in Dynamic Flow Systems.* Ph.D. Thesis, University of Warwick.
[7] Gamerman, D., Migon, H. (1993). Dynamic hierarchical models. *J. Roy. Statist. Soc.* (Ser. B) **55**, 629–642.
[8] Granger, C. W. J. (1988). Some recent developments in a concept of causality. *J. of Econometrics*, **39**, 199–211.
[9] Harrison, P. J., Stevens, C. (1976). Bayesian forecasting (with discussion). *J. Roy. Statist. Soc.* (Ser. B), **38**, 205–247.
[10] Jacquez, J. A. (1972). *Compartmental Analysis in Biology and Medicine.* Elsevier.
[11] Koeller, D., Lerner, U. (1999). Sampling in Factored Dynamic Systems. In *Sequential Monte Carlo Methods in Practice*, Doucet, A., de Freitas, N. and Gordon, N. (Eds.). Springer.
[12] Pearl, J. (1988). *Probabilistic Reasoning in Intelligent Systems.* San Mateo, CA: Morgan Kaufmann.
[13] Pearl, J. (2000). *Causality - Models, Reasoning and Inference.* Cambridge: Cambridge University Press.
[14] Queen, C. M., Smith, J. Q. (1993). Multiregression Dynamic Models. *Journal of the Royal Statistical Society (Ser. B)*, **55**, 849-870.
[15] Spirtes, P., Glymour, C., Scheines, R. (2000). *Causation, prediction and search.* New York: Springer-Verlag.
[16] Strotz, R. H., Wold, H. O. A. (1960). Causal models in the social sciences. *Econometrica*, **28**, 417–427.
[17] Tebaldi, C., West, M. (1998). Bayesian inference on network traffic using link count data. *Journal of the American Statistical Association*, **93**, 557–576.
[18] West, M., Harrison, P. J. (1986). Monitoring and Adaption in Bayesian forecasting models. *J. Amer. Statist. Assoc.*, **81**, 741–750.
[19] West, M. Harrison, P. J. (1989). Subjective intervention in formal models. *Journal of Forecasting,* **8**, 33–53.
[20] West, M. Harrison, P. J. (1997). *Bayesian Analysis of Time Series.* New York: Springer-Verlag.
[21] Whitlock, M. E., Queen, C. M. (2000). Modelling a Traffic Network with Missing Data. *Journal of Forecasting,* **19**, Issue 7, 561–574.
[22] Whittaker J., Garside, S., Lindveld, K. (1997). Tracking and predicting a network traffic process. *International Journal of Forecasting*, **13**, 51–61.

Graphical and Algebraic Representatives of Conditional Independence Models

Jiří Vomlel and Milan Studený

Institute of Information Theory and Automation,
Academy of Sciences of the Czech Republic,
Pod vodárenskou věží 4,
182 08 Prague 8, Czech Republic

Summary. The topic of this chapter is conditional independence models. We review mathematical objects that are used to generate conditional independence models in the area of probabilistic reasoning. More specifically, we mention undirected graphs, acyclic directed graphs, chain graphs, and an alternative algebraic approach that uses certain integer-valued vectors, named imsets. We compare the expressive power of these objects and discuss the problem of their uniqueness.

In learning Bayesian networks one meets the problem of non-unique graphical description of the respective statistical model. One way to avoid this problem is to use special chain graphs, named essential graphs. An alternative algebraic approach uses certain imsets, named standard imsets, instead. We present algorithms that make it possible to transform graphical representatives into algebraic ones and conversely. The algorithms were implemented in the R language.

1 Conditional Independence Models

The main motivation for this chapter is conditional independence models (CI models). The classical concept of independence of two variables has an interpretation of their mutual irrelevance, which means that knowing more about the state of the first variable does not have any impact on our knowledge of the state of the second variable. Similarly, the concept of conditional independence of two variables given a third variable means, that if we know the state of the third variable then knowing more about the state of the first variable does not have any impact on our knowledge of the state of the second variable. We will illustrate this concept using an example taken from Jensen [4].

Example 1 (CI model). Assume three variables:

- person's length of hair, denoted by h,
- person's stature, denoted by s, and
- person's gender, denoted by g.

We can describe relations between these three variables as follows:

- Seeing the *length of hair* of a person will tell us more about his/her *gender* and conversely. It means, the value of g is dependent on the value of h.
- Knowing more about the *gender* will focus our belief on his/her *stature*. It means that s is dependent on g and (through g) also on h.
- Nevertheless, if we know the *gender* of a person then *length of hair* of that person gives us no extra clue on his/her *stature*, that is h is independent of s given g.

Thus, we have indicated one (conditional) independence relation. Note that all the observations are implicitly understood as symmetric claims. That is, for example, the dependence between g and h is the same as the dependence between h and g.

The conditional independence relations of the above-mentioned type can formally be specified as conditional independence statements over a non-empty finite set of variables N.

Definition 1 (Disjoint triplet over N). *Let $A, B, C \subseteq N$ be pairwise disjoint subsets of a set of variables N. This disjoint triplet over N will be denoted by $\langle A, B \mid C \rangle$. The symbol $\mathcal{T}(N)$ will be used to denote the class of all possible disjoint triplets over N.*

Definition 2 (CI statement). *Let $\langle A, B \mid C \rangle$ be a disjoint triplet over N. Then the statement "A is conditionally independent of B given C" is a CI statement (over N), written as $A \perp\!\!\!\perp B \mid C$. The negation of this statement, the respective conditional dependence statement, will be denoted by $A \not\!\perp\!\!\!\perp B \mid C$.*

If any of the sets A, B, or C will be a one-element set we will omit the curly brackets and write a to denote $\{a\}$.

Example 2 (CI statement). In Example 1 we have indicated only one CI statement, $h \perp\!\!\!\perp s \mid g$. On the other hand, we have indicated two dependence statements, namely $g \not\!\perp\!\!\!\perp h$ and $s \not\!\perp\!\!\!\perp g$. These are viewed as conditional dependence statements, namely $g \not\!\perp\!\!\!\perp h \mid \emptyset$ and $s \not\!\perp\!\!\!\perp g \mid \emptyset$.

Using CI statements we can create a model of a real domain/system that describes conditional independence relations between all variables in the modeled domain.

Definition 3 (CI model). *A CI model is a set of CI statements.*

Example 3 (CI model). The situation in Example 1 can be modelled by a CI model containing just one non-trivial CI statement $h \perp\!\!\!\perp s \mid g$. However, one should also include trivial CI statements, which have the form $A \perp\!\!\!\perp \emptyset \mid C$, where $A, C \subseteq \{s, h, g\}$ are disjoint. They correspond to an intuitively evident statement that there cannot be any dependency on an empty set of variables. It is implicitly understood that the disjoint triplets that are not present in the list of CI statements are interpreted as conditional dependence statements.

One way to describe an independence model of a real system is to provide the list of all valid CI statements. However, for large domains of interest this list might be terribly long. This means, humans would hardly be able to to create, maintain, or verify the correctness of a list of this kind.

The list of CI statements can be shortened if we take into consideration certain properties of probabilistic CI models. The CI models that satisfy five basic axioms listed below are called *semi-graphoids* [8]. These axioms can be used as rules that generate from a certain set of CI statements called the *base* all valid CI statements in the modeled system. Thus, instead of storing the whole list of CI statements we just keep the CI statements in the base.

Definition 4 (Semi-Graphoid Axioms). *For each collection $A, B, C, D \subseteq N$ of pairwise disjoint sets the following axioms are assumed:*

$$A \perp\!\!\!\perp \emptyset \mid C,$$
$$A \perp\!\!\!\perp B \mid C \implies B \perp\!\!\!\perp A \mid C,$$
$$A \perp\!\!\!\perp B \cup D \mid C \implies A \perp\!\!\!\perp B \mid C,$$
$$A \perp\!\!\!\perp B \cup D \mid C \implies A \perp\!\!\!\perp B \mid C \cup D,$$
$$A \perp\!\!\!\perp B \mid C \cup D \ \land \ A \perp\!\!\!\perp D \mid C \implies A \perp\!\!\!\perp B \cup D \mid C.$$

Even if lists of CI statements are shortened so that they only contain CI statements in a base they still might be hardly understandable by humans. Therefore various auxiliary mathematical objects were proposed that can be used to generate CI models.

In Section 2 we review some of these mathematical objects that are traditionally used in the area of probabilistic reasoning. More specifically, we discuss *undirected graphs*, *acyclic directed graphs*, *chain graphs*, and an alternative algebraic approach that uses certain integer-valued vectors, named *imsets*. We compare their expressive power and discuss the problem of their uniqueness. The discussion is a starting point for introducing imsets since they meet two requirements:

- so-called standard imsets can represent each CI model generated by a discrete probability distribution, and
- imsets from a special class of standard imsets are unique representatives of CI models generated by acyclic directed graphs.

These properties play an important role in learning CI models.

In Section 3 we introduce two representatives of an equivalence class of acyclic directed graphs – *essential graphs* and *standard imsets*. The core of this chapter is the transition between these two representatives of Bayesian network models. Therefore, in Section 4, we explain in detail why an algorithm for this transition is desired in the area of learning Bayesian networks. Then, in Section 5, certain graphical characteristics of chain graphs are introduced that play a crucial role in the reconstruction of the essential graph on basis of the standard imset. In Section 6 we formulate lemmas on which the reconstruction algorithm is based and the algorithm is given in Section 7. In Section 8 we discuss the relation to hierarchical junction trees, which provide a third method for representing a Bayesian network model. We implemented all mentioned algorithms in the R language developed by R Development Core Team [10]. In Conclusions we discuss future perspectives.

2 Objects Generating Conditional Independence Models

The main motivation for using mathematical objects to generate CI models is that they often have much more compact form when compared with a list of CI statements. They are also better readable by humans. By a *CI object* over N we will understand a mathematical object defined over a finite set (of variables) N that can be used to generate a CI model.

In this section we review the most popular classes of CI objects: discrete probability distributions, undirected graphs, acyclic directed graphs, chain graphs, and imsets. We will describe how they generate CI models and compare their expressive power with respect to the collection of all CI models generated by discrete probability distributions. All CI models discussed in this section are semi-graphoids.

2.1 Probability Distributions

Every discrete probability distribution (PD) defined over variables from N can be used to induce a CI model over N. In Definition 5 the concept of conditional independence for discrete PDs is recalled.

Definition 5 (CI in PDs). *Let P be a discrete PD over N. Given any $A \subseteq N$, let v_A denote a configuration of values of variables from A and for $B \subseteq N \setminus A$ let $P(v_A \mid v_B)$ denote the conditional probability for $A = v_A$ given $B = v_B$. Given $\langle A, B \mid C \rangle \in \mathcal{T}(N)$, the CI statement $A \perp\!\!\!\perp B \mid C$ is induced by probability distribution P over N if for all v_A, v_B, v_C such that $P(v_C) > 0$*

$$P(v_A, v_B \mid v_C) = P(v_A \mid v_C) \cdot P(v_B \mid v_C) \ . \tag{1}$$

Example 4. Assume that h, s, g are variables from Example 1. For simplicity, further assume that they are all discrete and binary:

- h taking states *short* and *long*,
- s taking states *more than 164 cm* and *less than 164 cm*, and
- g taking states *male* and *female*.

Let $A = \{h\}$, $B = \{s\}$, and $C = \{g\}$. If a probability distribution $P(A, B, C)$ satisfies equation (1) for all respective configuration of values (v_A, v_B, v_C) from

$$\{short, long\} \times \{more\ than\ 164\ cm, less\ than\ 164\ cm\} \times \{male, female\}$$

then $A \perp\!\!\!\perp B \mid C$ (which is $h \perp\!\!\!\perp s \mid g$) is a CI statement induced by P.

2.2 Undirected Graphs

Probabilistic graphical models modelled by undirected graphs are also known as *Markov networks* [8].

Definition 6 (UG). *An* undirected graph *(UG) over N is a pair $(N, \mathcal{E})$, where N is a set of nodes and $\mathcal{E}$ a set of undirected edges, i.e., a set of unordered pairs $a - b$ where $a, b \in N$ and $a \neq b$.*

Definition 7 (Undirected path). *An* undirected path *between nodes a and b in an UG $G = (N, \mathcal{E})$ is a sequence $a \equiv c_1, \ldots, c_n \equiv b$, $n \geq 2$ of nodes such that $c_1, \ldots, c_n$ are distinct and $c_i - c_{i+1} \in \mathcal{E}$ for $i = 1, \ldots, n-1$.*

Definition 8 (Separation criterion for UGs). *$A \perp\!\!\!\perp B \mid C$ is represented in an UG G if every path in G between a node in A and a node in B contains a node from C.*

Example 5. The UG in Figure 1 represents the CI statement $h \perp\!\!\!\perp s \mid g$.

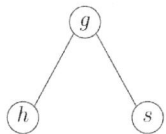

Fig. 1. The UG generating $h \perp\!\!\!\perp s \mid g$

Definition 9 (Clique). *A set of nodes $K \subseteq N$ of an UG $G = (N, \mathcal{E})$ is* complete *if $a - b \in \mathcal{E}$ for all $a, b \in K, a \neq b$. A maximal complete set of G with respect to set inclusion is called a* clique.

An important concept is that of a *decomposable* (undirected) graph. There are several equivalent definitions of a decomposable graph (see § 2.1.2 of Lauritzen [6]), one of them is the following.

Definition 10 (Decomposable graph). *An undirected graph $G = (N, \mathcal{E})$ is* decomposable *if its cliques can be ordered into a sequence $K_1, \ldots, K_m$, $m \geq 1$ satisfying the* running intersection property *(cf. Proposition 2.17(ii) in [6]):*

$$\forall i \geq 2 \ \exists k < i \quad S_i \equiv K_i \cap (\bigcup_{j<i} K_j) \subseteq K_k. \qquad (2)$$

It is a well-known fact that the collection of sets S_i, $2 \leq i \leq m$ does not depend on the choice of an ordering satisfying (2) – see Lemma 7.2 in [13]. We will call these sets *separators* of the graph. Moreover, the multiplicity $\nu(S)$ of a separator S, that is, the number of indices i for which $S = S_i$ also does not depend on the choice of an ordering satisfying (2). Note that the definition implies that the class of cliques is disjoint with the class of separators.

Example 6. The UG in Figure 1 has two cliques $C_1 = \{h, g\}$ and $C_2 = \{g, s\}$. It is decomposable and it has one separator $S = \{g\}$.

2.3 Acyclic Directed Graphs

Acyclic directed graphs are a very popular class of CI objects in the area of probabilistic reasoning. They constitute the structural part of probabilistic models known as *Bayesian network models* [8].

Definition 11 (Directed graph). *A* directed graph *is a pair* $(N, \mathcal{F})$, *where N is the set of nodes and $\mathcal{F}$ is a set of directed edges, i.e., a set of ordered pairs $a \to b$ where $a, b \in N$ and $a \neq b$ such that if $a \to b \in \mathcal{F}$ then $b \to a \notin \mathcal{F}$.*

Definition 12 (Directed path). *A* directed path *between nodes a and b in a directed graph $G = (N, \mathcal{F})$ is a sequence $a \equiv c_1, \ldots, c_n \equiv b, n \geq 2$ such that $c_1, \ldots, c_n$ are distinct and for $i = 1, \ldots, n-1$: $c_i \to c_{i+1} \in \mathcal{F}$.*

Definition 13 (Directed cycle). *A* directed cycle *in a graph $G = (N, \mathcal{F})$ is a sequence $c_1, \ldots, c_n, c_{n+1} \equiv c_1, n \geq 3$ of nodes in G such that $c_1, \ldots, c_n$ are distinct and for $i = 1, \ldots, n-1$: $c_i \to c_{i+1} \in \mathcal{F}$.*

Definition 14 (DAG). *An* acyclic directed graph *(DAG) is a directed graph that has no directed cycle.*

Remark 1. Researchers in the area of artificial intelligence became accustomed to the abbreviation DAG. It is based on the phrase *directed acyclic graph*, which is, however, imprecise from the grammatical point of view.

Definition 15 (Underlying graph of a DAG). Underlying graph *of a DAG $G = (N, \mathcal{F})$ is undirected graph $G' = (N, \mathcal{E})$, where*

$$\mathcal{E} = \{a - b : a \to b \in \mathcal{F} \vee a \to b \in \mathcal{F}\} \ .$$

Definition 16 (Set of parents in a DAG). *Let $G = (N, \mathcal{F})$ be a DAG. The set of* parents *of a node $b \in N$ is the set $\{a \in N : a \to b \in \mathcal{F}\}$, denoted by $pa_G(b)$.*

To introduce so-called moralization criterion for reading CI statements from a DAG we will use graphical concepts of *ancestral set* and *moral graph*.

Definition 17 (Ancestral set in a DAG). *Let $G = (N, \mathcal{F})$ be a DAG. Node a is an* ancestor *of node b in G if there exists a directed path from a to b. Given $B \subseteq N$, the set of ancestors of nodes in a set of nodes B is called the (least)* ancestral set *for B. It will be denoted $an_G(B)$.*

Definition 18 (Induced subgraph for a DAG). *Assume a DAG $G = (N, \mathcal{F})$ and $\emptyset \neq M \subseteq N$. The* induced subgraph *of G for M is graph $G_M = (M, \mathcal{F}_M)$, where $\mathcal{F}_M = \{a \to b \in \mathcal{F} : a \in M, b \in M\}$.*

Note that every induced subgraph of a DAG is again a DAG.

Definition 19 (Moralization for DAGs). *The* moral graph *of a DAG $G = (N, \mathcal{F})$ is graph G^{mor}, which is an undirected graph $(N, \mathcal{E})$, where*

$$\mathcal{E} = \{a - b : (a \to b \in \mathcal{F}) \vee (b \to a \in \mathcal{F}) \vee \\ (\exists c \in N \setminus \{a, b\} : a \to c, b \to c \in \mathcal{F})\}$$

Now, we can formulate the *moralization criterion*. To check whether a CI statement is represented in a DAG, we first create an ancestral graph, moralize it, and finally check the validity of the CI statement using the separation criterion for UGs as defined in Definition 8.

Definition 20 (Moralization criterion for DAGs). *$A \perp\!\!\!\perp B \mid C$ is represented in a DAG G if $A \perp\!\!\!\perp B \mid C$ is represented in the moral graph of induced graph $G_{an_G(A \cup B \cup C)}$.*

Example 7. In Figure 2 three DAGs generating the CI statement $h \perp\!\!\!\perp s \mid g$ are shown. They all have the same moral graph, which is in Figure 1. Note that there are three different DAGs that generate the same CI model.

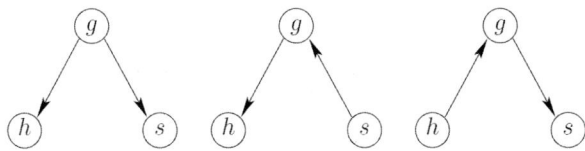

Fig. 2. Three DAGs generating $h \perp\!\!\!\perp s \mid g$

2.4 Chain Graphs

Chain graphs [7] are a common generalization of both UGs and DAGs. They form a special subclass of graphs that contain both directed and undirected edges, so called *hybrid graphs*.

Definition 21 (Hybrid graph). *A hybrid graph is a triplet $(N, \mathcal{E}, \mathcal{F})$, where N is the set of nodes, $\mathcal{E}$ is a set of undirected edges, i.e., a set of unordered pairs $a - b$ where $a, b \in N$ and $a \neq b$ and $\mathcal{F}$ is a set of directed edges, i.e., a set of ordered pairs $a \to b$ where $a, b \in N$ and $a \neq b$. Moreover, it is required that if $a \to b \in \mathcal{F}$ then $b \to a \notin \mathcal{F}$ and $a - b \notin \mathcal{E}$.*

Definition 22 (Semi-directed cycle). *A semi-directed cycle in a hybrid graph H is a sequence $c_1, \ldots, c_n, c_{n+1} \equiv c_1, n \geq 3$ of nodes in G such that $c_1, \ldots, c_n$ are distinct, $c_1 \to c_2 \in \mathcal{F}$, and either $c_i \to c_{i+1} \in \mathcal{F}$ or $c_i - c_{i+1} \in \mathcal{E}$ for $i = 2, \ldots, n$.*

Definition 23 (Chain graph). *A* chain graph *(CG) is a hybrid graph that has no semi-directed cycle.*

Example 8. In Figure 3 we give an example of a chain graph.

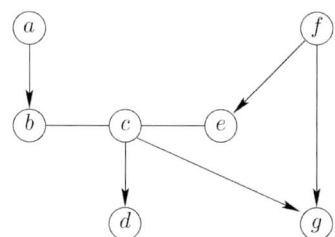

Fig. 3. A chain graph

The following definitions generalize definitions of undirected and directed paths.

Definition 24 (Path). *Path in a hybrid graph* $H = (N, \mathcal{E}, \mathcal{F})$ *between nodes a and b is a sequence $a \equiv c_1, \ldots, c_n \equiv b, n > 1$ such that $c_1, \ldots, c_n$ are distinct and for $i = 1, \ldots, n-1$ one of the following three conditions holds:*

(1) $c_i - c_{i+1} \in \mathcal{E}$,
(2) $c_i \rightarrow c_{i+1} \in \mathcal{F}$, *or*
(3) $c_{i+1} \rightarrow c_i \in \mathcal{F}$.

If for $i = 1, \ldots, n-1$ either the condition (1) or (2) holds the path is called descending.

Example 9. An example of a descending path in the chain graph in Figure 3 is the path $a \rightarrow b - c \rightarrow d$.

Next, we generalize the concepts of set of parents and ancestral set for CGs.

Definition 25 (Set of parents in a CG). *Let* $H = (N, \mathcal{E}, \mathcal{F})$ *be a CG. The set of parents of nodes in a set B is the set* $\{a \in N : a \rightarrow b \in \mathcal{F}, b \in B\}$. *It will be denoted* $pa_H(B)$.

Definition 26 (Ancestral set in a CG). *Let* $H = (N, \mathcal{E}, \mathcal{F})$ *be a CG. Node a is an* ancestor *of node b in H if there exists a descending path from a to b. The set of ancestors of nodes in a set of nodes B is called the (least)* ancestral set *for B. It will be denoted* $an_H(B)$.

Example 10. Let H be the chain graph from Figure 3. For example, the set of parents of $\{d\}$ is $\{c\}$, the ancestral set of $\{d\}$ is $\{a, b, c, e, f\}$. The ancestral set of $\{d, g\}$ is the set of all nodes N.

Next, we generalize special concepts from DAGs to CGs.

Definition 27 (Induced subgraph for a CG). *Let $H = (N, \mathcal{E}, \mathcal{F})$ be a CG and $\emptyset \neq M \subseteq N$. The* induced subgraph *of H for M is graph $H_M = (M, \mathcal{E}_M, \mathcal{F}_M)$, where $\mathcal{E}_M = \{a - b \in \mathcal{E} : a, b \in M\}$ and $\mathcal{F}_M = \{a \to b \in \mathcal{F} : a, b \in M\}$.*

Note that every induced subgraph of a CG is a CG.

Definition 28 (Components). *A set of nodes $C \subseteq N$ is* connected *in a CG $H = (N, \mathcal{E}, \mathcal{F})$ if for all $a, b \in C$ there exists a path in the undirected graph $G = (N, \mathcal{E})$. Maximal connected subsets of N with respect to set inclusion are called* components *in H. The class of all components in H will be denoted $\mathcal{C}(H)$.*

Remark 2. Observe that every induced subgraph of a CG H for a component $C \in \mathcal{C}(H)$ is an UG. Given a $C \in \mathcal{C}(H)$, $pa_H(C)$ is disjoint with C. Components in a CG $H = (N, \mathcal{E}, \mathcal{F})$ form a partition of N. Components in a DAG are singletons.

Example 11. The chain graph H from Figure 3 has five components. Thus, $\mathcal{C}(H) = \{ \{a\}, \{f\}, \{b, c, e\}, \{d\}, \{g\} \}$.

Definition 29 (Moralization). *The* moral graph *of a hybrid graph $H = (N, \mathcal{E}, \mathcal{F})$ is graph H^{mor}, which is the undirected graph $(N, \bar{\mathcal{E}})$, where*

$$\bar{\mathcal{E}} = \{a - b : (a \to b \in \mathcal{F}) \vee (b \to a \in \mathcal{F}) \vee \\ (a - b \in \mathcal{E}) \vee (\exists C \in \mathcal{C}(H) : a \neq b, \ \{a, b\} \subseteq pa_H(C))\}$$

Example 12. In Figure 4 we give the moral graph of the graph from Figure 3.

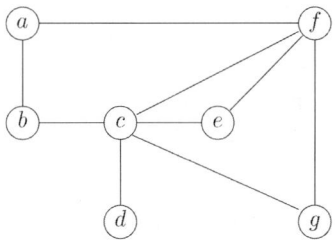

Fig. 4. The moral graph of the chain graph from Figure 3

Now, we are ready to introduce the moralization criterion for reading CI statements represented in a CG. To check a CI statement, we first create an ancestral graph, moralize it, and finally check the CI statement using the separation criterion for UGs as defined in Definition 8.

Definition 30 (Moralization criterion for CGs).
CI statement $A \perp\!\!\!\perp B \mid C$ is represented in a CG H if $A \perp\!\!\!\perp B \mid C$ is represented in the moral graph of induced graph $H_{an_H(A \cup B \cup C)}$.

Remark 3. Note that CGs are generalization of DAGs and UGs because the moralization criterion for CGs generalizes the separation criterion for UGs and the moralization criterion for DAGs.

Example 13. In order to verify whether the CI statement $a \perp\!\!\!\perp d \mid \{b, e, g\}$ is represented in the chain graph H in Figure 3 we check this CI statement using the separation criterion in the moral graph G of the ancestral graph for the set $\{a, b, d, e, g\}$, which is apparently the undirected graph in Figure 4. Since there exist a path from a to d that does not contain a node from $\{b, e, g\}$

$$a - f - c - d$$

the CI statement is not represented in the undirected graph G and, consequently, it is not represented in the CG H.

2.5 Imsets

An *imset* [13] is an algebraic object that can be used to describe a CI model.

Definition 31 (Imset). Let N be a finite set, $\mathcal{P}(N) = \{A : A \subseteq N\}$ the power set of N, and $\mathbb{Z}$ the set of all integers. An imset u is a function $u : \mathcal{P}(N) \mapsto \mathbb{Z}$.

Remark 4. Let $\mathbb{N}$ be the set of all natural numbers. Sometimes, function $m : \mathcal{P}(N) \mapsto \mathbb{N}$ is called multiset. The word *imset* is an abbreviation for Integer valued MultiSET.

Zero function on the power set of N will be denoted by 0. Note that every imset can be interpreted as a vector, whose components are integers indexed by subsets of N.

Example 14. Let $N = \{h, g, s\}$. An example of an imset over N is

B	$u(B)$
$\emptyset$	0
$\{h\}$	0
$\{g\}$	+1
$\{s\}$	0
$\{h, g\}$	−1
$\{g, s\}$	−1
$\{h, s\}$	0
$\{h, g, s\}$	+1

We will specify imsets using a special convention. Let us use Kronecker's symbol δ to denote an indicator function defined for $A, B \subseteq N$ as follows:

$$\delta_A(B) = \begin{cases} 1 & \text{if } A = B, \\ 0 & \text{otherwise.} \end{cases}$$

Then, given an imset u, one has

$$\forall B \subseteq N: \ u(B) = \sum_{A \subseteq N} u(A) \cdot \delta_A(B)$$

which can be abbreviated as

$$u = \sum_{A \subseteq N} c_A \cdot \delta_A$$

where $c_A = u(A) \in \mathbb{Z}$ is a the respective coefficient for every $A \subseteq N$.

Example 15. Using the convention we will write the imset from Example 14 as follows:

$$u = \delta_{\{g\}} - \delta_{\{h,g\}} - \delta_{\{g,s\}} + \delta_{\{h,g,s\}}$$

Definition 32 (Elementary imset). *Let $K \subseteq N$, $a, b \in N \setminus K$, and $a \neq b$. The* elementary imset *corresponding to $\langle a, b \mid K \rangle$ is given by the formula*

$$u_{\langle a,b|K \rangle} = \delta_{\{a,b\} \cup K} + \delta_K - \delta_{\{a\} \cup K} - \delta_{\{b\} \cup K} \ .$$

The symbol $\mathcal{E}(N)$ will denote the set of all elementary imsets over N.

Definition 33 (Structural imset). *An imset u is* structural *iff*

$$n \cdot u = \sum_{v \in \mathcal{E}(N)} k_v \cdot v \ ,$$

where $n \in \mathbb{N}$ and $v \in \mathcal{E}(N): k_v \in \mathbb{N} \cup \{0\}$.

As an analogy to graphical criteria in the case of UGs, DAGs, and CGs, we need a criterion that specifies how imsets generate CI models. This time we have an algebraic criterion.

Definition 34. *CI statement $A \perp\!\!\!\perp B \mid C$ is represented in a structural imset u over N if there exists $k \in \mathbb{N}$ such that $k \cdot u = u_{\langle A,B|C \rangle} + w$, where w is a structural imset.*

2.6 Comparison of CI Objects

The CI objects discussed above are often used to represent CI models induced by probability distributions. Thus, we may compare how well they can play this role. Let

- $M(P)$ be a CI model generated by a discrete probability distribution P over N,
- $\mathcal{M}$ denote the class of all CI models generated by discrete probability distributions over N,
- $\mathcal{O}$ denote a considered class of CI objects (e.g., UGs, DAGs, CGs, structural imsets).

Definition 35 (Perfectly Markovian). *A probability distribution P is perfectly Markovian with respect to an object $O \in \mathcal{O}$ if for every disjoint triplet $\langle A, B \mid C \rangle \in \mathcal{T}(N)$*

$$A \perp\!\!\!\perp B \mid C \text{ is represented in } P \iff A \perp\!\!\!\perp B \mid C \text{ is represented in } O\ .$$

In other words, perfect Markovness means that P and O generate the same CI model. Now, we can raise three basic questions about the relation between a class $\mathcal{O}$ of CI objects and the class $\mathcal{M}$ of CI models generated by discrete probability distributions.

Definition 36 (Faithfulness). *A class of CI objects $\mathcal{O}$ is faithfull if for every CI object $O \in \mathcal{O}$ there exists a CI model $M(P)$ from $\mathcal{M}$ such that P is perfectly Markovian with respect to O.*

Definition 37 (Completeness). *A class of CI objects $\mathcal{O}$ is complete if for every CI model $M(P)$ from $\mathcal{M}$ there exists a CI object $O \in \mathcal{O}$ such that P is perfectly Markovian with respect to O.*

Definition 38 (Uniqueness). *A class of CI objects $\mathcal{O}$ satisfies the uniqueness property if for every CI model $M(P)$ from $\mathcal{M}$ at most one CI object $O \in \mathcal{O}$ exists such that P is perfectly Markovian with respect to O.*

Table 1 compares properties of different classes of CI objects. In contrary to graphical probabilistic models (UGs, DAGs, and CGs) the class of structural imsets is complete, that is, it can describe all CI models generated by discrete PDs.

Table 2 shows the numbers of different CI models that can be generated by different classes of CI objects. We can see that, already for N having only four elements, only a small fraction of CI models generated by discrete PDs over N can be represented by DAGs, UGs, and CGs. Structural imsets are complete, which means that they can represent all CI models from $\mathcal{M}$.

Table 1. Properties of classes of CI objects

	Faithfulness	Completeness	Uniqueness
UGs	yes	no	yes
DAGs	yes	no	no
CGs	yes	no	no
structural imsets	no	yes	no

Table 2. Comparison of the number of CI models generated by different classes of CI objects for $|N| = 3, 4, 5$. The number of CI models generated by discrete PDs is not known for $|N| \geq 5$.

$|N|$	UGs	DAGs	CGs	$|\mathcal{M}|$
3	8	11	11	22
4	64	185	200	18300
5	1024	8782	11519	?

3 Representatives of Equivalence Classes of DAGs

We have already mentioned that several different DAGs may generate the same CI-model, that is, DAGs do not satisfy the uniqueness property. This unpleasant fact may cause some problems in learning Bayesian networks (see Section 4) and motivates the need for a uniquely determined representative of the respective CI model. In this section we present two unique representatives of an equivalence class of DAGs that generate the same CI model. First, we define the *essential graph* and note how it can be constructed from a DAG. Second, we introduce a uniquely determined algebraic representative of an equivalence class of DAGs – the *standard imset*.

Definition 39 (Independence equivalence). *Two CI-objects are called* independence equivalent *if they define the same CI-model. We will briefly say that they are equivalent.*

Verma and Pearl [16] gave a direct graphical characterization of equivalent DAGs. It uses the following concept.

Definition 40 (Immorality). *An* immorality *in a DAG $G = (N, \mathcal{F})$ is an induced subgraph of G for a set $\{a, b, c\} \subseteq N$ such that $a \to c$, $b \to c \in \mathcal{F}$ and $a \leftarrow b, a \to b \notin \mathcal{F}$.*

Lemma 1. *[16]*
Two acyclic directed graphs are equivalent iff they have the same underlying graph and immoralities.

3.1 Essential Graphs

Definition 41 (Essential graph). *The* essential graph *(EG)* G^* *of an equivalence class* $\mathcal{G}$ *of DAGs over* N *is a hybrid graph over* N *defined as follows:*

- $a \to b$ *in* G^* *if* $a \to b$ *in* G *for every* $G \in \mathcal{G}$,
- $a - b$ *in* G^* *if* $\exists G_1, G_2 \in \mathcal{G}$ *such that* $a \to b$ *in* G_1 *and* $a \leftarrow b$ *in* G_2.

We say that a hybrid graph H *over* N *is an EG over* N *if there exists an equivalence class* $\mathcal{G}$ *of DAGs over* N *such that* $H = G^*$.

Of course, this definition is in terms of the whole equivalence class of DAGs. Nevertheless, there exists an algorithm for getting the EG on basis of any $G \in \mathcal{G}$ – see Studený [11]. A graphical characterization of EGs was presented by Andersson et al. [1]. To recall it we need the following notion.

Definition 42 (Flag). *A* flag *in a CG* $H = (N, \mathcal{E}, \mathcal{F})$ *is an induced subgraph of* H *for a set* $\{a, b, c\} \subseteq N$ *such that* $a \to b \in \mathcal{F}$, $b - c \in \mathcal{E}$, *and* $a \leftarrow c, a \to c \notin \mathcal{F}, a - c \notin \mathcal{E}$.

Example 16. The chain graph in Figure 3 has two flags, $a \to b - c$ and $f \to e - c$.

Lemma 2. *[1, Theorem 4.1]*
A hybrid graph $H = (N, \mathcal{E}, \mathcal{F})$ *is an EG iff it is a CG without flags such that, for every component* $C \in \mathcal{C}(H)$, *the induced subgraph* H_C *is decomposable and, for every* $a \to b \in \mathcal{F}$, *at least one of the following conditions holds:*

- $\exists c \in N : (c \to a \in \mathcal{F}) \land (c \to b \notin \mathcal{F})$,
- $\exists c \in N : (c \to b \in \mathcal{F}) \land (c - a \notin \mathcal{E}) \land (c \to a \notin \mathcal{F})$,
- $\exists c_1, c_2 \in N : (c_1 \to b, c_2 \to b \in \mathcal{F}) \land (c_1 - a, c_2 - a \in \mathcal{E}) \land (c_1 \to c_2, c_2 \to c_1 \notin \mathcal{F}) \land (c_1 - c_2 \notin \mathcal{E})$

Example 17. An example of an EG is given in Figure 5.

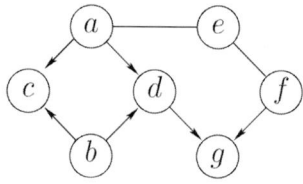

Fig. 5. Example of an EG.

Another useful result is a characterization of CGs equivalent to DAGs, which uses the following concept.

Definition 43 (Closure graph). *If C is a component in a CG H then by the closure graph for C we will understand the moral graph of the induced subgraph for the set $D = C \cup pa_H(C)$. It will be denoted by $\bar{H}(C)$.*

Lemma 3. *[2, Proposition 4.2]*
A chain graph H is equivalent to a DAG iff, for every component C of H, the closure graph $\bar{H}(C)$ is decomposable. In particular, the induced subgraph H_C is decomposable for any $C \in \mathcal{C}(H)$.

3.2 Standard Imsets

Another uniquely determined representative of an equivalence class of DAGs is the *standard imset* [13].

Definition 44 (Standard imset). *Given a DAG $G = (N, \mathcal{F})$, the standard imset for G is given by the formula*

$$u_G = \delta_N - \delta_\emptyset + \sum_{a \in N} \left\{ \delta_{pa_G(a)} - \delta_{\{a\} \cup pa_G(a)} \right\}. \tag{3}$$

Note that some terms in the formula (3) can cancel each other and some terms can be merged together. The basic observation is as follows.

Lemma 4. *[13, Corollary 7.1]*
Two DAGs G and H are independence equivalent iff $u_G = u_H$.

Thus, standard imsets can serve as unique representatives of the respective CI model. Another pleasant fact is that standard imsets, viewed as vectors, have many zero components. Therefore, they can effectively be kept in the memory of a computer.

Now, we give a formula for the standard imset on basis of any chain graph H over N which is equivalent to a DAG. It is based on Lemma 3. Let $\bar{\mathcal{K}}(C)$ denote the collection of cliques of $\bar{H}(C)$ and $\bar{\mathcal{S}}(C)$ the collection of separators in $\bar{H}(C)$. Further, let $\bar{\nu}_C(S)$ denote the multiplicity of a separator S in $\bar{H}(C)$. The standard imset for H is given by the following formula:

$$u_H = \delta_N - \delta_\emptyset + \sum_{C \in \mathcal{C}(H)} \left\{ \delta_{pa_H(C)} - \sum_{\bar{K} \in \bar{\mathcal{K}}(C)} \delta_{\bar{K}} + \sum_{\bar{S} \in \bar{\mathcal{S}}(C)} \bar{\nu}_C(\bar{S}) \cdot \delta_{\bar{S}} \right\}. \tag{4}$$

The point is that the formula (4) gives the same result for equivalent chain graphs.

Lemma 5. *[14, Proposition 20]*
Let G and H are equivalent chain graphs such that there exists a DAG equivalent to them. Then $u_G = u_H$.

Of course, if $H = G$ is a DAG, then (4) gives the same result as (3). On the other hand, since the EG G^* of an equivalence class of DAGs $\mathcal{G}$ is a CG equivalent to any $G \in \mathcal{G}$, we can conclude this:

Corollary 1. *Let G be a DAG and H the EG of the respective equivalence class. Then the formula (4) gives the standard imset for G.*

Example 18. In Table 3 we give the standard imset for the EG from Figure 5.

Table 3. The standard imset for the EG from Figure 5. The values for remaining subsets of N are zero.

B	$u(B)$
$\{a,b,c,d,e,f,g\}$	$+1$
$\emptyset$	$+1$
$\{a,b\}$	$+2$
$\{d,f\}$	$+1$
$\{a,b,c\}$	-1
$\{a,b,d\}$	-1
$\{d,f,g\}$	-1
$\{b\}$	-1
$\{a,e\}$	-1
$\{e,f\}$	-1
$\{e\}$	$+1$

4 Learning Bayesian Networks

Specific motivation for the transition between EGs and standard imsets is learning Bayesian networks. A *Bayesian network model* has two components:

- *structure*, determined by a DAG, whose nodes correspond to variables,
- *parameters*, namely the numbers in the collection of conditional probability tables, which correspond to the DAG.

We are interested in learning structure of a Bayesian network from data. Actually, our aim is to determine the respective *statistical model*, that is, the class of probability distributions with prescribed structure.

4.1 Quality Criterion

The basic division of methods for learning Bayesian networks is as follows:

- methods based on *statistical conditional independence tests*.
- methods based on the *maximization of a quality criterion*.

In this chapter, we are interested in maximization of a quality criterion

$$\mathcal{Q} : \mathsf{DAGS}\,(N) \times \mathsf{DATA}\,(N,d) \longrightarrow \mathbb{R}$$

where DAGS (N) is the class of DAGs over N and DATA (N, d) is the collection of databases over N of the length $d \geq 1$. In this context, the graph "represents" the respective statistical model. Because the direct maximization of a quality criterion is typically infeasible, the researchers in artificial intelligence developed various *methods of local search*, see Chickering [3].

The basic idea of these methods is to introduce the concept of *neighbourhood* for representatives of considered CI models (= graphs) and search for a local maximum of the criterion with respect to the neighbourhood structure. Typically, the change in the value of a (common reasonable) quality criterion is easy to compute. Natural neighbourhood concept from a mathematical point of view is so-called *inclusion neighbourhood*, see Kočka and Castelo [5].

4.2 Problem of Representative Choice

This topic is related to the question of internal computer representation of a Bayesian network model. There are two approaches:

- to use any DAG in the respective equivalence class (which need not be unique),
- to use a suitable uniquely determined representative.

We prefer using a unique representative. This is because we believe that non-uniqueness may lead to computational inefficiencies. Another reason is that using a unique representative is more elegant from a mathematical point of view. Two possible unique representatives, namely the EG and the standard imset, were already mentioned in Section 3.

A natural question is whether one can "translate" one to the other. Our special motivation is as follows. The inclusion neighbourhood of a given Bayesian network model is already characterized in terms of the EG – see Studený [12]. We would like to have its characterization in terms of the standard imset. Below we explain why we consider standard imsets particularly suitable for this purpose.

4.3 Algebraic Approach

The basic idea of an algebraic approach (to learning Bayesian networks) is to use the standard imset as a unique representative of the respective statistical model – se §8.4 in (Studený 2005). The advantage of this approach is that every imset can be interpreted as a vector and (reasonable) quality criteria appear to be affine (= shifted linear) functions of the standard imset.

More specifically, every criterion $\mathcal{Q}(G, D)$ depending on a DAG G and a database D, which is score-equivalent and decomposable Studený [13], has the form

$$\mathcal{Q}(G, D) = s_D^{\mathcal{Q}} + \sum_{S \subseteq N} t_D^{\mathcal{Q}}(S) \cdot u_G(S),$$

where $s_D^{\mathcal{Q}}$ is a constant depending on data and $[t_D^{\mathcal{Q}}(S)]_{S \subseteq N}$ is so-called *data vector* (relative to a criterion $\mathcal{Q}$).

Example 19. In the case of a well-known Bayesian information criterion (= BIC) one has the following formula for the respective data vector:

$$t_D^{\mathsf{BIC}}(S) = d \cdot H(\hat{P}_S | \prod_{i \in S} \hat{P}_i) - \frac{1}{2} \cdot \ln d \cdot \{ |S| - 1 + \prod_{i \in S} r(i) - \sum_{i \in S} r(i) \},$$

where $\emptyset \neq S \subseteq N$, d is the length of the database D, $H(*|*)$ denotes the relative entropy, $\hat{P}_S$ is the marginal of the empirical measure (based on D) for S and $r(i) = |\mathsf{X}_i|$, $i \in S$ are the cardinalities of the respective individual sample spaces. Moreover, one has $t_D^{\mathsf{BIC}}(\emptyset) = 0$.

Another pleasant fact is that, in the method of local search, the *move* between two neighbouring model in the sense of inclusion neighbourhood is characterized by a simple elementary imset, which is the difference of respective standard imsets. Therefore, the move can be interpreted in terms of a CI statement $a \perp\!\!\!\perp b \,|\, C$. In particular, the respective change in the value of $\mathcal{Q}$ takes a neat form of the scalar product of two vectors:

$$\langle t_D^{\mathcal{Q}}, u_{\langle a,b|C\rangle} \rangle = \sum_{S \subseteq N} t_D^{\mathcal{Q}}(S) \cdot u_{\langle a,b|C\rangle}(S).$$

5 Graphical Characteristics of Chain Graphs

The formula (4) can be simplified for chain graphs without flags. In this section, we introduce some characteristics of these graphs that will be used in a simplified formula given in Section 6. The proofs of claims from Sections 5.1 and 5.2 can be found in [14].

5.1 Initial Components

Definition 45 (Initial component). *A component $C \in \mathcal{C}(H)$ in a CG H such that $pa_H(C) = \emptyset$ will be called an* initial component *in H. Let us denote by $i(H)$ the number of initial components in H.*

Note that $i(H) \geq 1$ and this number appears to be the same for equivalent CGs.

Example 20. The chain graph H in Figure 5 has two initial components: $\{b\}$ and $\{a, e, f\}$. Thus, $i(H) = 2$.

5.2 Core

Definition 46 (Idle set). *We will say that a set B of nodes in a CG $H = (N, \mathcal{E}, \mathcal{F})$ is idle if the following two conditions hold:*

- $\forall b_1, b_2 \in B, b_1 \neq b_2, \quad (b_1 \to b_2 \in \mathcal{F}) \vee (b_2 \to b_1 \in \mathcal{F}) \vee (b_1 - b_2 \in \mathcal{E})$
- $\forall a \in N \setminus B, \forall b \in B, \quad a \to b \text{ in } H.$

The meaning of these conditions is that no non-trivial CI statement represented in H involves variables in B. The second condition implies that $\forall C \in \mathcal{C}$ if $C \cap B \neq \emptyset$ then $C \subseteq B$. Therefore, every idle set is the union of some components. One can easily show that every CG H over N has a unique maximal idle set of nodes, possibly empty. This set can be shown to be the same for equivalent chain graphs.

Definition 47 (Core, core-components). *The complement $N \setminus B$ of the maximal idle set B will be called the* core *of H and denoted by* $\text{core}(H)$. *The class of* core-components, *that is, components in H contained in the core, will be denoted by* $\mathcal{C}_{core}(H)$.

Observe that if the core is non-empty then every initial component is a core-component.

5.3 Cliques and Separators

If H is a CG without flags equivalent to a DAG then every its core-component C induces a decomposable graph H_C by Lemma 3. Let us denote by $\mathcal{K}(C)$ the class of its cliques, by $\mathcal{S}(C)$ the collection of its separators, and by $\nu_C(S)$ the multiplicity of $S \in \mathcal{S}(C)$ in H_C. Note that, the fact that C is connected implies that every $S \in \mathcal{S}(C)$ is a non-empty proper subset of $\mathcal{C}_{core}(H)$.

Example 21. The chain graph H in Figure 5 has an empty maximal idle set, i.e., the core is $\text{core}(H) = N$. Its core-components are $C_1 = \{a, e, f\}$, $C_2 = \{b\}$, $C_3 = \{c\}$, $C_4 = \{d\}$ and $C_5 = \{g\}$. All components except for C_1 have only one clique and no separator. The set of cliques of H_{C_1} is $\mathcal{K}(C_1) = \{\{a, e\}, \{e, f\}\}$ and the set of its separators is $\mathcal{S}(C_1) = \{\{e\}\}$.

5.4 Parent Sets

Definition 48 (Parent sets). *A set $P \subseteq N$ will be called a* parent set *in a CG H if it is non-empty and there exists a core-component $C \in \mathcal{C}_{core}(H)$ with $P = pa_H(C)$. The* multiplicity $\tau(P)$ *of a parent set P is the number of $C \in \mathcal{C}_{core}(H)$ with $P = pa_H(C)$. Let us denote the collection of parent sets in H by* $\mathcal{P}_{core}(H)$.

Evidently, every $P \in \mathcal{P}_{core}(H)$ is a proper subset of $\mathcal{C}_{core}(H)$.

Example 22. The parent sets in the chain graph H in Figure 5 are $\{a, b\}$ and $\{d, f\}$. The multiplicities are $\tau(\{a, b\}) = 2$ and $\tau(\{d, f\}) = 1$.

6 Simplified Formula for Imset of Essential Graph

For CGs without flags the formula (4) can be simplified as follows.

Lemma 6. *[15, Lemma 5.1].*
Let H be a CG without flags equivalent to a DAG with $\mathcal{C}_{core}(H) \neq \emptyset$. Then the standard imset for H is given by

$$u_H = \delta_{core(H)} - \sum_{C \in \mathcal{C}_{core}(H)} \sum_{K \in \mathcal{K}(C)} \delta_{K \cup pa_H(C)}$$
$$+ \sum_{S \in \mathcal{S}(C)} \nu_C(S) \cdot \delta_{S \cup pa_H(C)} + \sum_{P \in \mathcal{P}_{core}(H)} \tau(P) \cdot \delta_P + \{i(H) - 1\} \cdot \delta_\emptyset .$$

The point is that, in the case of a non-trivial EG H, none of the terms in the above formula cancel each other.

Lemma 7. *[15, Lemma 5.2].*
Let H be the EG of an equivalence class of DAGs over N such that $u_H \neq 0$. Then, for every $L \subseteq N$, exclusively one of the following six cases occurs:

(a) $L = core(H)$ and $u_H(L) = +1$,
(b) $L = K \cup pa_H(C)$ for $K \in \mathcal{K}(C)$, $C \in \mathcal{C}_{core}(H)$ and $u_H(L) = -1$,
(c) $L = S \cup pa_H(C)$ for $S \in \mathcal{S}(C)$, $C \in \mathcal{C}_{core}(H)$ and $u_H(L) = \nu_C(S) > 0$,
(d) $L = P$ for $P \in \mathcal{P}_{core}(H)$ and $u_H(L) = \tau(P) > 0$,
(e) $L = \emptyset$ and $u_H(L) = i(H) - 1 \geq 0$,
(f) none of the above cases occurs and $u_H(L) = 0$.

Lemma 7 implies that, given an EG H, the class of sets

$$\mathcal{K}_H \equiv \{K \cup pa_H(C); K \in \mathcal{K}(C), C \in \mathcal{C}_{core}(H)\}$$

and the class of sets

$$\mathcal{P}_{core}(H) \cup \mathcal{S}_H, \quad \text{where } \mathcal{S}_H \equiv \{S \cup pa_H(C); S \in \mathcal{S}(C), C \in \mathcal{C}_{core}(H)\},$$

can be determined on basis of u_H. Therefore, it follows from Lemma 7 that, given a non-zero standard imset, one can simply determine the core of the EG H, the number of its initial components, the collections of sets $\mathcal{K}_H$, and $\mathcal{P}_{core}(H) \cup \mathcal{S}_H$.

7 Reconstruction Algorithm

Lemma 7 is a basis of a two-stage reconstruction algorithm for the EG from the standard imset; the proof of its correctness is quite long – see (Studený, Vomlel 2005).

Table 4. Subroutine **Adapt** (u over $N \mid M$, w over M).

Input:	u ... a standard imset over a non-empty set of variables N
Output:	M ... a subset of N
	w ... an adapted standard imset over M

1	find a maximal set $M \subseteq N$ with respect to inclusion with $u(M) \neq 0$;
2	$w :=$ the restriction of u to the class of subsets of M;
3	return M, w;

The first stage of the algorithm is a *decomposition procedure*, whose output is an ordered sequence τ of subsets of the set of variables N. The procedure, described in Table 6, consists of repeated application of two subroutines until one gets a zero imset. The first subroutine is an adaptation subroutine for (a standard imset) u which is applied if $u(N) = 0$ – see Table 4.

The basic idea of the second subroutine, which is described in Table 5, is to reduce the set of variables N. Thus, the original imset u over N is "decomposed" into an imset w over a proper subset $M \subset N$ and a certain set of nodes T with $M \cup T = N$. Note the set T chosen in line 4 of Table 5 plays crucial role in the reconstruction phase and one can prove that is a clique $T = K \cup pa_H(C)$, $K \in \mathcal{K}(C)$ of the closure graph for a component $C \in \mathcal{C}(H)$ that is a leaf-clique of a junction tree for cliques of $\bar{H}(C)$ – for details see [15, Section 7]. The reduced imset w is obtained from the original imset u by subtracting a structural imset that corresponds to a CI statement $(M \setminus T) \perp\!\!\!\perp (T \setminus M) \mid (T \cap M)$ and by restricting to M – see lines 7-8 of Table 5.

Table 5. Subroutine **Reduce** (u over $N \mid T, M$, w over M).

Input:	u ... an adapted standard imset over a non-empty set of variables N
Output:	T ... a proper subset of the set of variables N
	M ... a proper subset of N such that $M \cup T = N$ and $T \setminus M \neq \emptyset$
	w ... a standard imset over M

1	$\mathcal{T} := \{L \subseteq N;\ u(L) < 0\}$;
2	$\mathcal{L} := \{L \subset N;\ u(L) > 0,\ L \neq \emptyset\}$;
3	$W := \bigcup \mathcal{L}$;
4	find $T \in \mathcal{T}$ such that $T \setminus W \neq \emptyset$ and $T \cap W \in \mathcal{L} \cup \{\emptyset\}$;
5	$R := T \setminus W$;
6	$M := N \setminus R$;
7	$\tilde{w} := u - \delta_N + \delta_M + \delta_T - \delta_{T \setminus R}$;
8	$w :=$ the restriction of $\tilde{w}$ to the class of subsets of M;
9	return T, M, w;

Table 6. The first stage **Decompose** (u over $N \,|\, \tau$ over N).

Input:	u ... a standard imset over a non-empty set of variables N
Output:	τ ... an ordered sequence of subsets of N
1	$Y := N$;
2	$\tau :=$ empty list;
3	if $u = 0$ then
4	append Y as the last item in τ;
5	return τ;
6	if $u \neq 0 = u(Y)$ then
7	**Adapt** (u over $Y \,|\, M, w$ over M);
8	append Y as the last item in τ;
9	go to 14;
10	if $u(Y) \neq 0$ then
11	**Reduce** (u over $Y \,|\, T, M, w$ over M);
12	append T as the last item in τ;
13	go to 14;
14	$Y := M$;
15	$u := w$;
16	go to 3;
17	exit;

The decomposition procedure is illustrated by Example 23.

Example 23. The ordered sequence of subsets that is the outcome of the first stage for the imset given in Table 3 is

$$\{a,b,c\}, \{d,f,g\}, \{a,b,d\}, \{a,e\}, \{b\}, \{e,f\}. \tag{5}$$

Note that, in this case, only the subroutine **Reduce** was applied.

The basis of the dual procedure is the extension subroutine, described in Table 7. It constructs the EG H over N on basis of its induced subgraph G for $M \subset N$ and a set $T \subseteq N$ with $M \cup T = N$. Of course, the set is assumed to have above-mentioned special form $T = K \cup pa_H(C)$. The crucial step to fully reconstruct H on basis of G and T is to decide which of two following cases occurs: either $T \cap M = pa_H(C)$ or $T \cap M = Z \cup pa_H(C)$, where $Z \in \mathcal{S}(C)$. The condition in line 6 of Table 7 is a necessary and sufficient condition for the second case.

Now, the second stage of the algorithm, the *composition procedure*, which consists of repeated application of the subroutine **Extend**. It is described in Table 8. Its input is the ordered sequence τ of sets obtained from the decomposition procedure. However, the sequence τ is processed in the reverse order. The procedure is illustrated by Example 24.

Example 24. The output of the second stage of the algorithm applied to the ordered sequence (5) of subsets of N is the EG in Figure 5.

Table 7. Subroutine **Extend** (G over M, $T \mid H$ over $M \cup T$).

Input:	G ... a chain graph over a non-empty set of variables M
	T ... a set of variables with $T \setminus M \neq \emptyset$
Output:	H ... a chain graph over $M \cup T$

1	$L := T \cap M$;
2	$R := T \setminus M$;
3	$Z := \emptyset$;
4	if $L \neq \emptyset$ then
5	choose a terminal component X in G_L;
6	if X is a complete component in G_L and $pa_G(X) = L \setminus X$
7	then put $Z := X$;
8	determine the edges in H as follows:
9	$H_M := G$;
10	$\forall\, x \in L \setminus Z, \forall\, z \in R$ include $x \to z$ in H;
11	$\forall\, y \in R \cup Z, z \in R, y \neq z$ include $y \text{---} z$ in H;
12	return H;

8 Construction of a Hierarchical Junction Tree

The sequence of sets that is the outcome of the first stage (Table 6) can also be used to construct a hierarchical junction tree similar to those introduced in [9].

Each set in the sequence defines a node of the hierarchical junction tree whose entering edges could be labeled by sets Z or L obtained during the second stage of the reconstruction algorithm (see Table 8). More specifically, each node T may or may not be ascribed an entering edge and the edge can be labeled either by a separator Z (if $Z \neq \emptyset$) or by a parent set L (if $Z = \emptyset$ and $L \equiv L \setminus Z \neq \emptyset$). These units can be used then to compose the whole

Table 8. The second stage **Compose** (τ over $N \mid H$ over N).

Input:	τ ... an ordered sequence $T_1, \ldots, T_n$, $n \geq 1$ of subsets of N
Output:	H ... a chain graph over M

1	$M := T_n$;
2	$H :=$ the complete undirected graph over M;
3	$G := H$
4	for $j = n-1, \ldots, 1$ do
5	**Extend** (G over M, $T_j \mid H$ over $M \cup T_j$)
6	$M := M \cup T_j$;
7	$G := H$;
8	return H;

hierarchical junction tree. Due to the lack of space we omit details of the construction and give two examples instead.

Example 25. The nodes of a hierarchical junction tree constructed from the sequence (5) are given in Figure 6, including their attributed separators and parent sets. The resulting hierarchical junction tree is given in Figure 7.

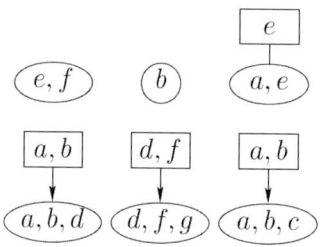

Fig. 6. Units of a hierarchical junction tree.

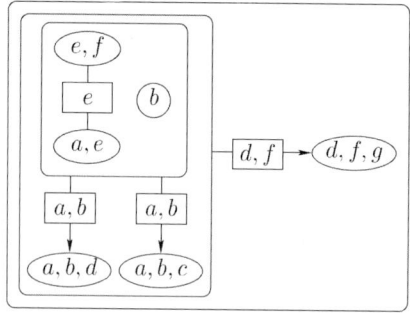

Fig. 7. A hierarchical junction tree.

Example 26. Figure 8 gives another example of an EG H. The first stage of the reconstruction algorithm applied to the standard imset u_H ends with the sequence of sets

$$\{a, b, c, d\}, \ \{a, b, d, e\}, \ \{a\}, \ \{b\}.$$

The nodes of the respective hierarchical junction tree are given in Figure 9, including their attributed separators and parent sets. The resulting hierarchical junction tree is given in Figure 10. In that picture, the parent set $pa_H(\{c, d, e\}) = \{a, b\}$ is attributed to just one node of the hierarchical junction tree. One can perhaps draw a picture in which every parent set is ascribed to every node of the respective component of the hierarchical junction tree.

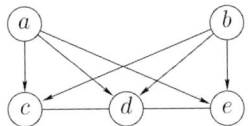

Fig. 8. An EG.

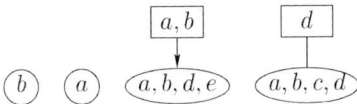

Fig. 9. Units of a hierarchical junction tree.

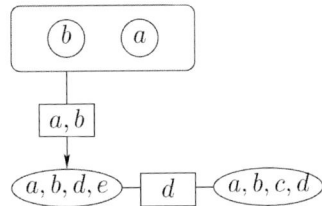

Fig. 10. A hierarchical junction tree.

Conclusions

The presented procedures for the transition between graphical and algebraic representatives of a CI model generated by DAG. These can be the first step on the way towards a fully algebraic method for learning structure of Bayesian networks. We hope that the procedures can be utilized to find a characterization of the inclusion neighborhood of a given DAG in terms of the standard imset. This will be a topic of a future research. We also plan to study the polytope generated by standard imsets over N hoping that linear programming maximization methods can be applied in learning Bayesian networks.

Acknowledgments

This research has been supported by the grant GAČR nr. 201/04/0393.

References

[1] Andersson SA, Madigan D, and Perlman MD. A characterization of Markov equivalence classes for acyclic digraphs. *Annals of Statistics*, 25:pp. 505–541 (1997).

[2] Andersson SA, Madigan D, and Perlman MD. On the Markov equivalence of chain graphs, undirected graphs and acyclic digraphs. *Scandinavian Journal of Statistics*, 24:pp. 81–102 (1997).
[3] Chickering DM. Optimal structure identification with greedy search. *Journal of Machine Learning Research*, 3:pp. 507–554 (2002).
[4] Jensen FV. *Bayesian Networks and Decision Graphs*. Springer Verlag (2001).
[5] Kočka T and Castelo R. Improved learning of Bayesian networks. In J Breese and D Koller, eds., *Uncertainty in Artificial Intelligence 17*, pp. 269–276. Morgan Kaufmann (2001).
[6] Lauritzen SL. *Graphical Models*. Clarendon Press (1996).
[7] Lauritzen SL and Wermuth N. Graphical models for associations between variables some of which are qulitative and some quantitative. *Annals of Statistics*, 17:pp. 31–57 (1989).
[8] Pearl J. *Probabilistic Reasoning in Intelligent Systems - Networks of Plausible Inference*. Morgan Kaufmann (1988).
[9] Puch RO, Smith JQ, and Bielza C. Hierarchical junction trees: conditional independence preservation and forecasting in dynamic Bayesian networks with heterogeneous evolution. In JA Gámez, S Moral, and A Salmerón, eds., *Advances in Bayesian Networks*, pp. 57–75. Springer-Verlag (2004).
[10] R Development Core Team. *R: A language and environment for statistical computing*. R Foundation for Statistical Computing, Vienna, Austria (2004). ISBN 3-900051-00-3.
[11] Studený M. Characterization of essential graphs by means of the operation of legal merging of components. *International Journal of Uncertainty, Fuzziness and Knowledge-Based Systems*, 12:pp. 43–62 (2004).
[12] Studený M. Characterization of inclusion neighbourhood in terms of the essential graph. *International Journal of Approximate Reasoning*, 38:pp. 283–309 (2005).
[13] Studený M. *On Probabilistic Conditional Independence Structures*. Springer-Verlag (2005).
[14] Studený M, Roverato A, and Štěpánová Š. Two operations of merging components in a chain graph. *Scandinavian Journal of Statistics* (2005). Submitted.
[15] Studený M and Vomlel J. A reconstruction algorithm for the essential graph. Tech. rep., ÚTIA AV ČR (2005). http://staff.utia.cz/vomlel/sv.ps.
[16] Verma T and Pearl J. Equivalence and synthesis of causal models. In PP Bonissone, M Henrion, LN Kanal, and JF Lemmer, eds., *Uncertainty in Artificial Intelligence 6*, pp. 220–227. Elsevier (1991).

Bayesian Network Models with Discrete and Continuous Variables

Barry R. Cobb[1], Rafael Rumí[2], and Antonio Salmerón[2]

[1] Department of Economics and Business,
 Virginia Military Institute,
 Lexington, VA 24450, USA
[2] Department of Statistics and Applied Mathematics
 University of Almería,
 Ctra. Sacramento s/n, E-04120 Almería, Spain

Summary. Bayesian networks are powerful tools for handling problems which are specified through a multivariate probability distribution. A broad background of theory and methods have been developed for the case in which all the variables are discrete. However, situations in which continuous and discrete variables coexist in the same problem are common in practice. In such cases, usually the continuous variables are discretized and therefore all the existing methods for discrete variables can be applied, but the price to pay is that the obtained model is just an approximation. In this chapter we study two frameworks where continuous and discrete variables can be handled simultaneously without using discretization. These models are based on the CG and MTE distributions.

1 Introduction

Bayesian networks provide a framework for efficiently dealing with multivariate models. One important feature of these networks is that they allow probabilistic inference which takes advantage of the independence relationships among the variables. Probabilistic inference, commonly known as *probability propagation*, consists of obtaining the marginal distribution on some variables of interest given that the values of some other variables are known.

Much attention has been paid to probability propagation in networks where the variables are qualitative. Several exact methods have been proposed in the literature for this task [11; 15; 21; 36], all of them based on *local computation*. Local computation means to calculate the marginals without actually computing the joint distribution, and is described in terms of a message passing scheme over a structure called *join tree*.

In *hybrid Bayesian networks*, where both discrete and continuous variables appear simultaneously, it is possible to apply local computation schemes similar to those for discrete variables. However, the correctness of exact inference depends on the model.

The first model that allowed exact inference through local computation in hybrid networks was based on the conditional Gaussian (CG) distribution [16; 17; 18; 28]. However, networks where discrete variables have continuous parents are not allowed. To avoid this restriction, Koller et al. [12] model the distribution of discrete nodes with continuous parents by a mixture of exponentials, but then inference is carried out by means of Monte Carlo methods, and therefore the results are approximate. In a more general setting, one way of using local computation is to discretize the continuous variables [13], and then treat them as if they were quantitative.

Recently, the so-called MTE (Mixture of Truncated Exponentials) model [22] has been introduced as a valid alternative for tackling problems where discrete and continuous variables appear simultaneously. The advantage with respect to CG is that discrete nodes with continuous parents are allowed and inference can be carried out by means of local computations [3; 32]. Another important feature of the MTE model is that it can approximate several known distributions accurately [2; 23; 33].

In this chapter we review the state-of-the-art in CG and MTE models. Section 2 describes the CG model, including issues about inference and learning from data. Section 3 is devoted to the MTE model. The application of the MTE model to a practical situation in which discrete and continuous variables appear is illustrated through an example in section 4. The chapter ends with the conclusions in section 5.

2 Conditional Gaussian (CG) Distributions

Bayesian networks where the variables are continuous are difficult to manage in a general way. The usual treatment is to discretize the variables and then use the known methods for discrete variables. This is an approximate approach since we introduce some error in the discretization procedure.

There are some kinds of continuous variables in which the computations can be done in an exact way, one of which is the Multivariate Gaussian distribution [1]. In this chapter we will study a generalization of that model, the *Conditional Gaussian (CG) distribution* [4; 17], in which discrete and continuous variables appear simultaneously.

In Section 2.1 the model is defined, specifying the probability distributions required in the model. Section 2.2 explains the probability propagation process, only affordable under some restrictions. These restrictions have been, up to some degree, overcome according to different criteria [18; 20; 12]. In Section 2.3 the problem of learning a CG network from data is studied.

2.1 Model Definition

Since it is a mixed model, there will be discrete and continuous variables. So, the set of variables $\mathbf{X}$ will be divided in two, $\mathbf{X} = \mathbf{Y} \cup \mathbf{Z}$, where $\mathbf{Y}$ represent the discrete variables and $\mathbf{Z}$ the continuous, and $|\mathbf{Y}| = d$ and $|\mathbf{Z}| = c$ define the number of continuous and discrete variables in the model respectively. So, an element in the joint state space is denoted by:

$$\mathbf{x} = (\mathbf{y}, \mathbf{z}) = (y_1, \ldots, y_d, z_1, \ldots, z_c) \,,$$

where y_i for $i = 1, \ldots, d$ are qualitative data and z_j for $j = 1, \ldots, c$ are real numbers.

In the CG model, the distribution of the continuous variables given the discrete ones is a Multivariate Gaussian distribution:

$$\mathbf{Z}|\mathbf{Y} = \mathbf{y} \to \mathcal{N}(\xi(\mathbf{y}), \Sigma(\mathbf{y})) \tag{1}$$

with Σ being positive definite, in the case that $p(\mathbf{Y} = \mathbf{y}) > 0$.

In general, the joint distribution of the variables in the network can be defined as follows:

Definition 1. *(CG Distribution) A mixed random variable $\mathbf{X} = (\mathbf{Y}, \mathbf{Z})$ is said to have a **conditional Gaussian** distribution if the joint distribution of the model variables (discrete and continuous) has a density*

$$f(\mathbf{x}) = f(\mathbf{y}, \mathbf{z}) = \chi(\mathbf{y}) \exp\{g(\mathbf{y}) + h(\mathbf{y})^T \mathbf{z} - \mathbf{z}^T K(\mathbf{y}) \mathbf{z}/2\} \,, \tag{2}$$

where $\chi(\mathbf{y}) \in \{0,1\}$ shows if f is positive on $\mathbf{y}$, g is a real function, h a function returning a vector of size c, and K a function returning a matrix $c \times c$.

The vector (g, h, K) is known as the *canonical characteristics* of the distribution, only well defined if $\xi(\mathbf{y}) > 0$. We can also refer to the distribution in terms of (p, ξ, Σ) which are the *moment characteristics*.

There are expressions that can be used to switch from one set of characteristics to the other [17]:

- From canonical to moment characteristics:

$$\xi(\mathbf{y}) = K(\mathbf{y})^{-1} h(\mathbf{y}) \,, \quad \Sigma(\mathbf{y}) = K(\mathbf{y})^{-1} \,,$$
$$p(\mathbf{y}) \propto \{det \Sigma(\mathbf{y})\}^{\frac{1}{2}} \exp\{g(\mathbf{y}) + h(\mathbf{y})^T \Sigma(\mathbf{y}) h(\mathbf{y})/2\} \,.$$

- From moment to canonical characteristics:

$$K(\mathbf{y}) = \Sigma(\mathbf{y})^{-1} \,,$$
$$h(\mathbf{y}) = K(\mathbf{y}) \xi(\mathbf{y}) \,,$$
$$g(\mathbf{y}) = \log p(\mathbf{y}) + \{\log det K(\mathbf{y}) - c \log 2\pi - \xi(\mathbf{y})^T K(\mathbf{y}) \xi(\mathbf{y})\}/2 \,.$$

The CG distribution can be easily extended to a *CG potential*, which is

$$\phi(\mathbf{x}) = \phi(\mathbf{y},\mathbf{z}) = \chi(\mathbf{y})\exp\{g(\mathbf{y}) + h(\mathbf{y})^T\mathbf{z} - \mathbf{z}^T K(\mathbf{y})\mathbf{z}/2\} ,$$

where $K(i)$ is only assumed to be a symmetric matrix.

2.2 Inference with CLG Model

The first probability propagation procedure developed [17] computes exact means and variances for continuous variables and exact probabilities for discrete variables. It is based on a local computation scheme over a *junction tree* [1].

The operations over CG potentials are defined in terms of the *canonical characteristics*, but only restriction and combination are closed; the outcome of the marginalization operation may not be a CG potential, but a *mixture of CG potentials*. Due to this problem, some restrictions have to be imposed to the network topology in order to make sure that the propagation can be carried out [17].

Marked graphs

A *marked graph* is a graph, $\mathcal{G} = (V, E)$ where V is a finite set of vertices, and E a set of edges. The vertices are *marked*, in the sense that they are divided in two sets, Δ for the discrete variables and Γ for the continuous variables.

Definition 2. *(Strong decomposition) A triple (A, B, C) of disjoint subsets of the vertex set V of an undirected, marked graph $\mathcal{G}$ is said to form a strong decomposition of $\mathcal{G}$ if $V = A \cup B \cup C$ and the following three conditions hold:*

1. *C separates A from B.*
2. *C is complete subset of V.*
3. *$C \subseteq \Delta$ or $B \subseteq \Gamma$.*

When this happens it is said that (A, B, C) decomposes $\mathcal{G}$ into the components $\mathcal{G}_{A \cup C}$ and $\mathcal{G}_{B \cup C}$.

A decomposable graph is one that can be successively decomposed into cliques.

Definition 3. *(Decomposable graph) An undirected, marked graph is said to be decomposable if it is complete, or if there exists a decomposition (A, B, C), with A, B, and C non-empty, into decomposable subgraphs $\mathcal{G}_{A \cup C}$ and $\mathcal{G}_{B \cup C}$.*

To get a decomposed graph, the marked graph must not have certain types of paths [19]:

Proposition 1. *An undirected marked graph is decomposable if and only if it is triangulated and does not contain any path between two discrete vertices passing through only continuous vertices, with the discrete vertices not being neighbors.*

A decomposable graph is obtained as follows. First, the moral graph is obtained, removing directions of the edges, and adding new edges between parents with common children. To get from this a decomposable graph, new edges created to to avoid paths shown in Proposition 1 need to be added. This can be done by a variable elimination procedure, in which continuous variables must be removed first [4].

Now it is possible to get a junction tree fulfilling the *runing intersection property* [1], but to make the propagation algorithm work properly, it must have a *strong root*:

Definition 4. *(Strong root) A clique R on a junction tree is a strong root if any pair A, B of neighbors of the tree, with A closer to R than B satisfies*

$$(B \setminus A) \subseteq \Gamma \quad or \quad (B \cap A) \subseteq \Delta. \tag{3}$$

Leimer in [19] shows that the cliques in a decomposable graph can always be arranged so that at least one of them is a strong root.

First and second moment exact inference

The original graph specifies the (in)dependence relations among the variables, and represents a factorization of the joint probability distribution of the variables in the network as the combination of the conditional distribution of each variable given its parents in the graph. But the CG model imposes the restriction that discrete nodes cannot have continuous parents. This restriction has been partially overcome in [12; 20; 26] where the *augmented CLG networks* were introduced, in which discrete nodes are allowed to have continuous parents, and the corresponding conditional distribution is modelled using *softmax* functions. The posterior distributions are computed by means of *Monte carlo* methods in [12] and approximating the combination of a Gaussian and a softmax function by a Gaussian distribution in [20; 26]. These new proposals return approximate posterior distributions, except for [20], where the first and second moment of the distributions are exact, like in the algorithm explained here.

Once checked that the network satisfies this restriction, the first thing to do is to specify the corresponding conditional distributions. For each discrete variable a probability table will be specified, since its parents will always be discrete. For each continuous variable, Z_i, the conditional distribution given its parents is

$$Z_i | \pi_i \to \mathcal{N}(\alpha(\mathbf{y}) + \beta(\mathbf{y})^T \mathbf{z}, \gamma(\mathbf{y})), \tag{4}$$

where $(\mathbf{y}, \mathbf{z})$ are the states of the parents, $\mathbf{y}$ the states of the discrete parents, $\mathbf{z}$ the states of the continuous parents, $\gamma(\mathbf{y}) > 0$, $\alpha(\mathbf{y})$ a real number, and $\beta(\mathbf{y})$ a vector of the same size as the continuous part of the parents. This assumes that the mean of a potential depends linearly on the continuous parent variables and that the variance does not depend on the continuous parent variables.

For each configuration of the discrete parents, a different linear function of the continuous parents is specified as the mean of the normal distribution, and a different real number is specified as the variance.

Up to this point, the rest of the probability propagation algorithm [17] is similar to the algorithms based on clustering [1]. They perform a message passing scheme between the cliques of the junction tree. This algorithm obtains correct mean and variance for every continuous variable, and correct probabilities for every discrete variable, whenever a strong root of the junction tree is selected as the root in the propagation algorithm.

A modification of this algorithm was presented in [18], where the numerical instability of the algorithm was improved by introducing some restrictions in the junction tree. Also, this new algorithm is capable of managing deterministic conditional probability distributions, i.e., distributions with null variance.

2.3 Learning a CG Model

Learning a CG model from a database D requires obtaining:

a) A structure representing the (in)dependece relations of the variables.
b) A conditional distribution for each variable given its parents.

Most of the learning algorithms search for a candidate network in the space of all possible networks. Once a network in this space has been selected, the corresponding parameters of the conditional distributions are first estimated, then the network is evaluated in terms of quality in the representation of D.

Estimating the conditional distributions for a candidate network

Given a candidate network, a conditional distribution for each variable is estimated. In the case of discrete variables, the distribution to learn is a probability table, since its parents have to be discrete as well (see Section 2.2). This probability table can be estimated using the *Maximum Likelihood Estimator* (MLE), ratio of appearance of the case, or using the Laplace correction, in the case of a small sample.

In the case of continuous variables, the distribution to learn is a different Gaussian distribution for each configuration of the discrete parents. First, the sufficient statistic of the joint distribution is computed, and afterwards the MLE of the parameters of the conditional distribution are obtained. Since the mean of the distribution is a linear combination of the parents, it can be obtained from a linear regression model, as explained in [25]. The *EM algorithm* can be applied in the case of an incomplete database.

Measuring the quality of a candidate network

Once the parameters of the candidate network are estimated, the quality of the network is scored, in terms of its posterior probability given a database.

Geiger and Heckerman [8] present a metric capable of scoring a CG network, under some restrictions in the topology of the network.

3 Mixtures of Truncated Exponentials (MTEs)

The CG model is useful in situations in which it is known that the joint distribution of the continuous variables for each configuration of the discrete ones follows a multivariate Gaussian. However, in practical applications it is possible to find scenarios where this hypothesis is violated, in which case another model, like discretization, should be used. Since discretization is equivalent to approximating a target density by a mixture of uniforms, the accuracy of the final model could be increased if, instead of uniforms, other distributions with higher fitting power were used.

This is the idea behind the so-called *mixture of truncated exponentials (MTE) model* [22]. The MTE model is formally defined as follows:

Definition 5. *(MTE potential) Let* $\mathbf{X}$ *be a mixed n-dimensional random vector. Let* $\mathbf{Y} = (Y_1, \ldots, Y_d)$ *and* $\mathbf{Z} = (Z_1, \ldots, Z_c)$ *be the discrete and continuous parts of* $\mathbf{X}$*, respectively, with* $c + d = n$. *We say that a function* $f : \Omega_{\mathbf{X}} \mapsto \mathbb{R}_0^+$ *is a Mixture of Truncated Exponentials potential (MTE potential) if one of the next conditions holds:*

i. $\mathbf{Y} = \emptyset$ *and* f *can be written as*

$$f(\mathbf{x}) = f(\mathbf{z}) = a_0 + \sum_{i=1}^{m} a_i \exp\left\{\sum_{j=1}^{c} b_i^{(j)} z_j\right\} \quad (5)$$

for all $\mathbf{z} \in \Omega_{\mathbf{Z}}$, *where* a_i, $i = 0, \ldots, m$ *and* $b_i^{(j)}$, $i = 1, \ldots, m$, $j = 1, \ldots, c$ *are real numbers.*

ii. $\mathbf{Y} = \emptyset$ *and there is a partition* $D_1, \ldots, D_k$ *of* $\Omega_{\mathbf{Z}}$ *into hypercubes such that* f *is defined as*

$$f(\mathbf{x}) = f(\mathbf{z}) = f_i(\mathbf{z}) \quad \text{if} \quad \mathbf{z} \in D_i ,$$

where each f_i, $i = 1, \ldots, k$ *can be written in the form of equation (5).*

iii. $\mathbf{Y} \neq \emptyset$ *and for each fixed value* $\mathbf{y} \in \Omega_{\mathbf{Y}}$, $f_{\mathbf{y}}(\mathbf{z}) = f(\mathbf{y}, \mathbf{z})$ *can be defined as in ii.*

Definition 6. *(MTE density) An MTE potential* f *is an MTE density if*

$$\sum_{\mathbf{y} \in \Omega_{\mathbf{Y}}} \int_{\Omega_{\mathbf{Z}}} f(\mathbf{y}, \mathbf{z}) d\mathbf{z} = 1 , \quad (6)$$

where $\mathbf{Y}$ *and* $\mathbf{Z}$ *are the discrete and continuous coordinates of* $\mathbf{X}$ *respectively.*

In a Bayesian network, two types of probability density functions can be found:

1. For each variable X which is a root of the network, a density $f(x)$ is given.
2. For each variable X with parents $\mathbf{Y}$, a conditional density $f(x|\mathbf{y})$ is given.

A *conditional MTE density* $f(x|\mathbf{y})$ is an MTE potential $f(x,\mathbf{y})$ such that after fixing $\mathbf{y}$ to each of its possible values, the resulting function is a density for X.

In [22] a data structure was proposed to represent MTE potentials, called *mixed probability trees* or mixed trees for short. Mixed trees can represent MTE potentials defined by parts. Each entire branch in the tree determines one sub-region of the space where the potential is defined, and the function stored in the leaf of a branch is the definition of the potential in the corresponding sub-region.

Example 1. Consider the following MTE potential, defined for a discrete variable (Y_1) and two continuous variables (Z_1 and Z_2).

$$\phi(y_1, z_1, z_2) = \begin{cases} 2 + e^{3z_1+z_2} & \text{if } y_1 = 0,\ 0 < z_1 \leq 1,\ 0 < z_2 < 2 \\ 1 + e^{z_1+z_2} & \text{if } y_1 = 0,\ 0 < z_1 \leq 1,\ 2 \leq z_2 < 3 \\ \frac{1}{4} + e^{2z_1+z_2} & \text{if } y_1 = 0,\ 1 < z_1 < 2,\ 0 < z_2 < 2 \\ \frac{1}{2} + 5e^{z_1+2z_2} & \text{if } y_1 = 0,\ 1 < z_1 < 2,\ 2 \leq z_2 < 3 \\ 1 + 2e^{2z_1+z_2} & \text{if } y_1 = 1,\ 0 < z_1 \leq 1,\ 0 < z_2 < 2 \\ 1 + 2e^{z_1+z_2} & \text{if } y_1 = 1,\ 0 < z_1 \leq 1,\ 2 \leq z_2 < 3 \\ \frac{1}{3} + e^{z_1+z_2} & \text{if } y_1 = 1,\ 1 < z_1 < 2,\ 0 < z_2 < 2 \\ \frac{1}{2} + e^{z_1-z_2} & \text{if } y_1 = 1,\ 1 < z_1 < 2,\ 2 \leq z_2 < 3 \end{cases}$$

A possible representation of this potential by means of a mixed probability tree is displayed in Figure 1.

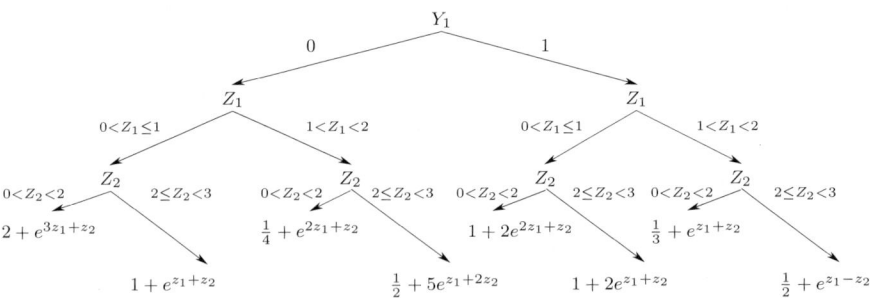

Fig. 1. An example of mixed probability tree.

3.1 Inference with MTEs

The operations required for probability propagation in Bayesian networks (restriction, marginalization and combination) can be carried out by means of algorithms very similar to those described for discrete probability trees in [13; 34]. Furthermore, it was shown in [22] that the class of MTE potentials is closed for the operations above mentioned. Hence, any exact propagation algorithm that is based in these three operations can be applied to networks with MTEs. The use of the Shenoy-Shafer and the Penniless propagation algorithms with MTEs is studied in [32].

3.2 Learning Bayesian Networks with MTEs

Consider a mixed random vector $\mathbf{X} = \{X_1, \ldots, X_n\}$, and a sample of $\mathbf{X}$,

$$D = \{\mathbf{x}^{(1)}, \ldots, \mathbf{x}^{(m)}\} \ .$$

We will describe how to obtain a Bayesian network with variables $\mathbf{X}$, that agrees with the data D, following the method proposed in [31]. Basically, the problem of learning Bayesian networks from data can be approached as repeating the next three steps until an optimal network is obtained:

1. Selection of a candidate structure G.
2. Estimation of the conditional distributions, $\hat{\theta}$, for G.
3. Determination of the quality of $(G, \hat{\theta})$.

The method proposed in [31] consists of exploring the space of possible networks for variables $\mathbf{X}$ using an optimization approach. The starting point is a network without arcs, and considering the movement operations of arc insertion, deletion and reversal. After each movement, the conditional distributions corresponding to the families involved in the change are estimated.

Estimating the conditional distributions for a candidate network

The problem of estimating the parameters of truncated distributions has been previously studied [38; 39], as in the case of the truncated Gamma [10; 27], but the number of parameters is usually equal to one, and the maximum likelihood estimator (that not always exists) or the UMVUE (Uniformly of Minimum Variance Unbiased Estimator) is obtained by means of numerical methods [6; 35]. In the case of the MTE model, no similar techniques have been applied so far, due to the high number of parameters involved in the MTE densities.

Another usual way to compute maximum likelihood estimates in mixture models is the *EM algorithm* [5; 30]. The difficulty in applying this method to learning MTE models lies in the fact that we may have negative coefficients for some of the densities we are combining and also in the computation of the conditional expectations in each iteration of the algorithm.

Due to the difficulties described above, the seminal paper on estimating MTEs from data [23], followed an approach based on regression techniques for the case of univariate densities. Besides the estimation of the parameters, the construction of an MTE density involves the determination of the number of terms and the splits into which its domain is partitioned. Heuristics to approach these issues are proposed in [23]. The estimation procedure can be summarized in the next algorithm.

Algorithm MTE-fitting
INPUT:

- A sample $x_1, \ldots, x_n$.

OUTPUT:

- Estimates of the parameters of the fitted model, $\hat{a}, \hat{b}, \hat{c}, \hat{d}$ and $\hat{k}$.

1. Using sample $x_1, \ldots, x_n$, obtain two vectors $(x_1^*, \ldots, x_n^*)$ and $(y_1, \ldots, y_n)$, where the first one contains values of the variable, and the second one contains their corresponding empirical density values.
2. Divide the range of the variable into subintervals in terms of concavity/convexity and increase/decrease of the curve determined by the points in vectors $(x_1^*, \ldots, x_n^*)$ and $(y_1, \ldots, y_n)$.
3. For each subinterval do:
    - Fit $y = f(x) = k + a \exp\{bx\} + c \exp\{dx\}$, using an iterative least squares procedure.
4. Normalize the whole function to integrate up to one.
5. Let $\hat{a}, \hat{b}, \hat{c}, \hat{d}$ and $\hat{k}$ be the coefficients of the normalized function.
6. RETURN$(\hat{a}, \hat{b}, \hat{c}, \hat{d}, \hat{k})$.

Although the core of this algorithm is Step 3, the results strongly depend on Step 1, in two ways:

1. *The accuracy of the estimation of the empirical density using the given sample.* If the estimation is poor, the result can be a density far away from the original one.
2. *The size of the vectors obtained.* Even if the empirical density is properly captured, if the exponential regression in Step 3 is computed from a scarce set of points, the accuracy of the approximation can be poor.

The method described in [23] used the empirical histogram as an approximation of the actual density of the sample points. In [31], with the aim of avoiding the two problems mentioned above, the estimation procedure is improved using kernel approximations to the empirical density.

First of all, a Gaussian kernel density [37] is fitted to the data corresponding to each leaf. The Gaussian kernel density provides a smooth approximation to the empirical density of the data, which is specially useful in situations in which the amount of data is scarce, since peaks in the density owing to the lack of data are not as rough as in the case of using histograms.

Out of the fitted Gaussian kernel density, $f(x)$, an artificial sample consisting of pairs $(x_1, f(x_1)), \ldots, (x_h, f(x_h))$, is drawn by taking equidistant points, and an MTE density is obtained from it using the regression-based algorithm described in [23; 33].

The performance of Gaussian kernels versus histograms or other types of kernels for fitting MTE densities is evaluated in Figures 2 and 3, which respectively show the MTE density obtained through histogram and kernel approximations out of a sample of 200 points randomly sampled from a standard normal distribution.

The method described so far for constructing estimators for the parameters of the univariate MTE density is not valid for the conditional case, since more restrictions should be imposed over the parameters in order to force the MTE potential to integrate up to 1 for each combination of values of the conditioning variables, i.e. to force the MTE potential to actually be a conditional density. This problem was approached in [24] by partitioning the domain of the conditioning variables and then fitting a univariate density in each one of the splits using the method described above. More precisely,

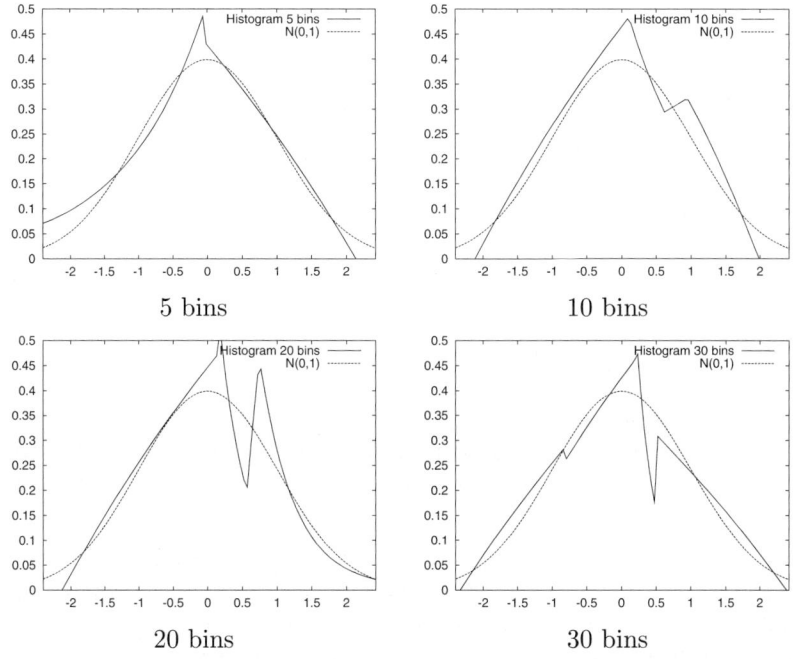

Fig. 2. Results of fitting an MTE from 5, 10, 20 and 30 bin histograms.

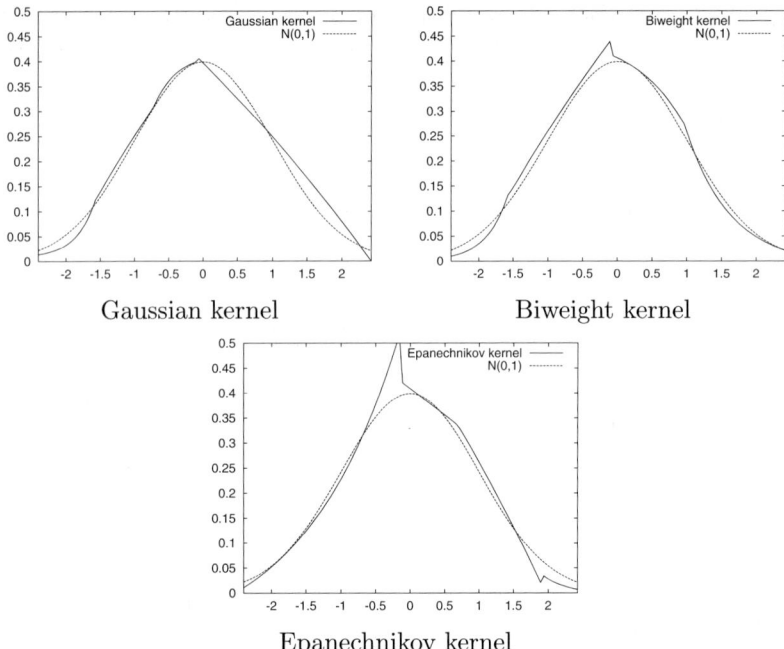

Fig. 3. Results of fitting an MTE from Gaussian, Biweight and Epanechnikov kernels.

the algorithm learns a mixed tree in which the leaves contain MTE densities that depend only on the child variable, and that represents the density of the child variable given by the values contained in the region determined by the corresponding branch of the mixed tree. The tree is learnt in such a way that the leaves discriminate as much as possible, following a schema similar to the construction of decision trees [29].

Measuring the quality of a candidate network

The metric used in [31] to measure the quality of a candidate networks is defined as:

$$Q(G|D,\hat{\theta}) = \log L(D;G,\hat{\theta}) - \frac{\log m}{2} \mathrm{Dim}(G), \qquad (7)$$

where $L(D;G,\hat{\theta})$ is the likelihood of the data given the current network and $\mathrm{Dim}(G)$ is the number of parameters needed to specify the network G that must be learnt from the data.

The number of parameters that are learnt from data for the conditional density of any variable given its parents in the network, is the result of adding the number of parameters in the leaves of the corresponding mixed tree that will be used to represent the density, and the number of points that determine the partition of the domain of each continuous variable.

Once the metric is defined, it can be used to guide the process of learning the network from data. The problem can be viewed as the optimization of the metric defined in equation (7) within the space of candidate networks. In this direction, hill climbing and simulated annealing algorithms have been used [31].

4 Example

This section presents an example of a Bayesian network with discrete and continuous variables where MTE potentials are used to model continuous distributions.

4.1 Description

The following example is adapted from a problem used by Cobb et al. [2]. A refinery has entered into a forward contract to sell 1,000 barrels of gasoline in five months at a price of $67 per barrel. In four months, the refinery will buy a futures contract for 1,000 barrels of crude oil to be delivered in one month in order to produce the contracted output. The futures price (X) is currently $59 and the price is expected to evolve for the next four months (90 business days) according to a geometric Brownian motion stochastic process if cetain market factors (M) are favorable ($M = 0$). If market factors are unfavorable ($M = 1$), the price is expected to evolve according to a jump diffusion stochastic process. Market factors are expected to be favorable with probability 0.80. The probabilities for the two states of M are denoted by $\eta(M = 0) = P(M = 0)$ and $\eta(M = 1) = P(M = 1)$.

Using 65 daily price observations, the mean (drift) and the standard deviation (volatility) of the both the GBM and jump diffusion process (the mean and standard deviation of the daily returns) are estimated as -0.00172 and 0.02089. The terminal distribution for the GBM process is a lognormal distribution with a mean of 50.465 and a variance of 102.056. The corresponding parameters of the lognormal distribution for the terminal price (X) given a favorable market ($M = 0$) are $\mu = 3.90$ and $\sigma^2 = 0.20$, i.e. $X \mid \{M = 0\} \sim LN(3.90, 0.20)$. A distribution for the terminal price (X) given an unfavorable market ($M = 1$) has no standard form. Using Monte Carlo simulation, a histogram for this distribution is simulated (see Figure 4). The jump diffusion process is simulated by using

$$X_{t+\tau} = X_t \cdot \exp\{(\mu - \sigma^2/2)\tau + \sigma\sqrt{\tau}Z_0 + \sum_{i=1}^{N(\tau)}(\gamma Z_i - \gamma^2/2)\} \ , \quad (8)$$

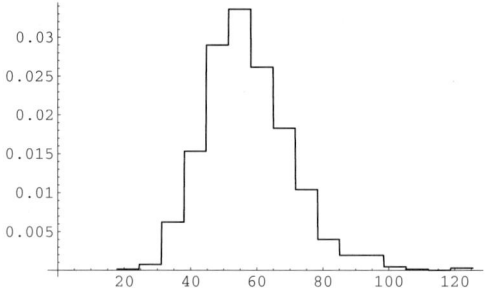

Fig. 4. A simulated histogram for the distribution of oil prices (X) given $M = 1$.

where $Z_i \sim N(0,1)$ i.i.d., $N(\tau) \sim Poisson(\lambda\tau)$, and the jump sizes are i.i.d. lognormally distributed $LN(\alpha_i, \beta_i^2)$, with $\alpha_i = -\gamma^2/2$ and $\beta_i = \gamma$ ($\gamma > 0$); in this example, we assume $\gamma = 0.20$. The Poisson random variable represents the number of jumps within each interval of length τ; in this case, we assume an average of four annual jumps, or $\lambda = 4$ and $\tau = 1/252$ (where there are an average of 252 business days per year).

The company incurs fixed costs per month of \$5,000 and additional variable costs of \$4 per barrel; additionally, the cost has a random component that can be modeled using a normal distribution with a mean of zero and a standard deviation of 8000. Because the random component is normally distributed, the conditional distribution of profits (Y) given a value for oil price is normally distributed with a mean that is a linear function of oil prices, $g(X) = 58000 - 1000X$ with an independent variance, i.e. $Y \mid x \sim N(g(x), 8000^2)$. The Bayesian network for this example is shown in Figure 5.

Fig. 5. The Bayesian network for the example problem.

4.2 Fitting MTE Potentials

Standard PDF's

Cobb et al. [2] describe a method for fitting an MTE potential to approximate a standard PDF with known parameters. Using this method with the lognormal distribution requires dividing the domain of the potential into pieces according to the absolute maximum and inflection points of the function. The absolute maximum, m, is defined where the first derivative of the lognormal PDF equals zero, or $m = \exp\{\mu - \sigma^2\}$. The inflection points, $d^{\pm}$, are defined where the second derivative of the lognormal PDF equals zero, or

$$d^{\pm} = \exp\left\{\frac{1}{2}(2\mu - 3\sigma^2 \pm \sigma\sqrt{4+\sigma^2})\right\}.$$

In this example, the lognormal PDF with $\mu = 3.90$ and $\sigma^2 = 0.20$ has an absolute maximum of $m = 40.4473$ and inflection points of $d^- = 23.1442$ and $d^+ = 57.8734$. The MTE approximation captures the same area under the lognormal PDF as is contained in the normal PDF within three standard deviations of the mean, or 0.9973. Define $x_{min}^0 = 12.9146$ and $x_{max}^0 = 188.9799$. The $LN(3.9, 0.2)$ PDF will have an MTE approximation defined with separate pieces for the intervals $[x_{min}^0, d^-), [d^-, m), [m, d^+),$ and $[d^+, x_{max}^0]$. Within each interval, we use a nonlinear optimization procedure to select parameters for the MTE potential that minimize the Kullback-Leibler (KL) divergence [14] between the PDF and the MTE approximation. For the $LN(3.9, 0.2)$ distribution, this results in the MTE approximation

$\phi_0(x) = P(X \mid \{M = 0\}) =$

$$\begin{cases} -0.005157 + 0.065997 \exp\{0.088333(x - 40.4473)\} \\ +0.101073 \exp\{1.419659(x - 40.4473)\} \\ \qquad\qquad\qquad\qquad\qquad \text{if } 12.9146 \leq x < 23.1442 \\[6pt] 0.110188 - 0.070500 \exp\{-0.016903(x - 40.4473)\} \\ -0.019732 \exp\{0.063467(x - 40.4473)\} \\ \qquad\qquad\qquad\qquad\qquad \text{if } 23.1442 \leq x < 40.4473 \\[6pt] 0.021151 + (1.221457E-14) \exp\{1.399310(x - 40.4473)\} \\ -0.001195 \exp\{0.102630(x - 40.4473)\} \\ \qquad\qquad\qquad\qquad\qquad \text{if } 40.4473 \leq x < 57.8734 \\[6pt] -0.100751 + 0.028812 \exp\{-0.368836(x - 40.4473)\} \\ +0.099982 \exp\{0.000049(x - 40.4473)\} \\ \qquad\qquad\qquad\qquad\qquad \text{if } 57.8734 \leq x \leq 188.9799 \, . \end{cases}$$
(9)

All MTE potentials in this section are equal to zero in unspecified regions. The MTE approximation is shown graphically in Figure 6 overlayed on a graph of the actual $LN(3.9, 0.2)$ distribution.

Estimation from Data

Using the simulated data from the jump diffusion process, the least squares regression estimation procedure [23] described in Section 3.2 is used to fit an MTE approximation (ϕ_1) for the distribution of oil prices (X) given $M = 1$.

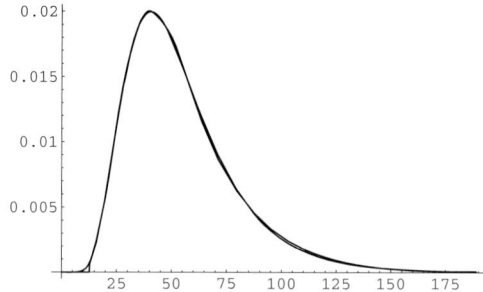

Fig. 6. The MTE approximation to the distribution for X given $M = 0$ overlayed on the actual $LN(3.9, 0.2)$ distribution.

Given a partition of the domain of the simulated data, the procedure is used to estimate the following function for the m-th piece of the MTE approximation (denoted ϕ_{1m}):

$$\phi_{1m} = K_m + a_m \exp\{b_m x\} + c_m \exp\{d_m x\} . \qquad (10)$$

The mid-points of the bins in the empirical density (estimated from a frequency distribution or histogram) in the domain of the m-th piece of the MTE approximation are denoted as $\mathbf{x} = (x_{m1}, x_{m2}, \ldots, x_{mn})$ and the values of the empirical density at the corresponding points are denoted as $\mathbf{y} = (f(x_{m1}), f(x_{m2}), \ldots, f(x_{mn}))$. Since two exponential terms are used in (10), a two-stage exponential regression is used to estimate the parameters for the MTE potential. First, the following model is estimated via least squares regression:

$$\ln\{y\} = \ln a_m \exp\{b_m x\} = \ln\{a_m\} + b_m x . \qquad (11)$$

Next, a similar regression is performed to fit the additional exponential term $c_m \exp\{d_m x\}$ to the differences between the empirical distribution and the values of $a_m \exp\{b_m x\}$. A constant term is then determined as the average of the remaining differences between the MTE approximation and the empirical density.

The MTE potential ϕ_1 for oil prices (X) given $M = 1$ is shown in Figure 7, overlayed on the empirical histogram constructed from the simulated data.

MTE approximation to the Normal PDF

Moral et al. [24] describe a mixed tree structure that can be used to define an MTE potential for a conditional distribution. The domain of the continuous parent variables are divided into hypercubes and an MTE potential is defined within each hypercube for the continuous child. Define $x_{max} = \text{Max}[x^0_{max}, x^1_{max}]$ and $x_{min} = \text{Min}[x^0_{min}, x^1_{min}]$, where x^1_{max} and

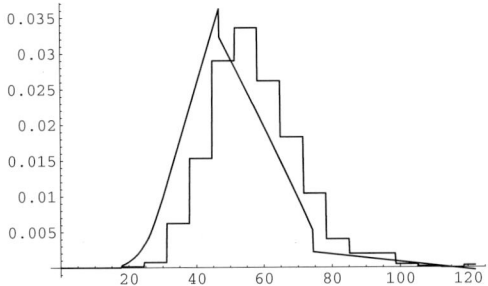

Fig. 7. The MTE potential for X given $M = 1$ fit to the simulated data from the jump diffusion process.

x_{min}^1 are the endpoints of the domain for the MTE potential fragment ϕ_1 for X given $M = 1$. The hypercube for X contains 12 intervals, with the endpoints of the intervals defined as $x_0^{lim}, x_1^{lim}, \ldots, x_{12}^{lim}$, where $x_i^{lim} = x_{min} + i \cdot (x_{max} - x_{min})/12$ for $i = 0, \ldots, 12$. An MTE approximation to the normal PDF is defined by Cobb and Shenoy [3] in terms of the parameters μ and σ^2. For this example, define $\mu_i = 58000 - 1000\left((x_i^{lim} + x_{i-1}^{lim})/2\right)$ for $i = 1, \ldots, 12$ and $\sigma^2 = 8000^2$.

The MTE potential for profits (Y) given oil price (X) is defined as

$$\varphi_i(x,y) = \begin{cases} 8000^{-1}\left(-0.010564 + 197.055720 \exp\{2.2568434(\frac{y-\mu_i}{8000})\}\right. \\ \quad -461.439251 \exp\{2.3434117(\frac{y-\mu_i}{8000})\} \\ \quad +264.793037 \exp\{2.4043270(\frac{y-\mu_i}{8000})\}\right) \\ \qquad \text{if } \left(x_{i-1}^{lim} \le x \le x_i^{lim}\right) \cap (\mu_i - 24000 \le y < \mu_i) \\ 8000^{-1}\left(-0.010564 + 197.055720 \exp\{-2.2568434(\frac{y-\mu_i}{8000})\}\right. \\ \quad -461.439251 \exp\{-2.3434117(\frac{y-\mu_i}{8000})\} \\ \quad +264.793037 \exp\{-2.4043270(\frac{y-\mu_i}{8000})\}\right) \\ \qquad \text{if } \left(x_{i-1}^{lim} \le x \le x_i^{lim}\right) \cap (\mu_i \le y \le \mu_i + 24000) \end{cases}$$
(12)

for $i = 1, \ldots, 12$.

4.3 Solution

Calculating marginal MTE potentials for X and Y requires using the operations of restriction (if evidence is available), combination and marginalization in a message passing scheme over a join tree structure. The Shenoy-Shafer architecture is used because the class of MTE potentials is closed under the operations of restriction, combination and marginalization, and these are the only three operations required in this architecture. Combination of MTE potentials is pointwise multiplication of functions and marginalization is integration over the continuous variables being removed and summation over the discrete variables being removed.

To calculate the marginal distribution for X, the potentials η and ϕ (which consists of the potential *fragments* ϕ_0 and ϕ_1) are combined via pointwise multiplication, then the variable M is removed by summation. This operation is denoted by $\xi = (\eta \otimes \phi)^{\downarrow X}$, with ξ calculated as

$$\xi(x) = \eta(M=0) \cdot \phi_0(x) + \eta(M=1) \cdot \phi_1(x) \ .$$

This results in the MTE potential shown graphically in Figure 8, which has a mean and variance of 53.8757 and 552.6200, respectively.

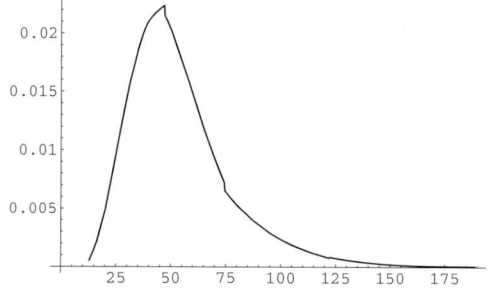

Fig. 8. The marginal distribution for X calculated by removing M from the combination of η and ϕ.

The marginal distribution for Y is calculated by combining the potentials ξ and φ via pointwise multiplication, then removing the variable X by integration. This operation is denoted by $\rho = (\xi \otimes \varphi)^{\downarrow Y}$, with ρ calculated as

$$\rho(y) = \int_{x_{min}}^{x_{max}} \xi(x) \cdot \varphi(x,y) \, dx \ .$$

This results in the MTE potential shown graphically in Figure 9, which has a mean and standard deviation of 4106.5735 and 25111.9326, respectively. The probability that the firm earns a positive profit is calculated as

$$P(Y > 0) = \int_0^\infty \rho(y) \, dy = 0.6737 \ .$$

5 Conclusions

In this chapter we have reviewed two frameworks for handling hybrid Bayesian networks, based on the CG and MTE distributions. In both cases, we have

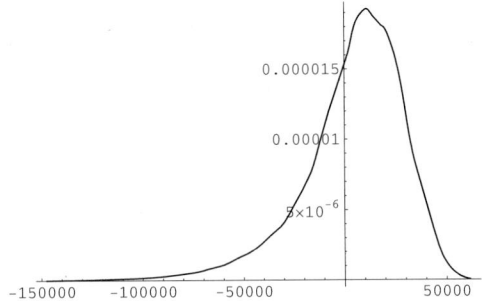

Fig. 9. The marginal distribution for Y calculated by removing X from the combination of ξ and φ.

studied how inference and learning from data can be carried out. Also, the behavior of the MTE model is illustrated through an extensive example.

The CG model relies on a very solid theoretical development and it allows efficient inference, but with the restriction that discrete nodes cannot have continuous parents. On the other hand, the MTE model fits more naturaly to local computation schemes for inference, since it is closed for the basic operations used in inference regardless the structure of the network. Furthermore, it is a good tool for approximating other distributions more accurately than discretization.

References

[1] E. Castillo, J. Gutiérrez, A. Hadi, Expert systems and probabilistic network models, Springer-Verlag, New York, 1997.
[2] B. Cobb, P. Shenoy, R. Rumí, Approximating probability density functions with mixtures of truncated exponentials, in: Proceedings of Information Processing and Management of Uncertainty in Knowledge-Based Systems (IPMU-2004), In press, Perugia, Italy, 2004.
[3] B. Cobb, P. Shenoy, Inference in hybrid Bayesian networks with mixtures of truncated exponentials, International Journal of Approximate Reasoning, 2006, In press.
[4] R. Cowell, A. Dawid, S. Lauritzen, D. Spiegelhalter, Probabilistic Networks and Expert Systems, Series: Statistics for Engineering and Information Science, Springer, 1999.
[5] A. Dempster, N. Laird, D. Rubin, Maximun likelihood from incomplete data via the EM algorithm, Journal of the Royal Statistical Society B 39 (1977) 1–38.
[6] M. El-Taha, W. Evans, A new estimation procedure for the right-truncated exponential distribution, in: Proceedings of the 23th Pittsburgh Conference on Modelling and Simulation, 1992, pp. 427–434.

[7] Elvira Consortium, Elvira: An environment for creating and using probabilistic graphical models, in: J. Gámez, A. Salmerón (Eds.), Proceedings of the First European Workshop on Probabilistic Graphical Models, 2002, pp. 222–230.

[8] D. Geiger, D. Heckerman, Learning Gaussian networks, Proceedings of the 10th Conference on Uncertainty in Artificial Intelligence, Morgan & Kaufmann, 1994, 235–243.

[9] W. Gilks, S. Richardson, D. Spiegelhalter, Markov chain Monte Carlo in practice, Chapman and Hall, London, UK, 1996.

[10] L. Hegde, R. Dahiya, Estimation of the parameters of a truncated Gamma distribution, Communications in Statistics. Theory and Methods 18 (1989) 561–577.

[11] F.V. Jensen, S.L. Lauritzen, K.G. Olesen, Bayesian updating in causal probabilistic networks by local computation, Computational Statistics Quarterly 4 (1990) 269–282.

[12] D. Koller, U. Lerner, D. Anguelov, A general algorithm for approximate inference and its application to hybrid Bayes nets, Proceedings of the 15th Conference on Uncertainty in Artificial Intelligence, Morgan & Kaufmann, 1999, 324–333.

[13] D. Kozlov, D. Koller, Nonuniform dynamic discretization in hybrid networks, in: D. Geiger, P. Shenoy (Eds.), Proceedings of the 13th Conference on Uncertainty in Artificial Intelligence, Morgan & Kaufmann, 1997, pp. 302–313.

[14] S. Kullback, S., R.A. Leibler, On information and sufficiency, Annals of Mathematical Statistics 22 (1951) 79–86.

[15] S.L. Lauritzen, D.J. Spiegelhalter, Local computations with probabilities on graphical structures and their application to expert systems, Journal of the Royal Statistical Society, Series B 50 (1988) 157–224.

[16] S.L. Lauritzen, N. Wermuth, Graphical models for associations between variables, some of which are qualitative and some quantitative, The Annals of Statistics 17 (1989) 31–57.

[17] S. Lauritzen, Propagation of probabilities, means and variances in mixed graphical association models, Journal of the American Statistical Association 87 (1992) 1098–1108.

[18] S. Lauritzen, F. Jensen, Stable local computation with conditional Gaussian distributions, Statistics and Computing 11 (2001) 191–203.

[19] H. Leimer, Triangulated graphs with marked vertices, Annals of Discrete Mathematics 41, 1989, pp. 311–324.

[20] U. Lerner, E. Segal, D. Koller, Exact inference in networks with discrete children of continuous parents, Proceedings of the 17th Conference on Uncertainty in Artificial Intelligence, Morgan & Kaufmann, 2001, pp. 319–328.

[21] A.L. Madsen, F.V. Jensen, Lazy propagation: a junction tree inference algorithm based on lazy evaluation, Artificial Intelligence 113 (1999) 203–245.

[22] S. Moral, R. Rumí, A. Salmerón, Mixtures of truncated exponentials in hybrid Bayesian networks. ECSQARU'01, Lecture Notes in Artificial Intelligence 2143 (2001) 135–143.

[23] S. Moral, R. Rumí, A. Salmerón, Estimating mixtures of truncated exponentials from data, in: J. Gámez, A. Salmerón (Eds.), Proceedings of the First European Workshop on Probabilistic Graphical Models, 2002, pp. 156–167.

[24] S. Moral, R. Rumí, A. Salmerón, Approximating conditional MTE distributions by means of mixed trees. ECSQARU'03, Lecture Notes in Artificial Intelligence 2711 (2003) 173–183.

[25] K. Murphy, Fitting a Conditional Linear Gaussian Distribution, Technical Report, 1998.

[26] K. Murphy, A variational approximation for Bayesian networks with discrete and continuous latent variables, Proceedings of the 15th Conference on Uncertainty in Artificial Intelligence, Morgan & Kaufmann, 1999, 467–475.

[27] G. Nath, Unbiased estimates of reliability for the truncated Gamma distribution, Scandinavian Actuarial Journal 3 (1975) 181–186.

[28] K.G. Olesen, Causal probabilistic networks with both discrete and continuous variables, IEEE Transactions on Pattern Analysis and Machine Intelligence 15 (1993) 275–279.

[29] J. Quinlan, Induction of decision trees, Machine Learning 1 (1986) 81–106.

[30] R. Redner, H. Walker, Mixture densities, maximum likelihood and the EM algorithm, SIAM Review 26 (1984) 195–239.

[31] V. Romero, R. Rumí, A. Salmerón, Learning hybrid Bayesian networks with mixtures of truncated exponentials, International Journal of Approximate Reasoning (2005) In press.

[32] R. Rumí, A. Salmerón, Penniless propagation with mixtures of truncated exponentials. ECSQARU'05. Lecture Notes in Computer Science 3571 (2005) 39–50.

[33] R. Rumí, A. Salmerón, S. Moral, Estimating mixtures of truncated exponentials in hybrid Bayesian network, Test (2005) In press.

[34] A. Salmerón, A. Cano, S. Moral, Importance sampling in Bayesian networks using probability trees, Computational Statistics and Data Analysis 34 (2000) 387–413.

[35] Y. Sathe, S. Varde, Minimum variance unbiased estimates of reliability for the truncated exponential distribution, Technometrics 11 (1969) 609–612.

[36] P.P. Shenoy, G. Shafer, Axioms for probability and belief function propagation, in: R.D. Shachter, T.S. Levitt, J.F. Lemmer and L.N. Kanal (eds.) Uncertainty in Artificial Intelligence, vol. 4. North Holland, Amsterdam, 1990, pp. 169–198.

[37] J. Simonoff, Smoothing methods in Statistics, Springer, 1996.

[38] W. Smith, A note on truncation and sufficient statistics, Annals of Mathematical Statistics 28 (1957) 247–252.
[39] J. Tukey, Sufficiency, truncation and selection, Annals of Mathematical Statistics 20 (1949) 309–311.

Sensitivity Analysis of Probabilistic Networks

Linda C. van der Gaag[1], Silja Renooij[1], and Veerle M.H. Coupé[2]

[1] Department of Information and Computing Sciences, Utrecht University, Utrecht, The Netherlands – {silja,linda}@cs.uu.nl
[2] Department of Clinical Epidemiology and Biostatistics, VU University Medical Centre, Amsterdam, The Netherlands – v.coupe@vumc.nl

Summary. Sensitivity analysis is a general technique for investigating the robustness of the output of a mathematical model and is performed for various different purposes. The practicability of conducting such an analysis of a probabilistic network has recently been studied extensively, resulting in a variety of new insights and effective methods, ranging from properties of the mathematical relation between a parameter and an output probability of interest, to methods for establishing the effects of parameter variation on decisions based on the output distribution computed from a network. In this paper, we present a survey of some of these research results and explain their significance.

1 Introduction

Sensitivity analysis is a general technique for investigating the effects of inaccuracies in the parameters of a mathematical model on the model's output. As a mathematical model, a probabilistic network can also be subjected to such an analysis. The basic idea of the analysis then is to systematically vary the assessments for the network's parameter probabilities over a plausible interval and study the effects on the output computed from the network.

Sensitivity analysis of a probabilistic network can be performed for various different purposes. The parameter probabilities for a network are generally estimated from statistical data or assessed by human experts in the domain of application. As a consequence of incompleteness of data and partial knowledge of the domain, the assessments obtained inevitably are inaccurate [11]. Since the output probabilities of a network are built from these assessments, they may be sensitive to the inaccuracies involved and may even be unreliable. In general, however, not every parameter will require the same level of accuracy to arrive at satisfactory behaviour of the network; some probabilities will typically have more impact on the network's output than others. During model construction, therefore, sensitivity analysis can be used to gain detailed insight into the level of accuracy that is required for the various parameters and to guide further knowledge elicitation efforts [7]. Moreover, if during an initial evaluation of the network with real-life data, the output of the network is different from what is expected based upon knowledge of the domain, then sensitivity analysis can be used to identify which parameters can be changed to arrive at the expected output [2; 3; 4]. Furthermore, upon using the network in a real-life setting, sensitivity analysis can be used to gain insight into the robustness of an output probability of interest as well as of a decision based upon this probability. The analysis thereby reveals the range of parameter values for which the network's output is valid.

For a probabilistic network, the simplest type of analysis is to systematically vary one of the network's parameter probabilities while keeping all other parameters fixed. Such an analysis serves to reveal the effect of just the parameter whose assessment is being varied, on the output probability of interest. A sensitivity analysis in which a single parameter is varied, is termed a *one-way sensitivity analysis*. In a *two-way sensitivity analysis* of a probabilistic network, two parameters are varied simultaneously. In addition to the separate effects of variation of the two parameters, a two-way sensitivity analysis reveals the joint effect of their variation on a probability of interest. In essence, it is also possible to systematically vary more than two parameters at the same time. The results of such an *n-way sensitivity analysis*, however, are often hard to interpret.

In recent years, one-way sensitivity analysis of probabilistic networks has been studied extensively, which has resulted in practicable methods for performing such an analysis. In this paper, we present a survey of some of the recent research results and explain their significance. The paper is organised as follows. In Sect. 2, we introduce some preliminaries on probabilistic networks and introduce our running example. Section 3 describes the functional relationship between a parameter being varied and the output probability computed from a network; this section further discusses the computation of these sensitivity functions and presents some bounds on the effects of parameter variation. In Sect. 4, we discuss the application of sensitivity analysis to study the robustness of output probabilities computed from a probabilistic network on the one hand, and the robustness of decisions based upon these probabilities on the other hand. The paper ends with our concluding remarks and some directions for further research in Sect. 5.

2 Preliminaries

A *probabilistic network* basically is a representation of a joint probability distribution Pr over a set of stochastic variables. It consists of a qualitative part and an associated quantitative part. The network's qualitative part takes the form of an acyclic directed graph. Each node in this digraph represents a variable that takes its value from a finite set of discrete values. The arcs in the digraph model the influential relationships among the represented variables;

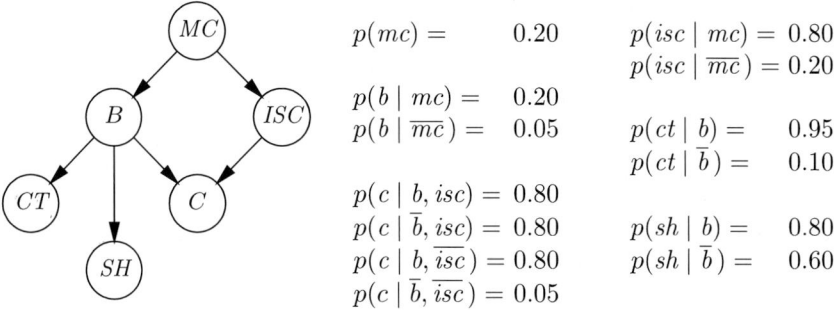

Fig. 1. The *Brain tumour* network.

more specifically, they capture independence by means of the d-separation criterion [20]. The strengths of the influential relationships are described by conditional probability distributions: for each variable V, conditional distributions $p(V \mid \pi(V))$ over its values are specified conditional on the various possible combinations of values for its set of parents $\pi(V)$ in the digraph. The specified probabilities are often referred to as the network's *parameters*. A probabilistic network in essence allows for the computation of any probability of interest over its variables. For this purpose, various efficient algorithms are available [20; 18; 15; 23], which we will refer to as the (*standard*) *propagation algorithms*. In the sequel, we will explicitly distinguish computed probabilities, written as Pr, from the parameter probabilities specified in the network, which are denoted by p.

For our running example we consider the *Brain tumour* network shown in Fig. 1. The network captures some (fictitious and incomplete) medical knowledge, adapted from [5]. It describes the problems associated with metastatic cancer for an arbitrary patient in oncology. Metastatic cancer (modelled by the variable MC) may lead to the development of a brain tumour (B) and typically gives rise to an increased level of serum calcium (ISC). The presence of a brain tumour can be established from a CT scan (CT). Severe headaches (SH) may also be indicative of the presence of a brain tumour. Either a brain tumour or an increased level of serum calcium are likely to ultimately cause a patient to fall into a coma (C). The digraph modelling the relationships among the six variables is shown on the left of the figure. In the network, all variables V are binary, taking one of the values *true* and *false*; we write v if V has the value *true* and $\bar{v}$ if it has the value *false*. The various parameter probabilities associated with the digraph are shown on the right of the figure. The probabilities specified for the variable ISC, for example, express that knowing whether or not metastatic cancer is present has a considerable influence on the probability of finding an increased level of serum calcium in an arbitrary patient. On the other hand, severe headaches are expressed as being quite common in both patients with and without a brain tumour.

3 Sensitivity Functions and Their Computation

Sensitivity analysis in general is a technique for investigating the effects of inaccuracies in the parameters of a mathematical model on its output. For a probabilistic network, more specifically, sensitivity analysis amounts to studying the effects of variation of the network's parameter probabilities on the output probabilities computed from the network. In this section, we will argue that any output probability of interest can be expressed as a simple mathematical function in the parameter under study. We further review the computational issues involved in establishing such a function.

3.1 Sensitivity Functions

Performing a sensitivity analysis of a probabilistic network in essence amounts to establishing, for each parameter and each output probability of interest, the *sensitivity function* that expresses this output probability in terms of the parameter being varied [6]. These sensitivity functions have a highly constrained functional form as a result of the graphical structure of a probabilistic network. Before elaborating on this functional form, we introduce some further notational conventions. We will denote a probability of interest by $\Pr(A = a \mid e)$, or $\Pr(a \mid e)$ for short, where a is a specific value of a variable A of interest and e denotes the available (possibly compound) evidence. A parameter under study will be denoted by $x = p(b_i \mid \pi)$, where b_i is a value of a variable B and π is a combination of values for the parents of B. We use $f_{\Pr(a\mid e)}(x)$ to denote the function that expresses the probability $\Pr(a \mid e)$ in terms of the parameter x; we often omit the subscript for the function symbol f, as long as ambiguity cannot occur.

In a one-way sensitivity analysis, a single parameter $x = p(b_i \mid \pi)$ for some variable B is varied. Upon varying this parameter, the other parameters $p(b_j \mid \pi)$, $j \neq i$, specified for the same variable need be *co-varied* to ensure that the parameters from the same distribution keep summing to one. Each such parameter $p(b_j \mid \pi)$ can thus be seen as a function $p(b_j \mid \pi)(x)$ of the parameter x under study. We assume that the parameters $p(b_j \mid \pi)$ are co-varied with $p(b_i \mid \pi)$ in such a way that their mutual proportional relationship is kept constant, that is,

$$p(b_j \mid \pi)(x) = p(b_j \mid \pi) \cdot \frac{1 - x}{1 - p(b_i \mid \pi)}$$

for $p(b_i \mid \pi) < 1$. This scheme of *proportional co-variation* has been shown to result in the smallest change, given the variation of the parameter x under study, in the output distribution [3].

Now, under the assumption of proportional co-variation, any sensitivity function $f_{\Pr(a\mid e)}(x)$ is a quotient of two functions that are linear in the parameter x under study [1; 9]. More formally, the function takes the form

$$f_{\Pr(a|e)}(x) = \frac{c_1 \cdot x + c_2}{c_3 \cdot x + c_4}$$

where the constants $c_1, \ldots, c_4$ are built from the original assessments for the parameters that are not being varied. The numerator of the function in essence describes the probability $\Pr(a, e)$ as a function of the parameter x; its denominator describes $\Pr(e)$ as a function of x. We observe that the probability distribution Pr defined by a probabilistic network can be written as a product of its parameter distributions. From the property of marginalisation, it then follows that both $\Pr(a, e)$ and $\Pr(e)$ can be written as a sum of products of parameters, one of which is the parameter under study.

We illustrate the form of a sensitivity function in general by studying, for the *Brain tumour* network, the sensitivity functions that describe the effects of varying the parameter probabilities $p(b \mid \overline{mc})$ and $p(sh \mid \overline{b})$ on the output probabilities $\Pr(b)$ and $\Pr(b \mid sh)$, respectively; the four functions are shown in Fig. 2. Fig. 2(c) and (d) show the sensitivity functions that express the output probability $\Pr(b \mid sh)$ in terms of the two separate parameters. Both functions are non-linear and monotone, and can in fact be written as a quotient of two linear functions. The sensitivity function expressing $\Pr(b \mid sh)$ in terms of

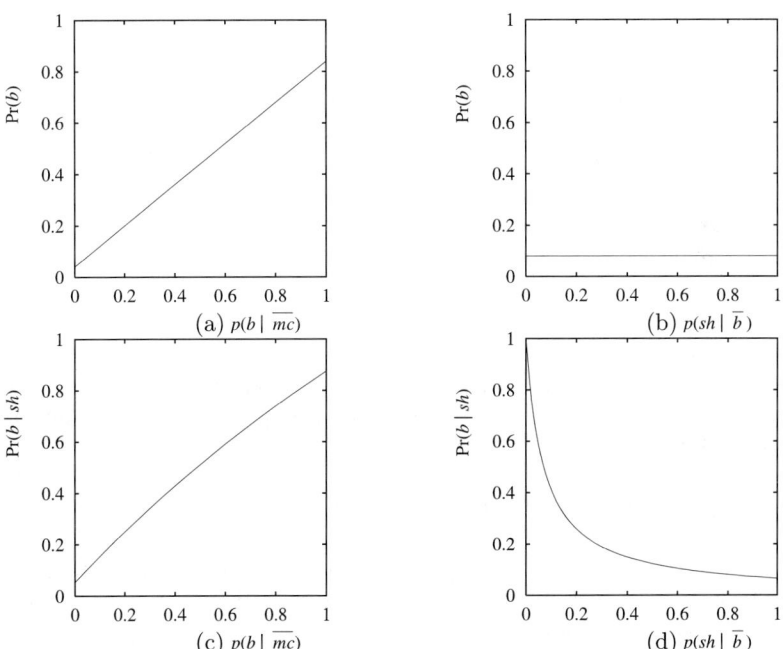

Fig. 2. Some sensitivity functions for the *Brain tumour* network; the effects of varying the parameters $p(b \mid \overline{mc})$ and $p(sh \mid \overline{b})$, respectively, on the prior probability $\Pr(b)$ (**a,b**) and on the posterior probability $\Pr(b \mid sh)$ (**c,d**) are shown.

$x = p(b \mid \overline{mc})$, for example, equals

$$f_{\Pr(b\mid sh)}(x) = \frac{0.640 \cdot x + 0.032}{0.160 \cdot x + 0.608} = \frac{4.0 \cdot x + 0.2}{x + 3.8}$$

Both sensitivity functions reveal that varying the parameter under study can have a considerable effect on the output probability of interest. We consider, for example, smaller values of the parameter $p(sh \mid \overline{b})$, meaning that it is becoming less likely to find severe headaches in patients without a brain tumour. If severe headaches then are found in a particular patient, this finding becomes more indicative of the presence of a brain tumour. This type of sensitivity is commonly found in real-life diagnostic networks [14].

Fig. 2(a) and (b) show the sensitivity functions that express the prior probability $\Pr(b)$ in terms of the two separate parameters. We observe that both functions are linear in the parameter under study. The sensitivity function expressing $\Pr(b)$ in terms of $x = p(b \mid \overline{mc})$, for example, equals

$$f_{\Pr(b)}(x) = 0.8 \cdot x + 0.04$$

In the absence of evidence, any probability of interest relates linearly to any network parameter. Linear functions are also found if the parameter under study pertains to an ancestor of the variable of interest and the parameter's variable has no observed descendants in the network's qualitative part. From Fig. 2(b) we further observe that variation of the parameter $p(sh \mid \overline{b})$ has no effect at all on our probability of interest; the associated sensitivity function is a constant function. If no information is available about the presence or absence of severe headaches, varying the diagnostic weight of headaches cannot have any impact on the output. In general, the sensitivity function is a constant function for any parameter associated with a variable that is not included in the sensitivity set of the variable of interest. This set of variables whose parameters may affect the probability of interest upon variation, is readily identified by simple inspection of the network's qualitative part [9].

From the four sensitivity functions in Fig. 2 we note that by entering evidence into a network, the exhibited sensitivities may change. While the observation of the presence of severe headaches hardly influences the effect of varying the parameter associated with the variable B, for example, we find that this same evidence changes the effect of varying the parameter associated with SH to a considerable extent. In general, quite different patterns of sensitivity may arise for different profiles of evidence.

For our discussion in the subsequent sections, it is convenient to observe that a sensitivity function is either a *linear* function or a fragment of a *rectangular hyperbola*. A rectangular hyperbola takes the general form

$$f(x) = \frac{r}{x - s} + t$$

where, for a hyperbolic sensitivity function, we have that

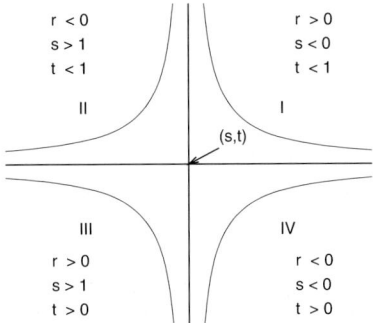

Fig. 3. Hyperbolas and their constants (specific for sensitivity functions).

$$s = -\frac{c_4}{c_3}, \quad t = \frac{c_1}{c_3}, \quad \text{and} \quad r = \frac{c_2}{c_3} + s \cdot t$$

with $c_1, \ldots, c_4$ as before. A rectangular hyperbola has two branches and two asymptotes. Fig. 3 illustrates the locations of the possible hyperbola branches relative to the two asymptotes. For $r < 0$, the branches lie in the second (II) and fourth (IV) quadrants relative to the asymptotes $x = s$ and $f(x) = t$; for $r > 0$, the branches are found in the first (I) and third (III) quadrants. Since any sensitivity function is well-defined for $x \in [0, 1]$, a hyperbolic sensitivity function is a fragment of one of the four possible hyperbola branches; the area with $0 \le x \le 1$ and $0 \le f(x) \le 1$ that defines the fragment, is called the *unit window* for the function. The vertical asymptote $x = s$ of the hyperbola lies either to the left of the unit window or to the right, that is, we have that $s < 0$ or $s > 1$. For first- and fourth-quadrant sensitivity functions more specifically we find $s < 0$, while for second- and third-quadrant functions we have that $s > 1$. In addition, the horizontal asymptote $f(x) = t$ of the hyperbola lies below $f(1)$ for first-quadrant functions and below $f(0)$ for second-quadrant functions; we then have that $t < 1$. The horizontal asymptote lies above $f(0)$ for third-quadrant and above $f(1)$ for fourth-quadrant functions, which implies that $t > 0$. Note that the horizontal asymptote of the hyperbola not necessarily lies within the unit window for the sensitivity function: for a first-quadrant function, for example, negative values of t are possible and for a fourth-quadrant function values of t larger than 1 can be found.

We illustrate the hyperbolic form by writing the sensitivity function for the posterior probability $\Pr(b \mid sh)$ and the parameter $x = p(b \mid \overline{mc})$ as a hyperbolic function:

$$f_{\Pr(b|sh)}(x) = -\frac{15.0}{x + 3.8} + 4.0$$

The three constants $s = -3.8, t = 4.0$, and $r = -15.0$ indicate that the sensitivity function is a fourth-quadrant function and, hence, is monotonically increasing. The vertical asymptote $x = -3.8$ is located at some distance to the left of the unit window; the horizontal asymptote $f(x) = 4.0$, moreover, is located at some distance above the unit window. These properties indicate that no major changes in the derivative of the function are expected within the unit window, as is confirmed by Fig. 2(c). For the first-quadrant function from Fig. 2(d) in contrast, the constants $s = -0.07$ and $t = 0$ are found.

3.2 Computing Sensitivity Functions

In the previous subsection, we argued that the effect of varying a single parameter upon an output probability computed from a probabilistic network can be described by a highly constrained mathematical function. The major advantage of this constrained form is that any sensitivity function can be established by computing just its constants.

The simplest method of determining the constants of a sensitivity function amounts to computing, from the network, the probability of interest for up to three values for the parameter under study; using the functional form of the function to be established, a system of linear equations is obtained, which can subsequently be solved [9]. For the network computations, any of the standard propagation algorithms can be used. With this method, the sensitivity function for a single parameter and a single output probability is established. A more efficient method determines the required constants by propagating information through a junction tree, similar to the standard junction-tree propagation algorithm [16]. This method requires a very small number of inward and outward propagations in the tree to determine either the constants of *all* sensitivity functions that relate the probability of interest to any one of the network parameters, or to determine the sensitivity functions for any output probability in terms of a single parameter.

In a full sensitivity analysis, the effects of varying all parameter probabilities of a network on all output probabilities of interest are investigated. From the example sensitivity functions for the *Brain tumour* network, we have seen that the effect of varying a particular parameter can change considerably for different output probabilities. For a full analysis, therefore, all sensitivity functions of interest need be established explicitly, for which purpose the two methods outlined above need be applied multiple times. Although the concept of sensitivity set can be used to forestall some of the computations involved, performing a full sensitivity analysis of a probabilistic network of realistic size is highly time consuming.

3.3 Bounding Sensitivity Functions

For real-life probabilistic networks, performing a full sensitivity analysis is infeasible in practice, mostly as a consequence of the large range of evidence profiles to be studied. Recent research therefore focused on the derivation of general bounds for sensitivity functions. These bounds provide for selecting, without actually performing the analysis for the full range of profiles, the sensitivity functions that are the most likely to reveal high sensitivities; we will return to this observation in Sect. 4.

Within a given network, we consider a parameter x for which the value x_0 is specified. We are interested in some prior or posterior probability p, for which the value p_0 is computed from the network, using the value x_0 for the parameter x. We will refer to the values x_0 and p_0 as the *original* values of the parameter and of the output probability, respectively; we assume that neither x_0 nor p_0 is equal to zero or one. Without any further knowledge of the network, we only know that the sensitivity function for x and p passes through the point (x_0, p_0). Now, any sensitivity function through this point is bounded by the two rectangular hyperbola's $i(x)$ and $d(x)$ [21], with

$$i(x) = \frac{p_0 \cdot (1 - x_0) \cdot x}{(p_0 - x_0) \cdot x + (1 - p_0) \cdot x_0} \text{ and } d(x) = \frac{p_0 \cdot x_0 \cdot (1 - x)}{(1 - p_0 - x_0) \cdot x + p_0 \cdot x_0}$$

These bounds follow from the observation that an increasing rectangular hyperbola can, in the most extreme case, pass through the points $(0, 0)$ and $(1, 1)$; a decreasing hyperbola may pass through $(0, 1)$ and $(1, 0)$.

For any sensitivity function $f(x)$ with $f(x_0) = p_0$, we now have that

$$\min\{i(x_j), d(x_j)\} \leq f(x_j) \leq \max\{i(x_j), d(x_j)\}$$

for all $x_j \in [0, 1]$. From the bounding hyperbolas, we can readily establish numerical bounds on the new value p_1 of the probability of interest that results from varying the parameter x from its original value x_0 to x_1. More specifically, these bounds are given by [3]:

$$\frac{p_0 \cdot e^{-\delta}}{p_0 \cdot (e^{-\delta} - 1) + 1} \leq p_1 \leq \frac{p_0 \cdot e^{\delta}}{p_0 \cdot (e^{\delta} - 1) + 1}$$

where

$$\delta = \left| \ln \frac{x_1}{1 - x_1} - \ln \frac{x_0}{1 - x_0} \right|$$

We would like to note that, if we know that a sensitivity function is linear in the parameter under study, we can also establish linear bounding functions that may lead to tighter numerical bounds. These linear bounding functions also pass through the point (x_0, p_0). Since both functions moreover should be well-defined within the unit window, the increasing bounding function further passes through either $(0, 0)$ or $(1, 1)$, and the decreasing bounding function goes through $(0, 1)$ or $(1, 0)$ [21].

To illustrate the various functions, Fig. 4 depicts two sets of bounding functions for the *Brain tumour* network. Given the original value $x_0 = 0.05$ for the parameter $x = p(b \mid \overline{mc})$, the posterior probability $p = \Pr(b \mid sh)$ equals $p_0 = 0.10$. Fig. 4(a) now depicts the increasing and decreasing bounding hyperbolas through the point $(x_0, p_0) = (0.05, 0.10)$; the true sensitivity function for x and p is shown in the figure by a dashed curve. For the prior

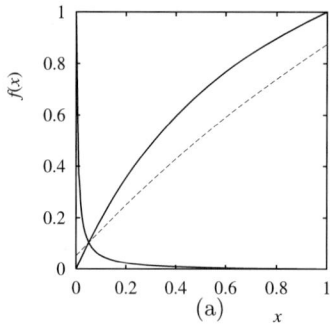 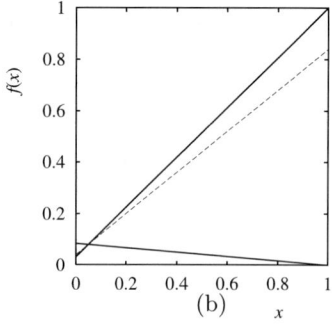

Fig. 4. Bounds on hyperbolic sensitivity functions through $(x_0, p_0) = (0.05, 0.10)$ (a) and on linear sensitivity functions through $(x_0, p_0) = (0.05, 0.08)$ (b); example sensitivity functions for the *Brain tumour* network are also shown (dashed).

probability $\Pr(b)$, we know that the sensitivity function that expresses this probability in terms of the same parameter x is a linear function. Fig. 4(b) now depicts the linear bounding functions established for this function; the true sensitivity function again is shown by a dashed line. Note that knowledge of the sensitivity function being linear leads to tighter bounds than the bounding hyperbolas would give.

The bounding functions introduced above all depend on the original values of the parameter and of the probability of interest, but are independent of any other aspect of the network under study. The bounds therefore apply to any pair (x_0, p_0) for any network. Their computation, moreover, does not require any network propagations, except for establishing the value p_0. For a given profile of evidence, the bounds on the sensitivity function can be further tightened. To this end, the probability of the evidence $\Pr(e)$ is expressed as a linear function $f_{\Pr(e)}(x) = c_3 \cdot x + c_4$ of the parameter under study; this function is readily computed using the junction-tree algorithm outlined in the previous subsection. The two constants c_3 and c_4 now determine the constant $s = -\frac{c_4}{c_3}$ in the hyperbolic form of any sensitivity function passing through the point (x_0, p_0) and thereby constrain the space of feasible functions [22].

4 Applications of Sensitivity Analysis

One of the purposes to which a sensitivity analysis is performed of a probabilistic network, is to study the robustness of the network's output. We thereby distinguish between the output probabilities that are computed from the network and the decisions based upon these probabilities.

4.1 Robustness of Output Probabilities

In the previous section, we argued that a sensitivity function $f(x)$ provides for establishing the change in the output probability of interest that is occasioned by a shift in the value of the parameter x under study. For the *Brain tumour* network, for example, we have from Fig. 2(c) that a shift in the value of the parameter $p(b \mid \overline{mc})$ occasions a more or less proportional shift in the output probability $\Pr(b \mid sh)$. A shift in the value of the parameter $p(sh \mid \overline{b})$ may have quite a considerable effect on this posterior probability; small deviations from the original value 0.60 for the parameter, however, will have little effect. The change in the output probability that is occasioned by a shift in the value of a parameter under study is indicative of the robustness of the output. Different concepts have been designed to measure this robustness. We discuss for this purpose the concepts of *sensitivity value*, which describes the change in an output probability for infinitesimally small shifts in a parameter under study, and *vertex proximity*, which indicates potential effects of larger shifts.

Sensitivity value

We consider a parameter x with an original value of x_0. We further consider an output probability p and the sensitivity function $f(x)$ that expresses p in terms of x. The *sensitivity value* of x and p now describes the effect of infinitesimally small shifts in the parameter's original value on the probability of interest. It is defined as $|f'(x_0)|$, that is, as the absolute value of the first derivative of the sensitivity function at the original value x_0 of the parameter. The sensitivity value of the parameter x and the output probability of interest can be established analytically by computing the sensitivity function from the network, taking its first derivative

$$f'(x) = \frac{c_1 \cdot c_4 - c_2 \cdot c_3}{(c_3 \cdot x + c_4)^2}$$

and filling in the original value x_0 for the parameter under study. Alternatively, the sensitivity value can be established directly from the network by performing two network propagations [17; 10].

To illustrate the concept of sensitivity value, we establish, for our *Brain tumour* network, the derivative of the sensitivity function that expresses the output probability $\Pr(b \mid sh)$ in terms of the parameter $x = p(b \mid \overline{mc})$. We find that

$$f'(x) = \frac{0.384}{(0.16 \cdot x + 0.608)^2}$$

Since the parameter x has an original value of 0.05, we find a sensitivity value of $|f'(0.05)| = 1.01$. This sensitivity value indicates that small shifts in the parameter under study will induce a more or less similar change in the output probability. For the parameter $x = p(sh \mid \overline{b})$ having an original value of 0.60, in contrast, we find a sensitivity value equal to

$$\left| \frac{-0.059}{(0.92 \cdot 0.60 + 0.064)^2} \right| = 0.16$$

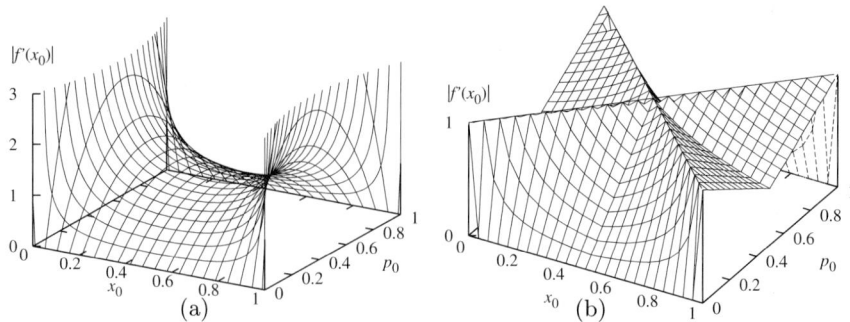

Fig. 5. The upper bound on the sensitivity value as a function of x_0 and p_0, for a hyperbolic sensitivity function (**a**) and for a linear sensitivity function (**b**).

which indicates that small shifts in the parameter's original value will hardly have any effect on the output probability of interest. We note that these observations are confirmed by Fig. 2(c) and (d).

In the previous section we argued that for a real-life probabilistic network it is generally infeasible to compute sensitivity functions given all profiles of evidence. By building upon the bounding functions for a sensitivity function, however, it is relatively straightforward to establish an upper bound on the sensitivity value [21; 2]. By taking the first derivatives of the hyperbolic bounding functions $i(x)$ and $d(x)$ for a parameter x with an original value of x_0 and an output probability with an original value of p_0, we find the following upper bound on the sensitivity value of x and p:

$$|f'(x_0)| \leq \frac{p_0 \cdot (1 - p_0)}{x_0 \cdot (1 - x_0)}$$

Fig 5(a) depicts this upper bound as a function of x_0 and p_0. The figure reveals that large sensitivity values are expected only for the more extreme values of x_0. For the sensitivity functions from Fig. 2(c) and (d), for example, with $p_0 = 0.10$, and $x_0 = 0.05$ for the parameter $p(b \mid \overline{mc})$ and $x_0 = 0.60$ for the parameter $p(sh \mid \overline{b})$, we find that the parameter with the smaller original value will have a sensitivity value of at most 1.96, whereas the less extreme value of the other parameter results in a bound of 0.39 on the sensitivity value. Note that the actual sensitivity values 1.01 and 0.16 for these parameters and the probability of interest are indeed below the established bounds.

We recall from the previous section that, if a sensitivity function is known to be linear in a parameter under study, then linear bounding functions can be established. From these linear bounding functions, also an upper bound on the sensitivity value can be computed. This bound will in general be tighter than the upper bound that is found from hyperbolic bounding functions [22]. Fig. 5(b) depicts the upper bound as a function of x_0 and p_0. As an example, we consider for our *Brain tumour* network the sensitivity function that describes the prior probability $\Pr(b)$ as a function of the parameter

$x = p(b \mid \overline{mc})$; recall that this function is depicted in Fig. 2(a). With $x_0 = 0.05$ and $p_0 = 0.08$, we find that the sensitivity value of this parameter and output probability is at most 0.97; the actual sensitivity value equals 0.80. Note that the sensitivity value for a linear function never exceeds 1.0, that is, a shift in the parameter under study can never result in a larger shift in the output probability of interest. Further note that the surface from Fig. 5(b) can be placed underneath that of Fig. 5(a), which confirms that knowledge of a sensitivity function being linear allows for tighter bounds.

In a full sensitivity analysis of a probabilistic network, typically a large range of different profiles of evidence need be studied, which renders the analysis infeasible in practice. The bounds on the sensitivity value established above can now be exploited for selecting the parameters for which a detailed analysis indeed is useful. We consider, as an example, a parameter with an original value of 0.5. From the bounds established above, we have that any output probability will be quite insensitive to small shifts in this parameter, regardless of the evidence profile under study. For such a parameter, therefore, a more detailed analysis is not required.

Vertex proximity

The concept of sensitivity value reviewed above has been designed to give insight in the effect of very small parameter variations. As the initial assessments for the parameters of a probabilistic network may be highly inaccurate, however, we are interested in the effects of larger parameter shifts as well. We observe that for linear sensitivity functions, the sensitivity value in essence is a constant function in the value of the parameter under study. The computed sensitivity value therefore remains valid also for larger parameter shifts. For hyperbolic sensitivity functions, this property does not hold. In fact, the sensitivity value can strongly differ for two values of the parameter under study. To illustrate this observation, we consider again, for our *Brain tumour* network, the sensitivity function from Fig. 2(d) which expresses the output probability $p = \Pr(b \mid sh)$ in terms of the parameter $x = p(sh \mid \overline{b})$. The sensitivity value of x and p, that is, the absolute value of the function's first derivative, is depicted in Fig. 6 as a function of the value of x. The figure reveals that the sensitivity value of x and p equals at most 1.0 if the parameter adopts a value within the interval $[0.2, 1]$. For smaller values of the parameter, the sensitivity value rapidly grows to infinity. Now, if the original value of the parameter would be slightly larger than $x_0 = 0.2$, we would find a relatively small sensitivity value which would be interpreted as indicating that the output probability is not very sensitive to variation of the parameter. Yet, if a more accurate value for the parameter would be slightly smaller than 0.2, we would conclude that the network's output is not very robust as a result of the larger sensitivity value found.

For a hyperbolic sensitivity function, we now take the point $(x_v, f(x_v))$ where the first derivative $|f'(x_v)|$ equals 1.0 as the point that marks the

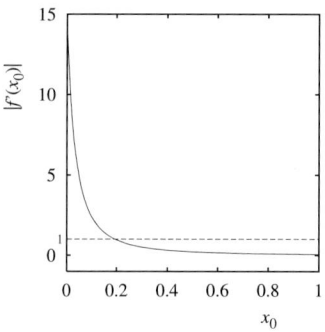

Fig. 6. The sensitivity value for the parameter $x = p(sh \mid \bar{b})$ and the output probability $\Pr(b \mid sh)$, as a function of x.

transition from large sensitivity values to small ones, and vice versa. This point is called the *vertex* of the hyperbola branch under study and can easily be computed from the constants of the sensitivity function using

$$x_v = \begin{cases} s + \sqrt{|r|}, \text{ if } s < 0 \\ s - \sqrt{|r|}, \text{ if } s > 1 \end{cases}$$

Now, if the original value x_0 of a parameter lies close to the x-value of the hyperbola's vertex, then the output probability of interest may be quite sensitive to variation of the parameter even if just a small sensitivity value is found. We say that the parameter's original value exhibits *vertex proximity*.

To illustrate the concept of vertex proximity, we consider again, for our *Brain tumour* network, the two sensitivity functions from Fig. 2(c) and (d). For the hyperbola branch that describes the output probability of interest as a function of the parameter $p(b \mid \overline{mc})$, with an original value of 0.05, we find that $x_v = -3.8 + \sqrt{15} = 0.07$. The vertex of the hyperbola branch thus is quite close to the parameter's original value. Since the sensitivity function is a fourth-quadrant hyperbolic function and $x_0 < x_v$, we conclude that the sensitivity value for values of x larger than x_v will be smaller than the sensitivity value of 1.01 computed for x_0; note that this observation is confirmed by Fig. 2(c). For the parameter $p(sh \mid \bar{b})$, the x-value of the vertex of the hyperbola branch is found around 0.20, which can be considered quite distant from the parameter's original value of 0.60. Both the sensitivity value of 0.16 and the lack of vertex proximity thus indicate that the output probability is not highly sensitivity to shifts in the parameter; note that this observation is confirmed by Fig. 2(d). If its original value had been 0.25, for example, then we would have found a sensitivity value of 0.68. The property of vertex proximity would then have been indicative of possibly significant effects of variation of the parameter to smaller values.

4.2 Robustness of Decisions

More and more, probabilistic networks are used within decision-support systems, where decisions are recommended based upon some probability distribution established from the network. In such a system, robustness of the recommended decision is often more important than robustness of the output probabilities themselves. Now, for some network parameters variation will have a considerable effect on an output probability and yet not induce a change in decision; for other parameters, variation will show little effect on the output probability and nonetheless result in a different decision. For studying the robustness of a recommended decision, therefore, studying the effects of parameter variation on an output probability no longer suffices: the effects on the decision itself need be taken into consideration explicitly. We discuss to this end the issue of output robustness in view of the *threshold model for decision making*, where a decision is based upon an output probability computed from the network, and in view of a model for decision making that is based upon the *most likely value* for a variable of interest.

Threshold decision making

The *threshold model* has been designed to support decision making for diagnostic problems in which the decision maker has to choose between gathering further evidence and acting without acquiring additional information. Although generally applicable, the model is used most notably for patient management in medical applications [19]. In such an application, the probability of disease is generally taken by an attending physician to decide upon management of a patient under consideration. The physician may decide to start treatment rightaway, or to withhold treatment altogether. Alternatively, if a diagnostic test can provide additional information which may affect the patient's probability of disease, then the physician may defer the treatment decision until this information has become available.

To support choosing among the various decision alternatives, the threshold model builds upon three threshold probabilities of disease. The *treatment threshold probability* $P^*(d)$ of a disease d being present, is the probability at which the physician is indifferent between giving treatment and withholding treatment; the *no treatment-test threshold probability* $P^-(d)$ is the probability at which the physician is indifferent between the decision to withhold treatment and the decision to obtain additional diagnostic information; and the *test-treatment threshold probability* $P^+(d)$ is the probability at which the physician is indifferent between obtaining further information and starting treatment rightaway. Now, as long as not all possible diagnostic tests have been performed, a physician has three decision alternatives at his disposal. Given the probability of disease $\Pr(d)$ for a patient, the model recommends the physician to withhold treatment if $\Pr(d) < P^-(d)$, to start treatment if $\Pr(d) > P^+(d)$, and to perform a diagnostic test if

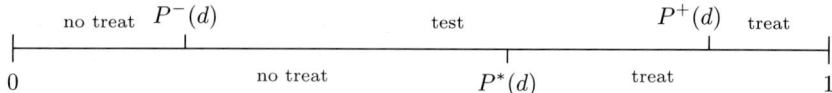

Fig. 7. The threshold model for patient management, indicating three threshold probabilities and the various decision alternatives at a physician's disposal.

$P^-(d) \leq \Pr(d) \leq P^+(d)$. If no further tests are available, the physician has to choose between only two alternative decisions. The model recommends to start treatment if $\Pr(d) > P^*(d)$ and to withhold treatment if $\Pr(d) \leq P^*(d)$. Fig. 7 summarises the basic idea of the threshold model.

For studying output robustness of a probabilistic network in view of the threshold model for decision making, the various threshold probabilities employed by the model should be taken into consideration [12]. More specifically, the effects of parameter variation on the recommended decision should be analysed. We consider an output probability of disease $\Pr(d \mid e)$ computed from the probabilistic network and suppose that $P^-(d) \leq \Pr(d \mid e) \leq P^+(d)$. Assuming that an appropriate diagnostic test is still available, the threshold model thus recommends to gather additional information. We further consider a parameter x from the network and the sensitivity function $f(x)$ that expresses the posterior probability of disease in terms of the parameter. We compute the values x^- and x^+ for the parameter x such that

$$f(x^-) = P^-(d) \text{ and } f(x^+) = P^+(d)$$

Note that these values can be readily established from the sensitivity function. We now have that the decision to gather additional information remains unaltered as long as the original value of the parameter is not varied beyond the interval $[x^-, x^+]$. Similar bounds on variation can be computed for an output probability of disease smaller than $P^-(d)$ or larger than $P^+(d)$.

We illustrate investigating the robustness of decisions for our *Brain tumour* network. We assume that the threshold probabilities of a brain tumour being present have been set at $P^-(b) = 0.045$ and $P^+(b) = 0.56$. The prior probability of the presence of a brain tumour in an arbitrary patient in oncology is computed from the network to be $\Pr(b) = 0.08$. For this probability, we have that $P^-(b) \leq \Pr(b) \leq P^+(b)$. For a patient about whom no further information is available, therefore, the threshold model recommends to gather additional evidence. We now study the robustness of this decision in terms of variation of the parameter $x = p(sh \mid \bar{b})$. The sensitivity function that expresses the prior probability $\Pr(b)$ in terms of x is a constant function; we recall that the function is depicted in Fig. 2(b). We thus find that the parameter can be varied within the entire $[0, 1]$-interval without inducing a change in decision. The decision to gather further evidence therefore is insensitive to variation of this parameter. Now suppose that the patient complains of severe headaches. From the network, the posterior probability of a brain tumour being present is computed to be $\Pr(b \mid sh) = 0.10$. For this probability, we find that $P^-(b) \leq \Pr(b \mid sh) \leq P^+(b)$ which again results in the recommendation to gather further information, for example by performing a CT scan. We again investigate the robustness of this decision in terms of variation of the parameter x. The sensitivity function that expresses the output probability $\Pr(b \mid sh)$ in terms of x no longer is a constant function; we recall that the function is shown in Fig. 2(d). The function equals

$$f_{\Pr(b \mid sh)}(x) = \frac{0.064}{0.92 \cdot x + 0.064}$$

From the intersections of the sensitivity function with the two threshold probabilities, we find that $x^+ = 0.055$ and $x^- = 1.476 > 1.0$. We thus find that the decision to gather additional information is robust to variation of the parameter x within the interval $[0.055, 1]$. Since the original value of the parameter equals $x_0 = 0.60$, we conclude that the decision is quite robust to variation of the parameter under study.

We would like to note that by building upon the bounding functions for a sensitivity function a *cautious* interval for variation of a parameter under study can be established. This interval provides boundaries between which the parameter can at least be safely varied. If the parameter is varied beyond these boundaries, however, robustness is no longer guaranteed and the recommended decision may change.

Most likely value

For classification problems, decision making with a probabilistic network often amounts to selecting the value that is the most likely for an output variable of interest, based upon the computed probability distribution. For studying output robustness of the network in view of this model for decision making, the effects of parameter variation on the most likely value of the output variable need be studied. For this purpose, the concept of *admissible deviation* has been designed. An admissible deviation captures the extent of the variation that can be applied to a parameter without changing the most likely value of the output variable [13]. It is a pair of real numbers (α, β) that describe the shifts to smaller values and to larger values, respectively, that are allowed in the parameter under study without inducing a change in the most likely value of the output variable; often the symbols $\leftarrow$ and $\rightarrow$ are used to express that the parameter can be varied as far as the boundaries of the probability interval. For a parameter with an original value of x_0, the admissible deviation (α, β) thus indicates that the parameter can be safely varied within the interval $[x_0 - \alpha, x_0 + \beta]$.

We consider a parameter x with an original value of x_0, and an output variable A; we suppose that, given the available evidence e, the most likely value of A is the value a_k. We observe that, if we compare just the two values a_k and a_i of A, then the intersection of the two sensitivity functions

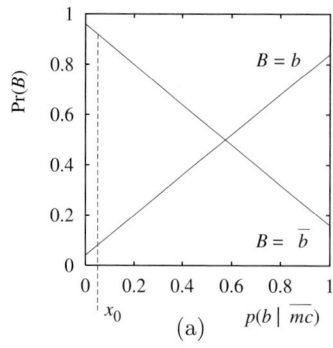
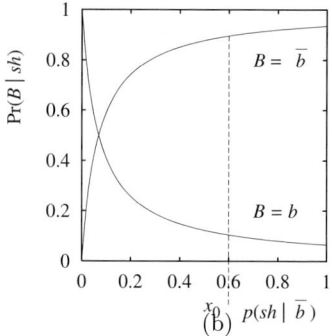

Fig. 8. Sensitivity functions for the possible values of the variable B, in terms of the parameters $p(b \mid \overline{mc})$ (a) and $p(sh \mid \overline{b})$ (b).

$f_{\Pr(a_k|e)}(x)$ and $f_{\Pr(a_i|e)}(x)$ marks the value of the parameter x at which the value a_i becomes more likely than the value a_k. Based upon this observation, the admissible deviation for the parameter x is now readily established from the intersections of the sensitivity function $f_{\Pr(a_k|e)}(x)$ for the most likely value a_k with the functions $f_{\Pr(a_i|e)}(x)$ pertaining to the other values a_i, $i \neq k$, of A. If the intersections are found at the values x_i for the parameter x with $x_1 \leq \ldots \leq x_j \leq x_0 \leq x_{j+1} \leq \ldots \leq x_n$, then we can conclude that the parameter can be varied between the values x_j and x_{j+1} without inducing a change in the most likely value of the output variable. The admissible deviation then equals $[x_0 - x_j, x_{j+1} - x_0]$.

To illustrate the concept of admissible deviation, we consider, in the *Brain tumour* network, the output variable B and its two values. For an arbitrary patient, the probability of a brain tumour being present equals $\Pr(b) = 0.08$. The absence of a brain tumour, therefore, is the most likely value of B. We now study the effects of varying the parameter $x = p(b \mid \overline{mc})$, with an original value of $x_0 = 0.05$, on this most likely value. The sensitivity functions that describe the prior probabilities of the two values of B in terms of x are

$$f_{\Pr(b)}(x) = 0.80 \cdot x + 0.04 \quad \text{and} \quad f_{\Pr(\overline{b})}(x) = -0.80 \cdot x + 0.96$$

These functions are depicted in Fig. 8(a). The two sensitivity functions intersect at $x = 0.575$. For the original value of x, we thus find an admissible deviation of $(\leftarrow, 0.525)$. We conclude that the diagnosis of this patient not having a brain tumour is quite robust to variation of the parameter x. We now consider a patient with severe headaches. For this patient, the probability of a brain tumour being present equals $\Pr(b \mid sh) = 0.10$. Again, the absence of such a tumour is the most likely value of the variable B. The effects of varying the parameter $x = p(sh \mid \overline{b})$, with an original value of $x_0 = 0.60$, on the posterior probabilities for B are described by the two sensitivity functions

$$f_{\Pr(b|sh)}(x) = \frac{0.064}{0.92 \cdot x + 0.064} \quad \text{and} \quad f_{\Pr(\overline{b}|sh)}(x) = \frac{0.92 \cdot x}{0.92 \cdot x + 0.064}$$

These functions are shown in Fig. 8(b). The two sensitivity functions intersect at $x = 0.07$, resulting in an admissible deviation of $(0.53, \rightarrow)$. We conclude that the most likely value of the output variable again is quite robust to variation of the parameter under study.

We would like to note that by building upon the bounding functions for a sensitivity function a *cautious* interval for variation of a parameter under study can be established. This interval provides boundaries between which the parameter can at least be safely varied. If the parameter is varied beyond these boundaries, however, robustness is no longer guaranteed and the recommended decision may change. From this interval, a *minimal admissible deviation* is established [13].

5 Concluding Observations

Recent research in sensitivity analysis of probabilistic networks has resulted in a variety of new insights and effective methods. The insight that has proved to be most significant, is that the mathematical properties of a probabilistic network strongly constrain the functional form of the relation between a parameter probability and an output probability of interest. In fact, this so-called sensitivity function is a fraction of two linear functions in the parameter under study, taking the form of either a linear function or a fragment of a hyperbola branch. Because of this constrained functional form, any sensitivity function can be established analytically from a probabilistic network by computing just its constants using a limited number of network propagations.

The significance of the sensitivity function lies in that it provides for easy computation of various sensitivity properties pertaining to the robustness of the output of a network. We detailed the concepts of sensitivity value and vertex proximity as giving insight into the effects of parameter variation on an output probability of interest. These concepts can be used during model construction, for example, to gain detailed insight into the level of accuracy that is required for the various parameters and to guide further elicitation efforts. We further discussed the issue of output robustness in view of two different models for decision making that build upon the output distribution computed from a network, and presented methods that give insight into the effects of parameter variation on a recommended decision. These methods reveal the range of parameter values for which a decision based upon the network's output distribution is valid when used in a real-life setting.

Another use of the sensitivity function is for *parameter tuning*. Parameter tuning is typically employed if the output of a probabilistic network differs from what is expected based upon knowledge of the domain and amounts to changing the values of one or more of the network's parameters. Sensitivity functions can then be used to identify which parameters had best be changed to arrive at the expected output. Parameter tuning, however, should be performed with care since changing even a single parameter may easily have unwanted effects over the large range of evidence profiles. We feel that further research into the global effects of local changes is still required.

To conclude, we focused in this paper on *one-way sensitivity analysis*. In essence it is possible to also perform a more general *n-way analysis* of a probabilistic network. Compared to one-way sensitivity analysis, however, higher-order analysis of probabilistic networks has so far received far less attention [8; 4]. Since a higher-order analysis is particularly useful for uncovering and studying synergistic effects of variation of the parameter probabilities of a network, we feel that it is worthwhile to direct further research efforts to gaining new insights and developing effective methods for conducting such analyses and interpreting their results.

References

[1] E. Castillo, J.M. Gutiérrez and A.S. Hadi (1997). Sensitivity analysis in discrete Bayesian networks. *IEEE Transactions on Systems, Man, and Cybernetics*, vol. 27, pp. 412–423.

[2] H. Chan and A. Darwiche (2002). When do numbers really matter? *Journal of Artificial Intelligence Research*, vol. 17, pp. 265–287.

[3] H. Chan and A. Darwiche (2002). A distance measure for bounding probabilistic belief change. *Proceedings of the Eighteenth National Conference on Artificial Intelligence*, AAAI Press, Menlo Park, pp. 539–545.

[4] H. Chan and A. Darwiche (2004). Sensitivity analysis in Bayesian networks: from single to multiple parameters. In: M. Chickering, J. Halpern (editors), *Proceedings of the Twentieth Conference on Uncertainty in Artificial Intelligence*, AUAI Press, Arlington, VA, pp. 67–75.

[5] G.F. Cooper (1984). *NESTOR: a Computer-based Medical Diagnostic Aid that Integrates Causal and Probabilistic Knowledge*, Report HPP-84-48, Stanford University.

[6] V.M.H. Coupé and L.C. van der Gaag (1998). Practicable sensitivity analysis of Bayesian belief networks. In: M. Hušková, P. Lachout, J.A. Víšek (editors), *Prague Stochastics '98*, Union of Czech Mathematicians and Physicists, Prague, pp. 81–86; also available as Report UU-CS-1998-10, Utrecht University.

[7] V.M.H. Coupé, L.C. van der Gaag and J.D.F. Habbema (2000). Sensitivity analysis: an aid for probability elicitation. *Knowledge Engineering Review*, vol. 15, pp. 1–18.

[8] V.M.H. Coupé, F.V. Jensen, U. Kjærulff and L.C. van der Gaag (2000). *A computational architecture for n-way sensitivity analysis of Bayesian networks*. Technical Report, Aalborg University.

[9] V.M.H. Coupé and L.C. van der Gaag (2002). Properties of sensitivity analysis of Bayesian belief networks. *Annals of Mathematics and Artificial Intelligence*, vol. 36, pp. 323–356.

[10] A. Darwiche (2003). A differential approach to inference in Bayesian networks. *Journal of the ACM*, vol. 50, pp. 280–305.
[11] M.J. Druzdzel and L.C. van der Gaag (1995). Elicitation of probabilities for belief networks: combining qualitative and quantitative information. In: P. Besnard, S. Hanks (editors), *Proceedings of the Eleventh Conference on Uncertainty in Artificial Intelligence*, Morgan Kaufmann, Palo Alto, pp. 141–148.
[12] L.C. van der Gaag and V.M.H. Coupé (1999). Sensitivity analysis for threshold decision making with Bayesian belief networks. In: E. Lamma, P. Mello (editors), *AI*IA 99: Advances in Artificial Intelligence*, LNAI, Springer-Verlag, Berlin, pp. 37–48.
[13] L.C. van der Gaag and S. Renooij (2001). Analysing sensitivity data from probabilistic networks. In: J. Breese, D. Koller (editors), *Proceedings of the Seventeenth Conference on Uncertainty in Artificial Intelligence*, Morgan Kaufmann, San Francisco, pp. 530–537.
[14] L.C. van der Gaag and S. Renooij (2006). On the sensitivity of probabilistic networks to reliability characteristics. In: B. Bouchon-Meunier, G. Coletti and R.R. Yager (editors), *Modern Information Processing: From Theory to Applications*, Elsevier B.V., pp. 395-405.
[15] F.V. Jensen, S.L. Lauritzen and K.G. Oleson (1990). Bayesian updating in causal probabilistic networks by local computations. *Computational Statistics Quarterly*, vol. 4, pp. 269–282.
[16] U. Kjærulff and L.C. van der Gaag (2000). Making sensitivity analysis computationally efficient. In: G. Boutilier, M. Goldszmidt (editors), *Proceedings of the Sixteenth Conference on Uncertainty in Artificial Intelligence*, Morgan Kaufmann, San Francisco, pp. 317–325.
[17] K.B. Laskey (1995). Sensitivity analysis for probability assessments in Bayesian networks. *IEEE Transactions on Systems, Man, and Cybernetics*, vol. 25, pp. 901–909.
[18] S.L. Lauritzen and D.J. Spiegelhalter (1988). Local computations with probabilities on graphical structures and their application to expert systems. *Journal of the Royal Statistical Society, Series B*, vol. 50, pp. 157–224.
[19] S.G. Pauker and J.P. Kassirer (1980). The threshold approach to clinical decision making. *New England Journal of Medicine*, vol. 302, pp. 1109–1117.
[20] J. Pearl (1988). *Probabilistic Reasoning in Intelligent Systems. Networks of Plausible Inference*, Morgan Kaufmann, Palo Alto.
[21] S. Renooij and L.C. van der Gaag (2004). Evidence-invariant sensitivity bounds. In: M. Chickering, J. Halpern (editors), *Proceedings of the Twentieth Conference on Uncertainty in Artificial Intelligence*, AUAI Press, Arlington, VA, pp. 479-486.
[22] S. Renooij and L.C. van der Gaag (2005). Exploiting evidence-dependent sensitivity bounds. In: F. Bacchus, T. Jaakkola (editors), *Proceedings*

of the Twenty-First Conference on Uncertainty in Artificial Intelligence, AUAI Press, Corvallis, OR, pp. 485-492.

[23] G. Shafer and P.P. Shenoy (1990) Probability propagation. *Annals of Mathematics in Artificial Intelligence*, vol. 2, pp. 327–352.

Part II

Inference

A Review on Distinct Methods and Approaches to Perform Triangulation for Bayesian Networks

M. Julia Flores and José A. Gámez

Departamento de Sistemas Informáticos & SIMD - i^3A
Universidad de Castilla-La Mancha
Campus Universitario s/n. Albacete. 02071
{julia,jgamez}@info-ab.uclm.es

Summary. Triangulation of a Bayesian network (BN) is somehow a necessary step in order to perform inference in a more efficient way, either if we use a secondary structure as the join tree (JT) or implicitly when we try to use other direct techniques on the network. If we focus on the first procedure, the goodness of the triangulation will affect on the simplicity of the join tree and therefore on a quicker and easier inference process.

The task of obtaining an optimal triangulation (in terms of producing the minimum number of triangulation links a.k.a. *fill-ins*) has been proved as an NP-hard problem. That is why many methods of distinct nature have been used with the purpose of getting as good as possible triangulations for any given network, especially important for big structures, that is, with a large number of variables and links.

In this chapter, we attempt to introduce the problem of triangulation, locating it in the compilation process and showing first its relevance for inference, and consequently for working with Bayesian networks. After this introduction, the most popular and used strategies to cope with the triangulation problem are reviewed, grouped into two main categories: heuristics and stochastic algorithms. Finally, another *family* of techniques could be understood as those based in decomposing the problem.

1 Introduction: the Compilation Process

If we consider that an expert system is composed of two main elements: *Knowlegde Base* (KB) + *Inference Engine* (IE), then a probabilistic expert system could be interpreted as a Bayesian network that models the particular problem (KB) and a secondary structure where inference is performed, normally an associated join tree[1] (IE).

---

[1] Also known as junction tree.

A Bayesian network[38] is made up of two elements:

- The directed graph $G = (\mathcal{V}, E)$, where $\mathcal{V}$ are the variables/nodes in the graph and E the set of edges present in the graph from which dependencies and independencies between domain variables can be extracted.
- The probability distribution, which is usually stored in the form of tables, including for every variable X_i its probability $P(X_i|pa(X_i))$ or just the prior probability $P(X_i)$ if the node has no parents[2]. This is because of the **factorisation rule** that states:

$$P(X_1,\ldots,X_n) = \prod_{i=1}^{n} P(X_i|pa(X_i)) \qquad (1)$$

When inference is performed for Bayesian networks we normally want to obtain the posterior probability for the problem variables given some facts or evidence e that we have previously observed. Then, we wish to obtain the value of $P(X_i|e)$ either for all variables or for a subset of them. Computing this marginal and posterior probability is not always a simple task. The structure of the JT is a representation of the network, and being free of cycles, it allows certain algorithms to propagate probabilities that result in general quite more efficient. JT is also used for other inference tasks as the search of explanations or abductive inference [19; 37].

There exist efficient algorithms to propagate evidence in acyclic networks (polytrees) [38], but unfortunately this is not common to have networks under this constraint. What has been broadly used in order to propagate in an exact manner for any kind of network is some form of the so-called clustering method, which basically entails grouping in a single node (cluster) variables that are strongly related together. Then, these groups/clusters are organised as a tree and sophisticated adaptations to the evidence propagation algorithm can also be done such as Lauritzen & Spiegelhalter method [32], Shenoy & Shafer propagation [43] or Hugin architecture [26]. A more recent technique is Lazy propagation [34].

We can find many other methods based on the previous ones that seek a more efficient evidence propagation and whose main feature is the possibility of an approximate inference to improve even more the speed-up, for instance Penniless propagation [9]. There is even a combination of two different techniques as Lazy Propagation with Penniless [10].

As a means to obtain this tree of clusters the BN has to be *compiled*. Compilation is the process of transforming a Bayesian network into a secondary structure called junction tree and it is a step to preprocess the network in order to make inference in a more efficient way on the whole. However, the resulting join tree from the compilation process is not unique for a given BN. Thus, a quite interesting feature is to have the ability of choosing the best

---

[2] $pa(X_i)$ is the set of parent nodes for variable X_i, i.e. the nodes which have a link pointing to X_i.

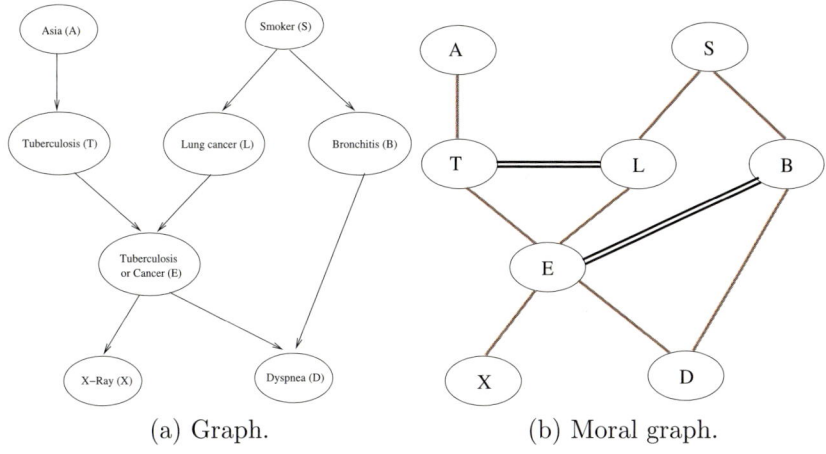

Fig. 1. Moralising graph for Asia network. Double lines indicate moral links.

tree among all possible ones for a particular BN. When using the join tree for inference, it is clear that the better this tree is the better our inference engine will work. Inference leads to a considerable number of operations, being our target to find a valid tree for our network, but also as simple as possible. We should note that a JT captures (in)dependence relations between variables. Nevertheless the groups of dependent variables might be bigger than necessary. Comparing two valid join trees, JT_1 and JT_2 related to the same network, the more complex the tree is, the more unnecessary dependencies it is actually including.

Let us indicate the four basic steps included in the compilation process[32]:

1. Obtain the *moral graph* from the original DAG that represented the *BN*.
2. Triangulate this *moral graph*.
3. Identify all the cliques.
4. Connect these cliques in order to form a valid join tree *JT*.

1.- Moralise the graph

The first step about moralisation takes the initial DAG G that forms the graphical part of the network and makes it undirected following these two rules: join those nodes with common parents by introducing a moral link[3], and drop directions of the directed edges. We show the graph for the *classical* network Asia or chest-clinic [32], originally directed (figure 1.(a)) and afterwards moralised (Figure 1.(b)).

---

[3] The origin of the term moral comes from *marrying* nodes with common children.

2.- Triangulate the moral graph

Second phase for compilation is the most problematic step: triangulation, since finding an optimal triangulation is an NP-hard problem [49]. To triangulate a graph it is needed to introduce a *chord* in those cycles of length greater than 3.

Normally, this process is done as the search of a deletion (or elimination) sequence σ which represents an ordering for all nodes in $\mathcal{V}$. Then, σ can also be seen as a function which relates every node $v_i \in \mathcal{V}$ with a unique number between 1 and $|\mathcal{V}|$. Therefore every node will have a position in the deletion sequence. Using this deletion sequence σ the necessary links to add (called *fill-ins*) will be obtained. If $adj(X_i)$ denotes the set of nodes adjacent to X_i in the undirected graph, then by *deleting* X_i we refer to the process of adding the necessary fill-ins in order to make $X_i \cup adj(X_i)$ a complete subgraph, and subsequently remove it and all its incident edges from the graph. The triangular graph G_T will be the result of adding to the moral graph the set ($\mathcal{F}$) of fill-ins added during the deletion process. That is, if $G_M = (\mathcal{V}, E_M)$ is the moral graph, then $G_T = (\mathcal{V}, E_M \cup \mathcal{F})$. Let us show one example for the moral graph of Asia network (Figure 2).

In Figures 2.(a)-(e) we proceed to use the deletion sequence σ showing the different steps for this ordering. So, as explained above, triangulation can be viewed as finding the deletion sequence. The method described in the previous paragraph is not complex, but the determination of a good deletion sequence is the most important step. For example, a sequence σ_2 as $\{D, S, L, B, E, T, A, X\}$ would produce the resulting triangulated graph shown in Figure 3.

As we can see in Figure 3 the graph is correctly triangulated, since there are no cycles of length 4 or greater without a chord. However, we have introduced 4 fill-ins instead of the only one needed with σ sequence. That introduces more *unnecessary* relations among nodes that will make a more dense triangulated graph, and will construct bigger clusters. The size of a cluster is crucial for the efficiency of join tree-based algorithms. Notice that the triangulation could still introduce many more fill-ins, for example if variable E is the first to be removed, 8 fill-ins in only one step will be introduced!!. And Asia network is a very simple one, since it presents only 8 nodes. It is obvious that the number of possible sequences is equal to all the possible permutations ($|\mathcal{V}|!$), that is, it increases more than exponentially in the number of nodes.

3.- Identify the cliques

Once the graph is triangulated it is time to determine which are the cliques (clusters) in this triangulated graph. Now we can give a proper definition:

Definition 1. *(Clique)* Let G be an undirected graph, then all the maximal complete subgraphs in G are called cliques.

In our particular case we will be interested in identifying the cliques corresponding to the triangulated graph, G_T. As we have already explained, these

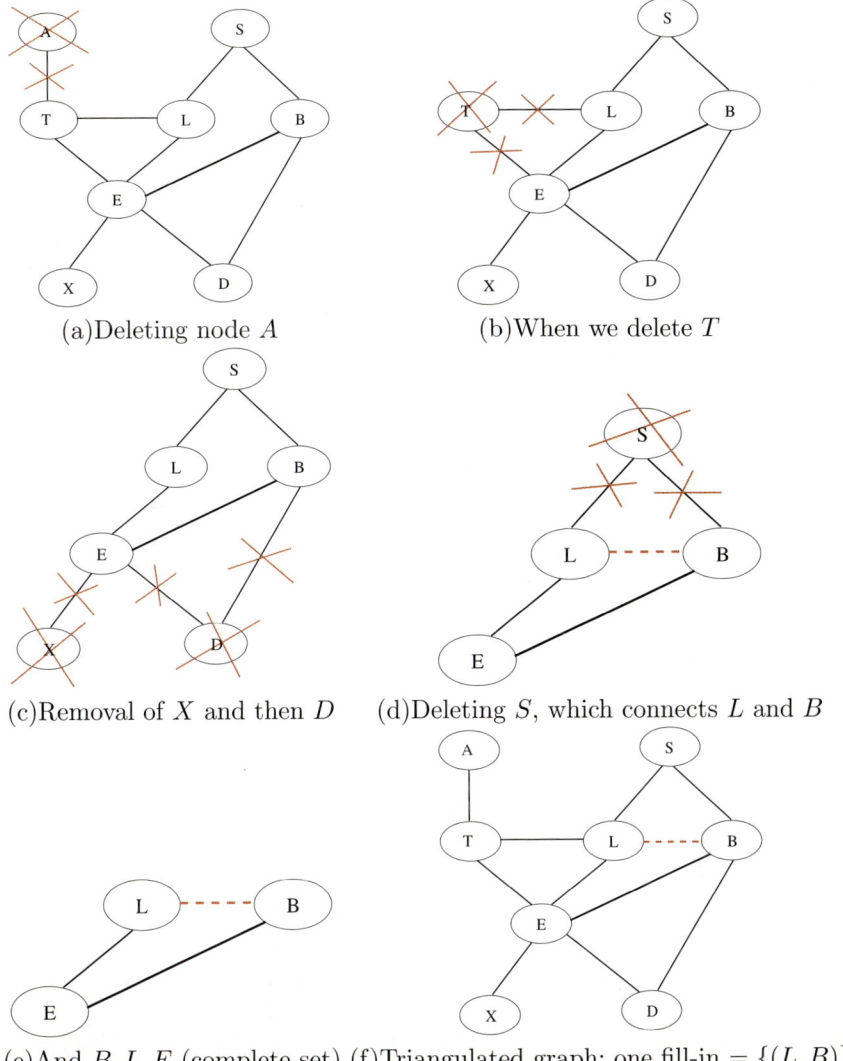

Fig. 2. Obtaining one possible triangulated graph for Asia. The used deletion sequence is $\sigma = \{A, T, X, D, S, B, L, E\}$.

cliques will be the nodes of the join tree. Since they are extracted from the triangulated graph they will also be dependent on the triangulation carried out, that is, on the introduced fill-ins.

Apart from determining the cliques we have to place them in a tree-shaped structure. And that leads us directly to the next step.

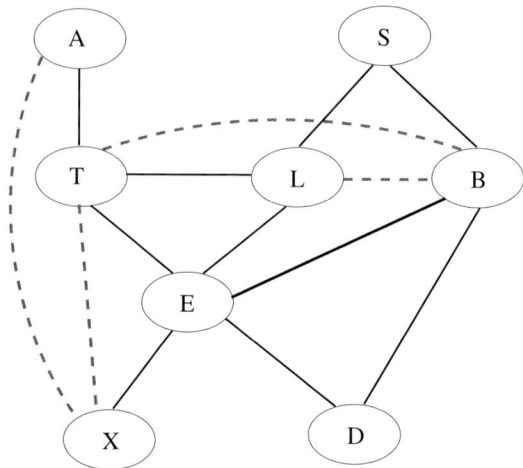

Fig. 3. Resulting triangulated graph when using deletion sequence $\sigma_2 = \{D, S, L, B, E, T, A, X\}$.

4.- Build the tree

It implies the establishment of the connections between cliques. From a triangulated graph there can be different possible join trees depending on the clique chosen as root, and sometimes a clique could be connected to different parents.

In order to guarantee that the running intersection property holds (see def. 2), we could use for example the maximum cardinality search [45] (MCS) with the aim of identifying the cliques and then connecting them in a tree.

Definition 2. *Running Intersection Property:* For every pair of clusters C_1 and C_2 whose intersection is not empty, that is, $V = C_1 \cap C_2 \neq \emptyset$, it is verified that V is contained in all nodes included in the path between C_1 and C_2.

Apart from MCS there are other alternative methods of ordering the cliques if no deletion sequence is available. And, on the other hand, it is possible to construct the tree from a deletion sequence if we know the cliques formed when deleting v_i and taking the reverse order of these. Figure 4 shows this second procedure for the Asia example with the previous deletion sequence σ. In any case, all these methods use he same idea: when identifying the cliques we need to have them ordered in a certain way that will assure the running intersection property. So that, this order will lead to an iterative way of constructing the tree.

i	v_i	$Clique_i$
1	A	{A,T}
2	T	{T,L,E}
3	X	{E,X}
4	D	{E,B,D}
5	S	{S,L,B}
6	B	{L,B,E}
7	L	
8	E	

(a) σ order

i:	v_i	$Clique_i$	Sep_i	parents
8:	E	-		
7:	L	-		
6:	B	{L,B,E}	∅	-
5:	S	{S,L,B}	[L,B]	6
4:	D	{B,E,D}	[B,E]	6
3:	X	{E,X}	[E]	6,4
2:	T	{T,L,E}	[L,E]	6
1:	A	{A,T}	[T]	2

(b) Inverse order

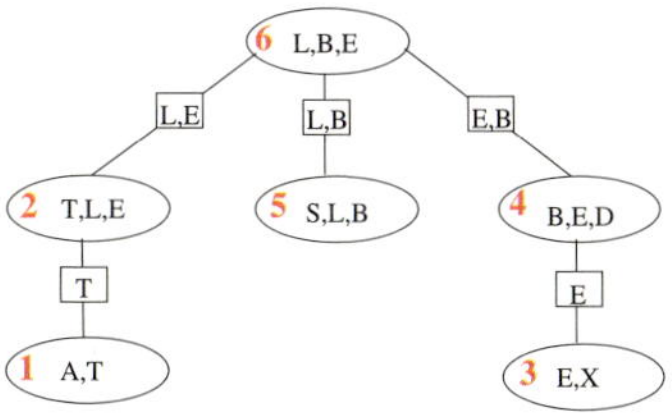

(c) Resulting join tree

Fig. 4. Ordering of the cliques from the triangulation $\sigma = \{A,T,X,D,S,B,L,E\}$ in Figure 2, identification of cliques and tree construction. Boldface when several options are possible indicates the randomly chosen one.

2 The Problem of Triangulation

As previously reviewed, the usual technique to triangulate a graph is selecting a *deletion/ elimination/ removing/ triangulation sequence* containing all the nodes in the graph. The method consists of an elimination process, following the sequence order, which will remove all the nodes. As pointed out before, finding an optimal deletion sequence is known to be an $\mathcal{N}P$-hard problem and coping with it involves a search over the space defined by all possible permutations of $|\mathcal{V}|$. Several approaches [40; 45; 28; 29; 8; 30; 24; 2; 1; 7; 20], most of them based on heuristics, have been proposed to search optimal solutions for this triangulation problem. Hence, these algorithms attempt to solve the problem of obtaining a good join tree from a BN, as next sections will show.

We should remark that these procedures to generate elimination sequences do not guarantee that we get an optimal triangulation either in terms of amount of added edges or in terms of the state space size when nodes are chosen randomly. Moreover, on average these measures in random sequences would normally be much larger than those corresponding to a minimum triangulation.

The question here is: *what we understand by an optimal triangulation?* If we refer to early work in graph triangulation what is understood by optimal triangulation is a *minimal triangulation*:

Definition 3. *(Minimal Triangulation)*
If we have a triangulation $\mathcal{F}$ for an undirected graph G, $G_T = (V, E \cup \mathcal{F})$, denoting the set of fill-ins adding during triangulation, $\mathcal{F}$ is said to be **minimal** if $\nexists \mathcal{F}'$ so that $\mathcal{F}' \subset \mathcal{F}$ and $\mathcal{F}'$ is a valid triangulation for G.

That is, a triangulation $\mathcal{F}$ is minimal if for each fill-in $f \in \mathcal{F}$, the graph, $(\mathcal{V}, E \cup \mathcal{F} - \{f\})$ is not any more triangulated. Strongly related with this concept are those deletion sequences that constitute *perfect orderings*:

Definition 4. *(Perfect ordering)*
Given an undirected graph $G = (\mathcal{V},E)$ and a sequence σ for $\mathcal{V}$, σ it is said to

be a *perfect ordering* if its use as deletion sequence to triangulate G does not produce any fill-in.

In fact, perfect orderings exist only for triangulated graphs and this is a way of checking if a given graph G is already triangulated.

In literature we can find many methods and studies for getting a minimal triangulation. The most known and first one is lexicographical search, LEX-M[40], providing a way of obtaining directly a minimal triangulation. The method consists of a particularly designed Breadth First Search (BFS), but labelling vertices (nodes) in a lexicographical way, LEX-BFS. LEX-M applies this labelling procedure along paths. More recent studies and (sometimes more efficient) methods have been designed [3; 6; 23; 39]. Among them, there is a recent successful technique [4] called MCS-M. This is a simplification of LEX-M where cardinality labels are used instead of lexicographical ones. In an analogical way, it applies the cardinality labelling of (neighbour) nodes along the path. MCS-M, as LEX-M, produces a minimal elimination ordering[4]. Even if both techniques could give different orderings it has been proved [47] that they create the same set of triangulations. LB-triang [5] is another recent algorithm that computes minimal triangulations with a computation complexity equal to the most efficient methods, and presenting certain properties that could make it especially interesting, such as it can also be implemented as an elimination scheme.

A different approach of obtaining a minimal triangulation is to follow an indirect path, that is, the method starts with a valid but non necessary minimal triangulation and then, it identifies the redundant fill-ins, so that eliminating them a minimal triangulation is obtained. If $\mathcal{F}$ is a set of fill-ins that make a graph G triangulated, $G = (\mathcal{V}, E \cup \mathcal{F})$, these methods identify a set of links $\mathcal{R}_{min} \subset \mathcal{F}$, so that $G_{min} = (\mathcal{V}, E \cup (\mathcal{F} \setminus \mathcal{R}_{min}))$ is minimally triangulated. The resulting minimal triangulation is therefore $\mathcal{F}_{min} = \mathcal{F} \setminus \mathcal{R}_{min}$. Among them, we find the method called *recursive thinning* designed by Kjærulf [28] and the algorithm proposed in [6].

The previous paragraphs discuss the problem of searching for minimal triangulations in the field of graph theory, however, when we move to the field of BNs triangulation things change. Now, the number of states of the variables plays a crucial role in the concept of optimal deletion sequences. Thus, in BNs triangulation a deletion sequence is optimal if it produces a triangulated graph whose associated join tree has minimal state space size. From this point of view a minimal triangulation does not need to be an optimal triangulation. As an example let us consider the Asia network previously introduced, where it is clear that triangulations $\mathcal{F}_1 = \{(L, B)\}$ and $\mathcal{F}_2 = \{(S, E)\}$ are minimal. If all the variables have two states (as it is the case) both triangulations are also optimal in the sense of state space size, but if we add an extra state to any of these variables, the two triangulations continue being minimal, but

---

[4] A deletion sequence that provides a minimal triangulation.

only one of them will be optimal with respect to the state space size criteria. In fact, Kjærulff [28] pointed out that most classical algorithms to look for a minimal triangulation are found to be highly ineffective when state space size is used as optimality criteria, being clearly surpassed by simple greedy heuristics (discussed in Section 3).

Although there is no general technique to perform always an optimal triangulation for any graph, there exist attempts to go as closer as possible as the algorithm QUICKTREE in [44], stated by the authors as the first algorithm that can optimally triangulate graphs with a hundred nodes in a reasonable time frame. In [21] we find a more modern *branch and bound* method, QUICKBB with similar purposes. In [12] graph triangulation is interestingly stated and solved as a constraint satisfaction problem. Another an more recent example is found in the commercial tool Hugin[5] where one technique for optimal triangulation has been implemented. This particular method, as indicated in [25] is a combined exact/heuristic method capable of producing an optimal triangulation, but only if sufficient computational resources (primarily storage) are available.

Finally, as a remark, several research works have shown that all existing methods for local computation will imply (maybe in a *hidden* way) a triangulation task. Besides, those methods not using a secondary structure like the junction tree either are less efficient or present another problem of NP-hardness [27].

3 Heuristic Greedy Methods

This group of techniques is characterised by establishing an ordering criterion based on the search rule *"the next node to be deleted is that one minimising $f()$"* where $f()$ is in function of one or several measures over the set of nodes within the graph $G = (\mathcal{V}, E)$. The most used measures [28] are based on:

1. Nodes $i \in \mathcal{V}$:
    - Size.- the number of variables: $s(i) = 1$.
    - Weight.- logarithm of the natural size: $w(i) = log_2 c(i)$, where natural size, $c(i) = |\Omega_{X_i}|$, i.e. the number of states of variable X_i. Depending on the author *Weight* is seen directly as $c(i)$[6].
    - Incident.- number of incident links in node/variable i within the moral graph: $|adj(X_i)|$.
2. Groups or clusters $C_i \in \mathcal{P}(V)$:
    - Size of the group: $V(C_i) = \sum_{j \in C_i} s(j) = |C_i|$. Then it refers to the number of variables in the group (or clique).

---

[5] http://www.hugin.com
[6] And that will be the approach when *minWeight* is referred in this chapter.

- Weight: $W(C_i) = \sum_{j \in C_i} w(j)$.
  As it happened with the nodes, sometimes this name is used for denoting the natural size: $S(C_i) = \prod_{j \in C_i} c(j)$.
- (*Fill-ins*).- number of introduced edges while the triangulation process: $F(C_i)$. That is, the number of edges necessary to make the group complete except those links already belonging to the moral graph.

We should indicate that other authors use the term size also for the *weight* measure. In this work we will try to write clearly which criterion we are referring to.

From these enumerated measures a set of criteria appear that give rise (among other) to the following heuristics[7]:

- **Minimum size**.- This criterion is based on selecting as the next node to be deleted that one which minimises the function $f(C_i) = V(C_i)$. That is, at each step, it chooses the next variable, among those not yet deleted, which produces a clique of minimum size.

As Rose[42] noted minimum size heuristics is fast[8], but it presents some drawbacks:

- It does not produce, in general, a perfect ordering (see def. 4) if the graph is already triangulated.
- It does not generally produce minimal triangulations.
- There exist examples for which the produced triangulation is arbitrarily greater than the triangulation obtained by *minimum fill* (see below).

- **Minimum weight**.- This criterion is based on selecting as the next node to be deleted that one which minimises the function $f(C_i) = S(C_i)$. This heuristics presents exactly the same advantages and disadvantages as *minimum size*. Note that when all nodes have the same weight both heuristics are identical.

This heuristics gives good results on the whole. It tries to minimise the total sum of the cliques sizes by minimising, at each step, the size of every clique which is being created. This does not guarantee that the total tree space state size (or weight) is optimal, since choosing one variable that produces a minimal clique could force us to produce bigger cliques when other variables are deleted later. However, in general, this method provides trees which are relatively manageable.

In [8] another particular heuristics based on the same idea arise, but attempting to avoid its weak points. The main underlying idea of these heuristics is that in the moment of deleting a variable it should be sought to minimise

---

[7] we will assume that $C_i = \{X_i\} \cup adj(X_i)$.
[8] It can be implemented in a computation time of order $O(|\mathcal{V}| + E')$, where $E' = E + |\mathcal{F}|$, being E the initial links and $\mathcal{F}$ those links added during triangulation.

the corresponding[9] $S(C_i)$. However, at the same time, the variable and all its corresponding links are deleted, which simplifies the resultant graph. Therefore, what they pursue is that this simplification for the resultant graph could also be taken into account.

Among the several heuristics that Cano and Moral [8] propose in their work, we find this approach called $H2$. This is very similar to *minimum weight*, at each case it chooses the variable X_i, among all the possible variables to be deleted, which minimises $S(i)/|\Omega(X_i)|$. With this feature when there are ties in (natural) size we remove first those variables of larger number of states, which leads to less complex cliques in the future formation of the tree.

- **Minimum fill**.- This criterion is based on selecting as the next node to be deleted that one which minimises the function $f(C_i) = F(C_i)$. In each case, it chooses the variable, among those not yet deleted, for which its elimination introduces a smaller number of fill-ins. This method presents the advantage of producing a perfect ordering when the graph is triangulated, but provokes the following drawbacks:

  - It is slightly slower than the minimum weight heuristics, that is because the adjacency set for every node has to be explored regarding edges.
  - In general it does not produce minimal triangulations.

There exist other heuristic techniques which attempt to tackle the problem of graph triangulation. In [24] they are classified in several groups:

1. Heuristics based on the relation between measures for nodes and clusters. They try to establish algebraic relationships between these two types of measures.
2. Heuristics based on measures for clusters and environments of nodes. They define the *k-neighbourhood* of a node by a distance k, which is determined as the minimum number of edges to go from one node to the other.
3. Compound heuristics. This sort of heuristics can be conceived as a hybridisation where the criterion to be used will vary on the different temporal stages of the triangulation process.
4. Iterative heuristics. Instead of using a single heuristic criterion to eliminate a node, they can make several iterations (each one with a different measure) in order to decide. They could be of k-iterations, where k could go from 1 (classical approach) until n ($n = |\mathcal{V}|$)). 2-iterations methods are studied in [24].

Since the complexity of finding a minimal triangulation grows as $n!$, it is not possible to carry out an exhaustive search directly, except when n is very small. Nevertheless, to construct an elimination order successively and to stop the execution when the total sum of the weights for the cliques (produced until this moment) exceeds the current smallest weight of a complete ordering could be of use to make an exhaustive search even for moderate-size graphs. Being

---

[9] Each deleted variable produces a group of variables, and when this is maximal it will therefore produce a clique.

triangulation an NP-complete problem, we can not generally expect that a *branch-and-bound* algorithm could find an optimal ordering within certain time limits. That is, the algorithm should finish either when the number of vertices exceeds a certain limit or when the number of the permutations left as discarded increases too slowly. Of course, the initial ordering will have a huge impact on the algorithm success. Thus, a *branch-and-bound* algorithm should be preferably used combining it with another quite faster algorithm (the first would be the last to apply) able of setting a "good" initial ordering for it with the goal of avoiding examining too many useless orderings and also with the goal of minimising the distance to some minimum ordering (we assume that low cost orderings are closer to a minimum one than a high cost ordering).

We could observe that the mentioned heuristics are only one-step lookahead, i.e., they just take into account that node which minimises a certain criterion if this node was deleted in the next step. We could then think of other heuristics able to look further than the next step. Unlike the heuristics above explained, about those looking beyond the next step, there is not much literature. This makes us think that, although they must produce better triangulations than the former, this improvement is not very significant in contrast to the complexity increase.

4 Methods Based on Stochastic Heuristics

The methods reviewed in section 3 present a good trade-off between the quality of the obtained deletion sequence (i.e. its associated join tree size) and the amount of computational resources (CPU time) required. Therefore, this kind of methods are suitable for *on-line* triangulation, that is, when there is a direct interaction between the user (knowledge engineer) and the compilation process and so a quick response is required. However, there are some occasions in which compilation can be carried out *off-line* and the time requirement can accordingly be relaxed. This is the case of compiling the final product (join tree or inference engine) to be given to the final user. At this stage, our goal should be to produce a junction tree as good as possible, because hundreds or thousands of propagations will be carried out over it. Thus, at this stage we can spend more time in the compilation process in order to achieve a better junction tree, and as a result algorithms requiring more CPU time are suitable.

When CPU time is not a strong constraint a family of algorithms arise as a good choice: *stochastic heuristic algorithms*. These algorithms are (in general) instances of metaheuristics that include stochastic behaviour so as to try to escape from local optima. Below we review some different approaches to the triangulation problem by using three outstanding representatives of this family of algorithms.

4.1 Simulated Annealing

Simulated annealing was (to our knowledge) the first stochastic heuristic used to solve the problem of Bayesian networks triangulation [29; 48].

Simulated annealing (SA) [50] is a stochastic optimisation algorithm used to look for global optima of NP-complete combinatorial problems having many local optima. SA is similar to a *hill climbing* algorithm, but sometimes it accepts to move to a worst solution in order to avoid to be trapped at local optima. The probability of accepting bad moves is controlled by a parameter t called *temperature*. Initially, during the exploration phase the temperature should be *high* in order to easily accept bad moves (exploration phase), but in successive iterations the temperature is decreased according to a cooling procedure and the probability of accepting cost-increases also decreases (exploitation phase).

When designing a SA algorithm for a given problem, different components have to be specified. Here we describe the algorithm proposed in [29].

- The search space is defined as all the possible elimination orderings for $\mathcal{V}$ (i.e., $|\mathcal{V}|!$ permutations).
- The neighbourhood of a deletion sequence σ is defined as all the deletion sequences $\{\sigma'\}$ obtained from σ by interchanging two of its nodes (positions).
- The cost/fitness of a deletion sequence is measured as the state space size of its associated join tree.

These three design decisions together with an appropriate cooling schedule are enough to have a SA algorithm that solves the triangulation problem, however Kjærulff [29] adds the following improvements in order to enhance the performance of the algorithm:

- Local computation of neighbour configurations (sequences). An efficient method is proposed that evaluates a new deletion sequence by only considering the cliques obtained when deleting the variables between the two interchanged positions.
- An additional parameter is introduced: the *radius*. The idea is to define a window w (length), such that, only positions inside this window can be interchanged. Initially a large window is set, so that free motion in the search space can be done (exploration). However, when the search process advances the window is reduced and so uniquely close neighbours are explored (exploitation). Because of the local evaluation proposed, this parameter is strongly related with the CPU time efficiency of the algorithm, since the smaller the window is the more efficiently the neighbours are evaluated.

The experiments carried out in [29] show that depending on the graph, on the average the state-space size of the join trees obtained by *minWeight* heuristics are 3 or 5.5 times larger than those obtained by SA.

4.2 Genetic Algorithms

SA carried out a local search that tries to escape from local optima by using a Monte Carlo method. On the contrary, Genetic Algorithms [22; 35] (GAs) do a *global* search by using a population of candidate solutions instead of a single one. In a GA we start with an initial population having solutions distributed over all the search space, then all the solutions are evaluated and a new population is obtained by: (1) selecting some of the individuals of the previous population (usually, higher fitness implies higher probability of being selected); (2) recombining some of the individuals of the previous population, that is, two individuals (parents) are selected and two off-spring are obtained by applying a crossover operator that mixes the representation of the parents; and (3) mutating some of the selected individuals (with low probability a small change is carried out over the individuals).

In the case of Bayesian networks triangulation the first contributions based on GAs are [24; 30]. The main features of the GA developed in [30] (that obtains similar results to the SA algorithm described in [29]) are:

- Individuals are represented by permutations (deletion sequences).
- The initial population is randomly generated in order to have initial points distributed uniformly through the search space.
- The selection mechanism is based on the rank of the individuals according to their fitness.
- A steady state GA is used. That is, in each generation instead of replacing the whole population, only a pair of off-spring are generated (selection+crossover+mutation) and (only if they are better) they replace the two worst individuals of the current population.
- Among the different specific operators for the case of permutations, Larrañaga et al. [30] found the combination of CX as crossover and ISM for mutation to be the best ones.

After these initial proposals different authors have used GAs to look for optimal deletion sequences. Concretely, Gámez and Puerta [20] slightly modified the algorithm proposed in [30] (where simplicial nodes are previously removed and informed initialisation of the population is used) obtaining better results (in terms of CPU time and join tree size).

4.3 Ant Colony Algorithms

The third stochastic heuristics we are going to review is *Ant Colony Optimisation* (ACO) [14].

Combinatorial optimisation based on *ant colony systems* is a recent metaheuristics that takes its basis on one aspect of ant behaviour, the ability to find shortest paths. Thus, in ACO a set of *artificial* ants (or agents) is used to look for the shortest paths in the same way as a *real* ant will do it: *following the pheromone track*. Concretely, when an artificial ant is located in a branch and has to take a decision, it makes a probabilistic decision biased by the amount of pheromone deposited on the different branches. Due to the fact that the shortest paths are more frequently visited, they receive a higher amount of pheromone and thereby become more attractive for the subsequent ants. In this way the amount of pheromone plays the role of memoristic information, but in ACO the decisions are based as well on heuristic information. Thus, in the initial Ant System when an ant k is located at node i, it chooses node j as the following node to be visited with a probability proportional to:

$$p_k(i,j) = \begin{cases} \frac{[\tau_{ij}]^\alpha \cdot [\nu_{ij}]^\beta}{\sum_{u \in J_k(i)} [\tau_{iu}]^\alpha \cdot [\nu_{iu}]^\beta} & \text{if } j \in J_k(i) \\ 0 & \text{otherwise} \end{cases} \qquad (2)$$

where τ_{ij} is the amount of pheromone in edge (i,j); ν_{ij} represents *heuristic information* (knowledge) about the problem; $J_k(i)$ is the set of nodes for which there is a direct path from node i and not yet visited by ant k; and α and β are two parameters used to control the relative importance of pheromone with respect to heuristic information.

Although this is not a complete description of ACO-based algorithms (see e.g. [14]), one of the main differences in regard to previous discussed metaheuristics is already evident: in ACO algorithms is very easy to integrate problem domain knowledge. The use of heuristic knowledge in ACO algorithms helps to focus upon the search process (and speed it up), and this is just the point studied in [20] where ACO algorithms are applied to the Bayesian networks triangulation problem. Below, we describe the main points of the approach presented in [20]:

- *Representation.* The first thing we need is a graph over which ants will walk. In [20] the complete graph defined over the network variables is used in such a way that it is always possible to reach a node i from a node j for every pair of nodes (i,j). In consequence, there is a graph-form representation equivalent to the one used for the TSP (Travelling Salesman Problem [13]), but in the asymmetrical case, on account of the fact that it is not generally the same deleting X_i before deleting X_j as in the reverse order.
- *Reduction.* In this work simplicial nodes are removed before starting the combinatorial optimisation problem. In this way the search space is (in general) drastically reduced and the search results faster.
- *Heuristic knowledge.* In ACO algorithms the heuristic knowledge is usually static, that is, it can be calculated before any ant is launched. This is not the case in the triangulation problem, because the knowledge associated to edge (i,j) does not only depend on itself, but on the nodes previously visited (deleted). In [20] each ant implements a greedy heuristics (minSize, minWeight, ...), that is, each ant carries out a triangulation over its own copy of the graph. In this way, the matrix of pheromone will be a global structure, while the heuristic knowledge will be local to each ant.

- *Origin nodes.* As solutions are permutations any node will be valid as the starting point. However in triangulation it has no sense to consider (equally) all the nodes as possible origins. Thus, in [20] the probability of a node to be chosen as origin is calculated as a function of its heuristic value (minFill, minSize, ...).
- *Transition rule.* A variant of the rule described in eq. 2 is used in [20]:

$$j = \begin{cases} \arg\max_{u \in J_k(i)} \{[\tau_{iu}] \cdot [\nu_{iu}]^\beta\} & \text{if } q \leq q_0 \\ J & \text{if } q > q_0 \end{cases} \quad (3)$$

where q is a random number uniformly distributed in [0,1], and $J \in J_k(i)$ is a node selected according to Eq. (2) with $\alpha = 1$. This is the rule proposed in Ant Colony Systems algorithms and it explicitly allows tuning (q_0) the amount of effort devoted to exploration/ exploitation.

Different experiments over a set of real and artificially generated networks are carried out in [20]. The obtained results turned out to be quite successful, regarding both accuracy and efficiency. Thus, ACO algorithms always obtain (on average) deletion sequences better that GAs, and due to the heuristic knowledge they use, the number of evaluated permutations (deletion sequences) is considerably smaller, making this approach faster than the one based on GAs. Furthermore, it presents also the advantage of having an ant-autonomy feature that could make them fit perfectly in a parallel environment with the aim of gaining efficiency.

5 Methods Based on Decomposition Techniques

Apart from the two previous approaches which are probably the most widely used, other triangulation techniques can be found in the literature, such as divide and conquer techniques based on the concept of *treewidth*[10] [1; 2]. The idea here is to use a different algorithm to triangulate in which the minimum vertex cut method is needed [15]. At each iteration it finds a minimum set of vertices X which being removed from graph G splits it into two disconnected components A and B such that $A \cup B \cup X = V$. This set X is then called the *minimum vertex cut*. This general algorithm proceeds in the two smaller problems $G[A \cup X]$ and $G[B \cup X]$, that is, those subgraphs obtained by projecting G on $A \cup X$ and $B \cup X$ respectively. And it goes on in this way so that each subgraph is triangulated such that X becomes a clique in it. As we will see MPSD is somehow based on this principle as well.

Within these techniques based on decomposition another related research line is the socalled *recursive hypergraph partitioning* (or simply hypergraph partitioning). They are quite broadly used in the context of VLSI design [41], but we can find some example of its application to join trees [11].

---
[10] Treewidth = number of variables, minus one, included in the biggest clique in the join tree

There exists another method capable of simplifying the triangulation task. In this case, it deals with a process to be performed prior to triangulate with the chosen method. In bibliography we can find it with different names, being *simplicial* (def. 5) the most broadly used. In [24] it is presented as *reduction*, and consists in eliminating all those nodes that, together with their neighbours, form a complete subgraph, i.e., no fill-in has to be added. This part of the network is then already triangulated and deleting them is not going to add any new fill-in. Another approach uses the application of preprocessing rules in order to reduce the graph [7]. In this approach the authors have developed a set of sophisticated *safe reduction rules* (being the first one removing simplicial nodes as well) to apply onto the graph before triangulation. The results are good, since a smaller (sub)graph has to be triangulated, but the technique requires more computation time than greedy heuristics.

Definition 5. *(Simplicial node)*
Let $G = (\mathcal{V}, E)$ be an undirected graph. A node $N \in \mathcal{V}$ is said to be **simplicial** if this node N together with its set of neighbours, $\{N \cup adj(N)\}$, form a complete node set.

5.1 A Recent Triangulation Approach Based on the *Divide & Conquer* Methodology

In this section we are going to describe the method *triangulation by retriangulation* which combines some of the philosophies previously noted. Firstly, as treewidth-oriented techniques, it uses a method for dividing the total graph in smaller components. Olesen and Madsen[36] launched the possibility of applying the Maximal Prime Subgraph (MPS) Decomposition to the problem of triangulation. So, the idea is to retriangulate separately each MPS, since it has been proved to be perfectly valid for the final result. And, secondly, for those portions it will apply some methods of triangulation based on the procedures to get a elimination sequence reviewed above. Then, the work in [16] exploited the previous idea by using both greedy heuristic algorithms and stochastic ones (genetic algorithms).

5.2 Maximal Prime Subgraph Decomposition

It is clear that the decomposition of an undirected graph can be used as a tool for the triangulation procedure. We can consider the problem as a set of solvable subgraphs, following *divide and conquer* philosophy (see Fig. 5).

In this particular case the <u>D</u>ecomposition using <u>M</u>aximal <u>P</u>rime <u>S</u>ubgraphs (MPSD)[11] of an undirected graph constitute an intermediate step in a new approach for triangulation. This idea [36] consists of working separately on different parts of the initial graph. The triangulation for each graph will be

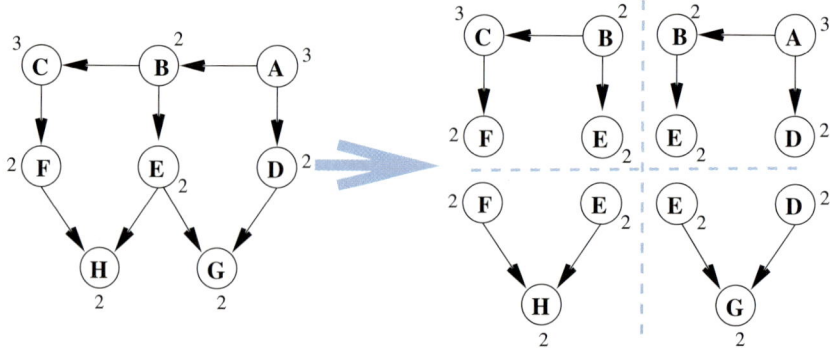

Fig. 5. Trying to reduce the problem of triangulating a network with n nodes to a set of k smaller subproblems: triangulate each subgraph S_k separately.

done separately and the global solution will then be the sum/combination of local solutions for smaller and independent graphs.

Let us just formalise the concept of maximal prime subgraph, for that, we also introduce the definition for **decomposition** (def. 6) of a graph and the characteristic for a graph of being **decomposable** (def. 7). Both of them can be easily related from previously presented ideas, since for constructing the JT we have made some kind of decomposition (MPS Tree will be the one which accomplishes the complete separators condition) whereas triangulated graphs are guaranteed to be decomposable [31] and that is somehow the justification for the necessity for a triangulation step. That is the reason why from here, we will refer to a decomposable graph as a triangulated graph.

Definition 6. *(Graph decomposition)*
Let $G = (\mathcal{V}, E)$ be an undirected graph, and let A and B be two sets of vertices in G, G can be decomposed in A and B if and only if the following conditions are satisfied:

- $A \cup B = \mathcal{V}$,
- $A \setminus B \neq \emptyset$,
- $B \setminus A \neq \emptyset$,
- Both $A \setminus B$ and $B \setminus A$ are separated by $A \cap B$ and
- And $A \cap B$ is a complete subset (called **clique separator**).

Definition 7. *(Decomposable graph)*
If a graph G and its subgraphs can be decomposed recursively until all the subgraphs are complete, then the graph is decomposable[12].

---

[11] Also known as decomposition by clique separators.
[12] Note that a graph can be decomposed without being decomposable.

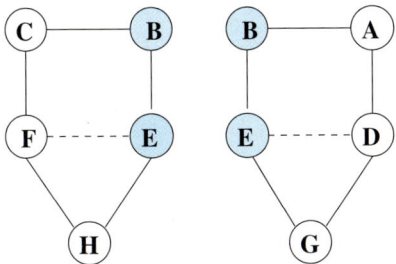

Fig. 6. A simple example of graph decomposition where $\{B, E\}$ is the clique separator for the BN in Fig. 8.a.

Then, it is said that a graph is *reducible* if it can be decomposed, that is, its set of nodes contains a clique separator, otherwise the graph is said to be *irreducible/**prime**/non-separable*. And this leads directly to def. 8:

Definition 8. *(Maximal Prime Subgraph)*
A subgraph $G(A) = (\mathcal{V}, E)^{\downarrow A}$ of a graph G is a Maximal Prime Subgraph of G if $G(A)$ is irreducible and $G(B)$ is not irreducible $\forall B$ so that $A \subset B \subseteq \mathcal{V}$.

Finally, from the previous concepts it just remains to indicate what the Maximal Prime Subgraph Decomposition is[13]:

Definition 9. *(Maximal Prime Subgraph Decomposition)*
Let $G = (\mathcal{V}, E)$ be an undirected graph. Its Maximal Prime Subgraph Decomposition is the set of induced maximal prime subgraphs of G resulting from a recursive decomposition of G.

$$G \longrightarrow G^M \longrightarrow G^{T_{min}} \longrightarrow \mathcal{T} \cdots\cdots\cdots\blacktriangleright \mathcal{T}_{MPD}$$

Fig. 7. Graphical process that indicates how to reach the MPST $\mathcal{T}_{MPD}$ from a Bayesian network $BN = (G, \mathcal{P})$, using as an intermediate step the join tree $\mathcal{T}$.

To obtain the MPSD of an undirected graph [45; 46; 33], the method in [36] is especially interesting for us, since it is based on the join tree constructed from a BN. The decomposition of the graph in MPSs is returned in a form of a tree: the <u>M</u>aximal <u>P</u>rime <u>S</u>ubgraph Decomposition <u>T</u>ree (MPST), sometimes denoted as $\mathcal{T}_{MPD}$. Figure 7 shows graphically this process to obtain the MPST in an schematic way. The MPST will express by itself a decomposition (every tree node will denote a group of variables belonging to the same MPS). We could say basically that once the triangulation from which a join tree is

---
[13] It can be proved that this decomposition is unique for an undirected graph, as it is the moral graph.

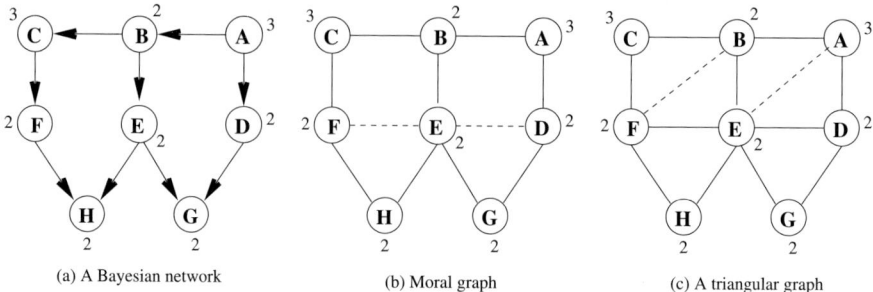

Fig. 8. Example of a Bayesian network (a), its associated moral graph (b) and a possible triangulation for it (c). Numbers next to each node indicate the number of states for the corresponding variable

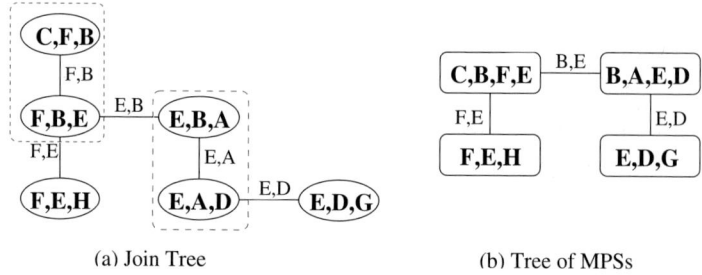

Fig. 9. Construction of the MPSs tree and the obtained result.

obtained is assured to be minimal, if we aggregate those cliques whose separators are not complete in G^M, we obtain the corresponding maximal prime subgraphs. If we have the network, moral graph and triangulated graph from Figure 8, then Figure 9 will show a corresponding join tree and the associated MPS tree. Since this is the necessary to guarantee that triangulations are minimal (def. 3), this can be achieved simply by using *recursive thinning*.

5.3 Triangulation of Bayesian networks by re-triangulation

It has been proved that is perfectly valid [18] to triangulate every subgraph in an independent way from the rest, and make a global triangulation of the graph by the combination of these *partial* triangulations. Retriangulating a graph can be worthy, even when the same triangulation method is used twice. That is to say, the same triangulation method is applied (first) when triangulating the moral graph, and (secondly) when triangulating each MPS separately.

The algorithm of RETRIANGULATION is as listed here:

1. Obtain moral graph G^M from BN.

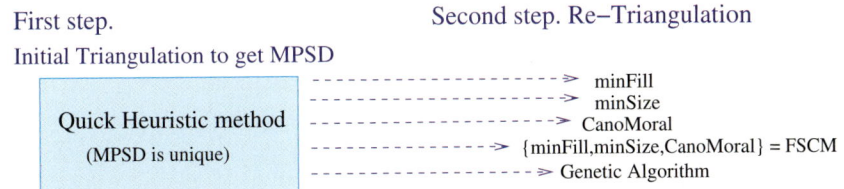

Fig. 10. Scheme for the retriangulation experiments: first step provides the decomposition and second step varies on the method M_i used to retriangulate the subgraphs S_i, giving place to the set of partial triangulations $\mathcal{F}_i$.

2. Using whichever triangulation method obtain the MPS decomposition $D = \{S_1, S_2, \ldots S_k\}$
3. Triangulate each S_i from D using a certain triangulation method M_i. Let $\mathcal{F}_i$ be the obtained triangulation for subgraph S_i.
4. Return the obtained result $\mathcal{F} = \cup_{i=1}^{k} \mathcal{F}_i$.

An experimental evaluation of this technique using it for real networks [16] was carried out, following the sketch in Figure 10. The obtained results have been quite satisfactory and should be regarded from two different points of view or goals:

1. Join Tree state space size:
   - Considering the heuristic techniques the tree size is generally better (smaller) when using the re-triangulation method, this difference is even bigger when we use a combined method called $FSCM^{14}$.
   - With respect to GA, the sizes of the obtained trees are quite similar.
2. CPU time:
   - Performing re-triangulation for heuristics implies a little more time, but this is due to the extra task of constructing the MPS Tree. This difference is only slightly noticeable for the heuristics because they are normally quicker.
   - FSCM is obviously three times slower than the rest (it tries the three methods), but since heuristic techniques are really quick and selecting the best one produces much better global results, it is a low price worth paying.
   - And the most important consequence related to time measuring is that a huge speed up is provided to GA. For example, in the case of Network Munin4 it can reduce (in this experiment settings) triangulation time from more than one day to less than 4 hours.

---

[14] It denotes a greedy technique that for every subgraph S_k tries the 3 different heuristics (min<u>F</u>ill, min<u>S</u>ize and <u>C</u>ano<u>M</u>oral) and it chooses the one that gives the best result, i.e., the smallest size. Since there is no an optimal heuristic for all cases, FSCM selects the best method M_i^* for every subgraph.

From these experiments and results we can mainly conclude that there exist some possibilities to optimise these results and to explore new combinations to get even better triangulations.

5.4 MPSD-based Incremental Compilation

In the last explained case, following the idea of *divide and conquer*, the natural decomposition of a graph into its prime subgraphs was exploited. In any manner, this decomposition tool is not reserved for triangulation itself, it can become even more powerful. The use of MPSD can be extended to the whole process of compilation. Since triangulation is the most *expensive* phase of compilation and this can be correctly and separately distributed among MPSs, we could sketch other techniques so that compilation could be less "dependent" on the global triangulation. For that, there exist a proposal to look more closely into the possibility of retriangulating some portions of the BN, and to use MPSD in order to perform incremental triangulations. This idea led directly to work on developing the approach of MPSD-Incremental Compilation of Bayesian networks [17]).

6 Main Conclusions

From this whole chapter and the analysed issues we can draw some main conclusions.

First, triangulation is still an unsolved problem, at least for a general case. There good techniques that might be *adjusted* depending on the problem, but not an optimal one (produced in a reasonable amount of time).

But on the other hand, triangulating has also been proved to be an unavoidable step in the computation of Bayesian networks. As a consequence, for solving queries and perform inference we must cope with this problem. This necessity for triangulating has brought about several endeavours to handle this problem, and the techniques found in literature are of distinct nature. So, we have shown most of the known approaches to tackle triangulation classifying them mainly in heuristic, stochastic algorithms and also techniques based on the division/decomposition of the problem. The last described algorithm (*ReTriangulation*) is of interest because it covers and integrates these three discussed manners of undertaking and solving the triangulation task.

Even though we find strong foundations for triangulation on the theory of graphs in literature, it is obvious that triangulation is still a quite open field to optimisation. It is illustrative to point out how this problem is already being studied in diverse mathematical and computing disciplines apart from Bayesian networks (probabilistic systems) such as the area of graph theory, VLSI (Very Large Scale Integration) circuits, data bases, constraint processing and graph algorithms.

References

[1] Amir, E. (2001) Efficient approximation for triangulation of minimum treewidth. In:*Proceedings of the 17th Conference on Uncertainty in Artificial Intelligene (UAI-01)*, 7–15
[2] Becker A. and Geiger D. (1996) A sufficiently fast algorithm for finding close to optimal junction trees. In: *Proceedings of the 12h Annual Conference on Uncertainty in Artificial Intelligence (UAI–96)*, 81–89
[3] Berry A., Blair J.R.S. and Heggernes P. (2002) Maximum Cardinality Search for Computing Minimal Triangulations. *WG '02: Revised Papers from the 28th International Workshop on Graph-Theoretic Concepts in Computer Science.* In: Lecture Notes in Computer Science, **2573**, 1–12, Springer Verlag
[4] Berry A., Blair J.R.S., Heggernes P. and Peyton B.W. (2004) Maximum Cardinality Search for Computing Minimal Triangulations of Graphs. *Algorithmica* **39**(4): 287–298
[5] Berry A. , Bordat J-P., Heggernes P., Simonet G. and Villanger Y. A wide-range algorithm for minimal triangulation from an arbitrary ordering. To appear in *Journal of Algorithms*
[6] Blair J.R.S, Heggernes P. and Telle J.A. (2001) A practical algorithm for making filled graphs minimal. *Theoretical Computer Science* **205**(1-2):125–141
[7] Bodlaender H.L., Koster A.M. et al (2001) Pre-processing for triangulation of probabilistic networks. In: *Proceedings of the 17th Conference on Uncertainty in Artificial Intelligene (UAI-01)*, 32–39
[8] Cano A. and Moral S. (1994) Heuristic algorithms for the triangulation of graphs. In: *Proceedings of the 5th International Conference on Information Processing and Management of Uncertainty in Knowledge Based Systems (IPMU)*, **1**, 166–171, Paris (France)
[9] Cano A., Moral S. and Salmerón A. (2000) Penniless propagation. *International Journal of Intelligent Systems*, **15**:1027–1059
[10] Cano A., Moral S. and Salmern A. (2002) Lazy evaluation in Penniless propagation over join trees. *Networks*, **39**:175–185
[11] Darwiche A. and Hopkins M. (2001) Using Recursive Decomposition to Construct Elimination Orders, Jointrees, and Dtrees. *Lecture Notes in Computer Science*, **2143**, 180–190, Springer Verlag.
[12] Dechter R. (2003) *Constraint Processing*. Morgan Kaufmann, 2003
[13] Dorigo M. and Gambardella L. (1997) Ant Colony System: a Cooperative Learning Approach to the Traveling Salesman Problem. *IEEE Transactions on Evolutionary Computation*, **1**:53–66
[14] Dorigo M. and Stützle T. *Ant colony optimization*. MIT Press, 2004.
[15] Even S. and Tarjan R.E. (1975) Network flow and testing graph connectivity. *SIAM Journal on Computing*, **4** :507–518

[16] Flores M.J. and Gámez J.A. (2003) Triangulation of Bayesian networks by retriangulation. *International Journal of Intelligent Systems* **18**(2):153–164
[17] Flores M.J., Gámez J.A. and Olesen K.G. (2003) Incremental Compilation of Bayesian networks. In: *Proceedings of the 19th Annual Conference on Uncertainty in Artificial Intelligence (UAI–03)*, Morgan Kaufmann, 233–240
[18] Flores M.J. (2005) Bayesian networks inference: Advanced algorithms for triangulation and partial abduction. PhD thesis, Departamento de Sistemas Informaticos (Computing Systems Department), University of Castilla - La Mancha UCLM, Spain
[19] de Campos L.M., Gámez J.A., and Moral S. (2002) On the problem of performing exact partial abductive inference in Bayesian belief networks using junction trees. In: B. Bouchon-Meunier , J. Gutierrez, L. Magdalena, and R.R. Yager, editors, *Technologies for Constructing Intelligent Systems 2: Tools*, 289–302 Springer Verlag
[20] Gámez J.A. and Puerta J.M. (2002) Searching for the best elimination sequence in Bayesian networks by using ant colony optimization. *Pattern Recognition Letters.*, **23**:261–277
[21] Gogate V. and Dechter R. (2004) A Complete Anytime Algorithm for Treewidth. In: *Proceedings of the Twentieth Annual Conference on Uncertainty in Artificial Intelligence (UAI–04)*, pp. 201–208, AUAI Press
[22] Goldberg D.E. (1989) *Genetic algorithms in search, optimization, and machine learning.* Addison-Wesley
[23] Heggernes P. and Villanger Y. (2002) Efficient Implementation of a Minimal Triangulation Algorithm. In: *ESA '02: Proceedings of the 10th Annual European Symposium on Algorithms.* Lecture Notes in Computer Science, **2461**, 550–561, Springer Verlag
[24] Hernández L.D. (1995) Diseño y validación de nuevos algoritmos para el tratamiento de grafos de dependencias (Validation and design of new algorithms to dependency graph processing.) Doctoral thesis, Dpto. de Ciencias de la Computación e I.A. Universidad de Granada, Spain
[25] HUGIN Expert A/S. *API manual for the Hugin Decision Engine V6.3.* http://developer.hugin.com/documentation/API_Manuals/
[26] Jensen F.V., Lauritzen S.L. and Olesen K.G. (1990) Bayesian updating in causal probabilistic networks by local computation. *Computational Statistics Quarterly*, **4**:269–282
[27] Jensen F.V. and Jensen F. (1994) Optimal junction trees. In: *Proceedings of the Tenth Annual Conference on Uncertainty in Artificial Intelligence (UAI–94)*, 360–366, Morgan-Kaufmann
[28] Kjærulff U. (1990) Triangulation of graphs - algorithms giving small total space. Technical Report R 90-09, Department of Mathematics and Computer Science. Institute of Electronic Systems. Aalborg University
[29] Kjærulff U. (1992) Optimal decomposition of probabilistic networks by simulated annealing. *Statistics and Computing*, **2**:7–17

[30] Larrañaga P., Kuijpers C.M., Poza M. and Murga R.H. (1997) Decomposing Bayesian networks: triangulation of the moral graph with genetic algorithms. *Statistics and Computing*, **7**:19–34

[31] Lauritzen S.L., Speed T.P. and Vijayan K. (1984) Decomposable graphs and hypergraphs. *Journal of the Australian Mathematical Society* Series A **36**: 12–29

[32] Lauritzen S.L. and Spiegelhalter D.J. (1988) Local computations with probabilities on graphical structures and their application to expert systems. *J.R. Statistics Society. Series B*, **50**(2):157–224

[33] Leimer H.G. (1993) Optimal decomposition by clique separators. *Discrete Mathematics*, **113**:99-123

[34] Madsen A.L. and Jensen F.V. (1999) Lazy Propagation: A Junction Tree Inference Algorithm based on Lazy Evaluation. *Artificial Intelligence*, **113** (1-2):203–245, Elsevier Science Publishers, North-Holland

[35] Michalewicz Z. (1996) *Genectic Algorithms + Data Structures = Evolution Programs*. Springer-Verlag

[36] Olesen K.G. and Madsen A.L. (2002) Maximal prime subgraph decomposition of bayesian networks. *IEEE Transactions on Systems, Man and Cybernetics*, Part **B**:(32), 21–31

[37] Park J.D. and Darwiche A. (2003): Solving MAP Exactly using Systematic Search. In: *Proceedings of the 19th Annual Conference on Uncertainty in Artificial Intelligence (UAI–03)*, Morgan Kaufmann, 459–468

[38] Pearl, J. (1988) *Probabilistic Reasoning in Intelligent Systems: Networks of Plausible Inference*. Morgan Kaufmann, San Mateo

[39] Peyton B.W. (2001) Minimal Orderings Revisited. *SIAM Journal on Matrix Analysis and Applications*, **23**(1):271–294

[40] Rose D., Tarjan R.E. and Lueker G.S. (1976) Algorithmic aspects of vertex elimination graphs. *SIAM Journal on Computing*, **5**:266–283

[41] Selvakkumaran N. and Karypis G. Multi-Objective Hypergraph Partitioning Algorithms for Cut and Maximum Subdomain Degree Minimization *IEEE Transactions on Computer Aided Design* (in press)

[42] Rose D. (1972) A graph theoretic study of the numerical solution of sparse positive definite systems of linear equations. In: R. Reed ed. *Graph Theory and Computing*, 183–217, Academic Press, New York

[43] Shafer G.R. and Shenoy P.P. (1990) Axioms for probability and belief-function propagation. In: R.D. Shachter, T.S. Levitt, L.N. Kanal and J.F. Lemmer (eds.), *Uncertainty in Artificial Intelligence*, **4**, 169–198, Elsevier Science Publishers B.V. (North-Holland)

[44] Shoikhet K. and Geiger D. (1997) Practical algorithm for finding optimal triangulations. In: *Proceedings of the National Conference on Artificial Intelligence*, AAAI, USA, 185–190

[45] Tarjan R.E. and Yannakakis M. (1984) Simple linear-time algorithms to test chordality of graps, text acyclicity of hypergraphs and selectively reduce acyclic hypergraphs. *SIAM Journal on Computing*, **13**:566–579

[46] Tarjan R.E. (1985) Decomposition by clique separators. *Discrete Mathematics*, **55**:221–232
[47] Villanger Y. (2004) LEX M versus MCS-M. Technical Report Reports in Informatics 261, University of Bergen, Norway
[48] Wen, W. X. (1990) Decomposing belief networks by simulated annealing. In: (C. P. Tsang, ed.) Proceedings of the Australian Joint Conference on Artificial Intelligence, 103–118
[49] Wen, W.X. (1991) Optimal decomposition of belief networks. In: P.P. Bonissone, M. Henrion, L.N. Kanal and J.F. Lemmer (eds.), *Uncertainty in Artificial Intelligence*, **6**, 209–224 North-Holland
[50] Kirkpatrick S., Gelatt C.D. and Vecchi M.P. (1983) Optimization by simulated annealing. *Science*, **220**:671–680

Decisiveness in Loopy Propagation

Janneke H. Bolt and Linda C. van der Gaag

Department of Information and Computing Sciences, Utrecht University
P.O. Box 80.089, 3508 TB Utrecht, The Netherlands

Summary. When Pearl's algorithm for reasoning with singly connected Bayesian networks is applied to a network with loops, the algorithm is no longer guaranteed to result in exact probabilities. We identify the two types of error that can arise in the probabilities yielded by the algorithm: the cycling error and the convergence error. We then focus on the cycling error and analyse its effect on the decisiveness of the approximations that are computed for the inner nodes of simple loops. More specifically, we detail the factors that induce the cycling error to push the exact probabilities towards over- or underconfident approximations.

1 Introduction

Bayesian networks [1] by now are being applied for a range of problems in a variety of domains. Successful applications are being realised for example for medical diagnosis, for traffic prediction, for technical troubleshooting and for information retrieval. In these applications, probabilistic inference plays an important role. Probabilistic inference with a Bayesian network amounts to computing (posterior) probability distributions for the variables involved. For networks without any topological restrictions, inference is known to be NP-hard [2]. For various classes of networks of restricted topology, however, efficient algorithms are available, such as Pearl's propagation algorithm for singly connected networks. The availability of these algorithms accounts to a large extent for the success of current Bayesian-network applications.

For Bayesian networks of complex topology for which exact inference is infeasible, the question arises whether good approximations can be computed in reasonable time. Unfortunately, also the problem of establishing approximate probabilities with guaranteed error bounds is NP-hard in general [3]. Although their results are not guaranteed to lie within specific error bounds, various approximation algorithms have been designed for which good performance has been reported. One of these algorithms is the *loopy-propagation algorithm*. The basic idea of this algorithm is to apply Pearl's propagation algorithm to a Bayesian network regardless of its topological structure. From an experimental point of view, Murphy et al. [4] reported good approximation behaviour of the loopy-propagation algorithm used on Bayesian networks whenever there was rapid convergence. Excellent performance has also been reported for algorithms equivalent to the loopy-propagation algorithm [5; 6; 7].

Several researchers have analysed the approximation behaviour of the loopy-propagation algorithm from a more fundamental point of view. Weiss and Freeman, more specifically, studied the performance of an equivalent algorithm on Markov networks [8; 9]; their use of Markov networks was motivated by the relatively easier analysis of these networks and justified by the observation that any Bayesian network can be converted into a pairwise Markov network. For pairwise Markov networks with a single loop, Weiss in fact derived an analytical relationship between the exact probabilities and the approximate probabilities computed for the nodes in the loop [8].

In this paper we study the performance properties of the loopy-propagation algorithm on Bayesian networks directly, and thereby provide further insights in the errors that it generates. We argue that two different types of error are introduced in the computed approximate probabilities, which we term the convergence error and the cycling error. A convergence error arises whenever messages that originate from dependent variables within a loop are combined as if they were independent. Such an error emerges in a convergence node only, that is, in a node with two or more incoming arcs on the loop under study. A cycling error arises when messages are being passed on within a loop repetitively and old information is mistaken for new by the variables involved. Cycling of information can occur as soon as for all the convergence nodes of a loop, either the convergence node itself or one of its descendants is observed. A cycling error arises in all nodes of the loop.

Weiss notes that the approximate probabilities found upon loopy propagation are overconfident as a result of double counting of evidence [8]. We observe, however, that overconfident as well as underconfident approximations can result. We use the term *decisiveness* to refer to the over- or underconfidence of an approximation. Decisiveness is an important concept as knowledge of the over- or underconfidence of an approximate probability provides an indication of where the exact probability lies. In this paper, we study the effect of the cycling error on the decisiveness of the approximations found for the inner nodes of a simple loop in a binary Bayesian network. We show that the effect depends on the qualitative influence between the parents of the loop's convergence node and the additional intercausal influence that is induced between these parents by the entered evidence. If the two influences have equal signs, the cycling error pushes the exact probabilities to overconfident approximations; otherwise, the approximations are pushed towards underconfidence.

The paper is organised as follows. In Sect. 2, we provide some preliminaries on Bayesian networks and on Pearl's propagation algorithm. In Sect. 3, we describe the two types of error that may be introduced by loopy propagation. In Sect. 4, we derive the relationship between the exact and approximate probabilities for the inner loop nodes and in Sect. 5 we investigate the decisiveness of the approximations. The paper is rounded off with some conclusions and directions for further research in Sect. 6.

2 Preliminaries

In Sect. 2.1 we provide some preliminaries on Bayesian networks; in Sect. 2.2 we review Pearl's algorithm for probabilistic inference with singly connected networks.

2.1 Bayesian Networks

A *Bayesian network* is a model of a joint probability distribution Pr over a set of stochastic variables, consisting of a directed acyclic graph and a set of conditional probability distributions. In this paper we assume all variables of the network to be binary, taking one of the values *true* and *false*. We will write a for $A = true$ and $\bar{a}$ for $A = false$. We use a_i to denote any value assignment to A, that is $a_i \in \{a, \bar{a}\}$. Each variable is represented by a node in the network's digraph; from now on, we will use the terms node and variable interchangeably. The probabilistic relationships between the variables are captured by the digraph's set of arcs according to the d-separation criterion [1]. Associated with the graphical structure are numerical quantities from the modelled distribution: for each variable A, conditional probability distributions $\Pr(A \mid p(A))$ are specified, where $p(A)$ denotes the set of parents of A in the digraph. Fig. 1 depicts a small example Bayesian network.

In the sequel, we distinguish between singly connected and multiply connected Bayesian networks. A network is *singly connected* if there is at most one trail between any two variables in its digraph. If there are multiple trails

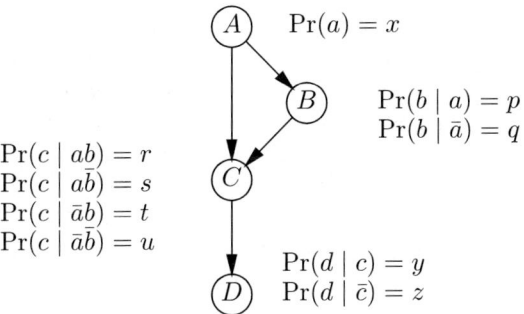

Fig. 1. A multiply connected Bayesian network with a convergence node C having the dependent parents A and B and the child D.

between variables, then the network is *multiply connected*. A multiply connected network includes one or more loops, that is, one or more cycles in its underlying undirected graph. We say that a loop is simple if none of its nodes are shared by another loop. A node that has two or more incoming arcs on a loop will be called a *convergence node* of this loop; the other nodes of the loop will be termed *inner nodes*. The network from Fig. 1 is an example of a multiply connected network. The trail $A \to B \to C \leftarrow A$ constitutes a simple loop in the network's digraph. Node C is the only convergence node of this loop; nodes A and B are the loop's inner nodes.

2.2 Pearl's Propagation Algorithm

We briefly review Pearl's propagation algorithm [1]. This algorithm was designed for exact inference with singly connected Bayesian networks. The term *loopy propagation* used throughout the literature, refers to the application of this algorithm to networks with loops. In the algorithm, each node X is provided with a limited set of rules that enable the node to calculate its probability distribution $\Pr(X \mid \mathbf{e})$ given the available evidence $\mathbf{e}$, from messages it receives from its neighbours. These rules are applied in parallel by the various nodes at each time step. The rule used by node X for establishing the probability distribution $\Pr(X \mid \mathbf{e})$ at time t is

$$\Pr{}^t(X \mid \mathbf{e}) = cst \cdot \lambda^t(X) \cdot \pi^t(X)$$

where the compound diagnostic parameter $\lambda^t(X)$ is computed from the diagnostic messages $\lambda^t_{Y^j}(X)$ it receives from each of its children Y^j:

$$\lambda^t(X) = \prod_{Y^j} \lambda^t_{Y^j}(X)$$

and the compound causal parameter $\pi^t(X)$ is computed from the causal messages $\pi^t_X(U^i)$ it receives from each of its parents U^i:

$$\pi^t(X) = \sum_{\mathbf{U}} \Pr(X \mid \mathbf{U}) \cdot \prod_{U^i} \pi^t_X(U^i)$$

where $\mathbf{U}$ denotes the set of all parents of node X. The rule for computing the diagnostic messages to be sent to its parent U^i is

$$\lambda^{t+1}_X(U^i) = cst \cdot \sum_X \lambda^t(X) \cdot \sum_{\mathbf{U}/U^i} \Pr(X \mid \mathbf{U}) \cdot \prod_{U^k,\, k \neq i} \pi^t_X(U^k)$$

and the rule for computing the causal messages to be sent to its child Y^j is

$$\pi^{t+1}_{Y^j}(X) = cst \cdot \pi^t(X) \cdot \prod_{Y^k,\, k \neq j} \lambda^t_{Y^k}(X)$$

where *cst* denotes a normalisation constant. Note that, in general, the number of messages that a node sends in each time step to a child equals the number of its own values; the number of messages that it sends to a parent equals the number of values of this parent. From here on, we will denote a diagnostic parameter $\lambda_X(U^i)$ and a causal parameter $\pi_{Y^j}(X)$ by the term *message vector*. For a binary parent U^i, we write the diagnostic messages as $(\lambda_X(u^i), \lambda_X(\bar{u}^i))$; for a binary child X, we write the causal messages as $(\pi_{Y^j}(x), \pi_{Y^j}(\bar{x}))$. All message vectors are initialised to contain just 1s. An observation for a node X now is entered into the network by multiplying the components of $\lambda^0(X)$ and $\pi^1_{Y^j}(X)$ by 1 for the observed value of X and by 0 for the other value(s).

3 Errors in Loopy Propagation

Pearl's propagation algorithm results in exact probabilities whenever it is applied to a singly connected Bayesian network. After a finite number of time steps, proportional to the diameter of the network's digraph, the probabilistic information present in the network will have been passed on to all nodes. The network then reaches an equilibrium state in which the computed probabilities and messages do no longer change upon further message passing. When applied to a multiply connected Bayesian network, that is, upon performing loopy propagation, the algorithm will often converge as well; we will consider the algorithm to have converged as soon as all causal and diagnostic messages and all computed probabilities change by less than a prespecified threshold value in the next time step. The resulting probabilities may then deviate from the probabilities of the modelled distribution, however.

The probabilities that result from performing loopy propagation on a multiply connected network, may include two different types of error. The first type of error originates at the convergence node(s) of a loop. In Pearl's algorithm, a node with two or more parents combines the messages from its parents as if these messages come from independent variables. In a singly connected network the parents of a node indeed are always independent. The parents of a convergence node, however, may be dependent. By assuming independence upon combining the causal messages from dependent variables, an error is introduced. A convergence error may be propagated to neighbours outside the loop. Given compound loops, a convergence error may enter the loop; for simple loops, however, this error does not affect the probabilities computed for the inner nodes of the loop. In our previous work [10], we studied the convergence error in detail; in this paper, we focus on the effect of the second type of error, called the *cycling error*. This type of error arises when messages are being passed on repetitively in the loop, where old information is mistaken for new by the nodes involved. This cycling of information can occur as soon as for each convergence node of the loop either the node itself or one of its descendants is observed. A cycling error arises in all loop nodes and is propagated to nodes outside the loop. Note that at the convergence node both types of error emerge upon loopy propagation, whereas at the inner nodes of simple loops only the cycling error originates. In the sequel we will denote the probabilities that result upon loopy propagation with $\widetilde{\Pr}$ to distinguish them from the exact probabilities which are denoted by Pr.

4 The Relationship Between the Exact and Approximate Probabilities

Upon performing loopy propagation on a Bayesian network with a simple loop, a cycling error may arise in the probabilities computed for all the nodes of the loop, as argued in the previous section. In this section we will derive, for the network from Fig. 1, an expression that relates the exact probabilities for the inner loop nodes to the computed approximate probabilities. Our derivation is analogous to the one constructed by Weiss [8] for an equivalent algorithm applied to binary Markov networks with a single loop. From the relationship between the exact and approximate probabilities, we then identify the factors, in terms of a network's specification, that determine whether the exact probabilities for the inner loop nodes are pushed towards overconfident or underconfident approximations. We will study these factors in Sect. 5.

We consider the Bayesian network from Fig. 1 and suppose that the evidence d has been entered. We now build upon the observation that the updating of a message vector during propagation can be captured by a transition matrix. We begin by deriving the matrices that describe the information that is included into a message vector during one clockwise cycle and during one counterclockwise cycle respectively, from node A back to itself. We will then use the eigenvalues of these matrices to express the relationship between the exact and approximate probabilities found at node A.

To derive the transition matrix that captures the information that is added during one clockwise cycle from node A back to itself, we consider the updating of the message vector $(1, 1)$ during the first cycle of the algorithm; we recall that node A initially receives this vector. In the first step of the algorithm, node A sends the vector

$$\pi_B(A) = \begin{bmatrix} x \\ 1 - x \end{bmatrix}$$

to node B, which subsequently sends the message vector

$$\pi_C(B) = \begin{bmatrix} p \cdot x + q \cdot (1 - x) \\ (1 - p) \cdot x + (1 - q) \cdot (1 - x) \end{bmatrix}$$

to node C. The diagnostic message that C receives from node D is

$$\lambda_D(C) = \begin{bmatrix} y \\ z \end{bmatrix}$$

Since node C does not have any other children, its compound diagnostic parameter also equals

$$\lambda(C) = \begin{bmatrix} y \\ z \end{bmatrix}$$

This compound diagnostic parameter and the causal message that node C receives from node B are combined with the information that node C has about its own conditional probabilities into the following diagnostic message from node C to node A:

$$\lambda_C(A) = \begin{bmatrix} \lambda(c) \cdot [r \cdot \pi_C(b) + s \cdot \pi_C(\bar{b})] + \lambda(\bar{c}) \cdot [(1-r) \cdot \pi_C(b) + (1-s) \cdot \pi_C(\bar{b})] \\ \lambda(c) \cdot [t \cdot \pi_C(b) + u \cdot \pi_C(\bar{b})] + \lambda(\bar{c}) \cdot [(1-t) \cdot \pi_C(b) + (1-u) \cdot \pi_C(\bar{b})] \end{bmatrix}$$

After the first clockwise cycle of the algorithm, therefore, the initial message vector $(1,1)$ has been updated to the vector given above. We now derive the transition matrix

$$M^{\circlearrowright A,d} = \begin{bmatrix} l & m \\ n & o \end{bmatrix}$$

that captures this update. We observe that the entries l, m, n and o of this matrix should adhere to

$$\begin{bmatrix} l & m \\ n & o \end{bmatrix} \cdot \begin{bmatrix} 1 \\ 1 \end{bmatrix} = \lambda_C(A)$$

from which we find that

$$l + m = \lambda_C(a)$$
$$n + o = \lambda_C(\bar{a})$$

We now split the expression for $\lambda_C(a)$ into separate terms for l and m and the expression for $\lambda_C(\bar{a})$ into separate terms for n and o. To this end, we observe that in the analysis above, the first component of the message vector from node A to node B pertains to a and the second component pertains to $\bar{a}$. Roughly speaking, now, since the first component is multiplied by l and n, these entries have to collect all information at node B concerning a. As $p = \Pr(b \mid a)$ pertains to a, therefore, all terms containing p are assigned to l and n. Likewise, all terms containing q are assigned to m and o. After rearranging the various terms in the expressions for $\lambda_C(a)$ and $\lambda_C(\bar{a})$ accordingly, we find that

$$l = [(y \cdot r + z \cdot (1-r)) \cdot p + (y \cdot s + z \cdot (1-s)) \cdot (1-p)] \cdot x$$
$$m = [(y \cdot r + z \cdot (1-r)) \cdot q + (y \cdot s + z \cdot (1-s)) \cdot (1-q)] \cdot (1-x)$$
$$n = [(y \cdot t + z \cdot (1-t)) \cdot p + (y \cdot u + z \cdot (1-u)) \cdot (1-p)] \cdot x$$
$$o = [(y \cdot t + z \cdot (1-t)) \cdot q + (y \cdot u + z \cdot (1-u)) \cdot (1-q)] \cdot (1-x)$$

The matrix that captures the information that is included during a single counterclockwise cycle into the messages from node A back to itself, is found to be

$$M^{\circlearrowleft A,d} = \begin{bmatrix} l & n \cdot \frac{1-x}{x} \\ m \cdot \frac{x}{1-x} & o \end{bmatrix}$$

Upon loopy propagation, the information captured by the above two transition matrices is included repeatedly in every cycle. We would like to observe that all message vectors are normalised, as described in Sect. 2.2. As a consequence, repeated multiplication by the transition matrices will not result in convergence to $(0,0)$.

Example We consider the example Bayesian network from Fig. 2 and address the approximate probabilities found for the inner node A upon performing loopy propagation. We consider the probabilities after the evidence d and $\bar{d}$ have been entered, respectively, during the first five cycles of the algorithm. The subsequent approximations are shown in Figs. 3 and 4; for comparison, the exact probabilities are depicted as well. We observe that the approximate probabilities asymptotically approach a particular value. The approximate probabilities given d oscillate around the final value, whereas given $\bar{d}$ the approximations go steadily towards the final value. In Sect. 5, we will return to this difference in approximation behaviour. □

The example demonstrates that upon loopy propagation the computed probabilities converge towards an equilibrium value. We now will exploit the eigenvalues of the transition matrices to relate the approximate probabilities in the equilibrium state to the exact probabilities. The eigenvalues λ_1 and λ_2 of the matrix $M^{\circlearrowleft A,d}$ are the solutions of

$$\lambda = \frac{1}{2} \cdot [(l+o) \pm \sqrt{(l+o)^2 - 4 \cdot (l \cdot o - m \cdot n)}]$$

where λ_1 is the largest of the two values. For $M^{\circlearrowleft A,\bar{d}}$ the same eigenvalues are found. Note that since the entries of the two matrices are positive, the eigenvalues λ_1 and λ_2 are real numbers and both λ_1 and $\lambda_1 + \lambda_2$ are positive.

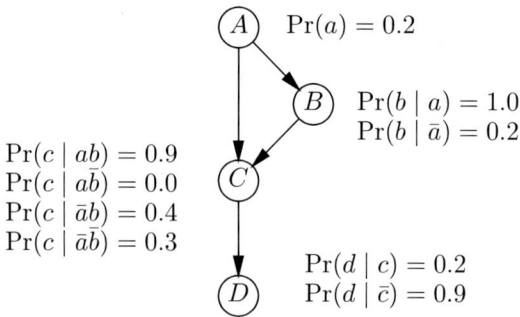

Fig. 2. An example Bayesian network.

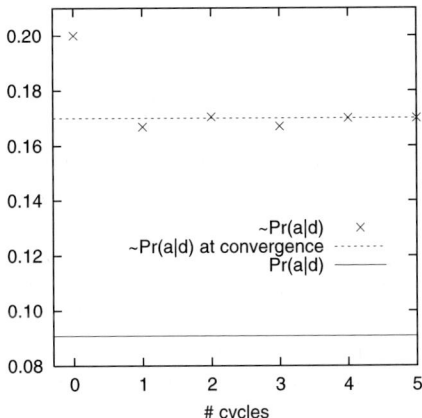

Fig. 3. The approximate probabilities given d found for node A in the network of Fig. 2 during the first five cycles of the loopy propagation algorithm.

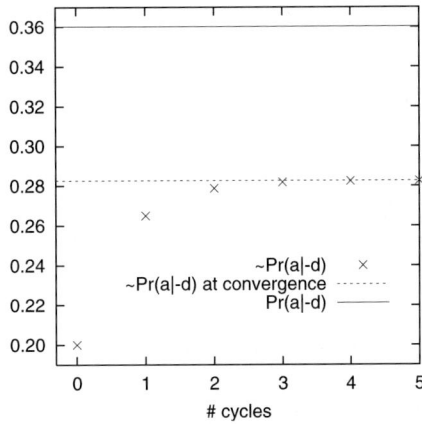

Fig. 4. The approximate probabilities given $\bar{d}$ found for node A in the network of Fig. 2 during the first five cycles of the loopy propagation algorithm.

The relationship between the exact and the approximate probability of a_i given d now is expressed by

$$\Pr(a_i \mid d) = \widetilde{\Pr}(a_i \mid d) - \frac{\lambda_2}{\lambda_1 + \lambda_2} \cdot (2 \cdot \widetilde{\Pr}(a_i \mid d) - 1)$$

To prove this property, we begin by observing that for each eigenvalue of $M^{\circlearrowleft A,d}$ and for each eigenvalue of $M^{\circlearrowleft A,d}$, an eigenvector direction is found. For $M^{\circlearrowleft A,d}$, we denote the normalised principal eigenvector by (α_1, β_1) and we denote a fixed arbitrary vector in the second eigenvector direction by (γ_1, δ_1);

for $M^{\circlearrowleft A,d}$ we denote the normalised principal eigenvector by (α_2, β_2). In the equilibrium state, now, node A receives the message $\lambda_C(A) = (\alpha_1, \beta_1)$ from node C and the message $\lambda_B(A) = (\alpha_2, \beta_2)$ from node B, cf. [11]. These two messages are combined by node A with its own knowledge $\pi(A) = (x, 1-x)$ of the prior distribution over its values, which results in the approximate probabilities

$$\widetilde{\Pr}(A \mid d) = cst_1 \cdot \begin{bmatrix} \alpha_1 \cdot \alpha_2 \cdot x \\ \beta_1 \cdot \beta_2 \cdot (1-x) \end{bmatrix}$$

where cst_1 is a normalisation constant.

We now relate the computed approximate probability $\widetilde{\Pr}(a_i \mid d)$ to the exact probability $\Pr(a_i \mid d)$. We begin by observing that for the entries l and o of the transition matrices $M^{\circlearrowleft A,d}$ and $M^{\circlearrowleft A,d}$, we have that $l = \Pr(d \mid a) \cdot \Pr(a)$ and $o = \Pr(d \mid \bar{a}) \cdot \Pr(\bar{a})$. For the exact probabilities $\Pr(a_i \mid d)$ we thus find that $\Pr(a \mid d) = l/(l+o)$ and $\Pr(\bar{a} \mid d) = o/(l+o)$. To express $\Pr(a_i \mid d)$ and $\widetilde{\Pr}(a_i \mid d)$ in similar terms, we now relate the entries l and o to the expressions $\alpha_1 \cdot \alpha_2 \cdot x$ and $\beta_1 \cdot \beta_2 \cdot (1-x)$. To this end, we diagonalise the matrix $M^{\circlearrowleft A,d}$ into

$$M^{\circlearrowleft A,d} = \begin{bmatrix} \alpha_1 & \gamma_1 \\ \beta_1 & \delta_1 \end{bmatrix} \cdot \begin{bmatrix} \lambda_1 & 0 \\ 0 & \lambda_2 \end{bmatrix} \cdot \begin{bmatrix} \mathcal{A} & \mathcal{B} \\ \mathcal{C} & \mathcal{D} \end{bmatrix}$$
$$= \begin{bmatrix} \alpha_1 \cdot \mathcal{A} \cdot \lambda_1 + \gamma_1 \cdot \mathcal{C} \cdot \lambda_2 & \alpha_1 \cdot \mathcal{B} \cdot \lambda_1 + \gamma_1 \cdot \mathcal{D} \cdot \lambda_2 \\ \beta_1 \cdot \mathcal{A} \cdot \lambda_1 + \delta_1 \cdot \mathcal{C} \cdot \lambda_2 & \beta_1 \cdot \mathcal{B} \cdot \lambda_1 + \delta_1 \cdot \mathcal{D} \cdot \lambda_2 \end{bmatrix}$$

where $\begin{bmatrix} \mathcal{A} & \mathcal{B} \\ \mathcal{C} & \mathcal{D} \end{bmatrix} = \begin{bmatrix} \alpha_1 & \gamma_1 \\ \beta_1 & \delta_1 \end{bmatrix}^{-1}$. We thus have that

$$l = \alpha_1 \cdot \mathcal{A} \cdot \lambda_1 + \gamma_1 \cdot \mathcal{C} \cdot \lambda_2$$
$$o = \beta_1 \cdot \mathcal{B} \cdot \lambda_1 + \delta_1 \cdot \mathcal{D} \cdot \lambda_2$$

From $\begin{bmatrix} \alpha_1 & \gamma_1 \\ \beta_1 & \delta_1 \end{bmatrix} \cdot \begin{bmatrix} \mathcal{A} & \mathcal{B} \\ \mathcal{C} & \mathcal{D} \end{bmatrix} = \begin{bmatrix} 1 & 0 \\ 0 & 1 \end{bmatrix}$ we find that $\gamma_1 \cdot \mathcal{C} = \beta_1 \cdot \mathcal{B}$ and $\delta_1 \cdot \mathcal{D} = \alpha_1 \cdot \mathcal{A}$. The entries l and o can therefore also be written as

$$l = \alpha_1 \cdot \mathcal{A} \cdot \lambda_1 + \beta_1 \cdot \mathcal{B} \cdot \lambda_2$$
$$o = \beta_1 \cdot \mathcal{B} \cdot \lambda_1 + \alpha_1 \cdot \mathcal{A} \cdot \lambda_2$$

To express $\mathcal{A}$ and $\mathcal{B}$ in terms of α_2, β_2 and x, we now rewrite the matrix $M^{\circlearrowleft A,d}$ as

$$M^{\circlearrowleft A,d} = \begin{bmatrix} \frac{1}{x} & 0 \\ 0 & \frac{1}{1-x} \end{bmatrix} \cdot (M^{\circlearrowleft A,d})^T \cdot \begin{bmatrix} x & 0 \\ 0 & 1-x \end{bmatrix}$$
$$= \begin{bmatrix} \frac{1}{x} & 0 \\ 0 & \frac{1}{1-x} \end{bmatrix} \cdot \begin{bmatrix} \mathcal{A} & \mathcal{C} \\ \mathcal{B} & \mathcal{D} \end{bmatrix} \cdot \begin{bmatrix} \lambda_1 & 0 \\ 0 & \lambda_2 \end{bmatrix} \cdot \begin{bmatrix} \alpha_1 & \beta_1 \\ \gamma_1 & \delta_1 \end{bmatrix} \cdot \begin{bmatrix} x & 0 \\ 0 & 1-x \end{bmatrix}$$

The first column of the product

$$\begin{bmatrix} \frac{1}{x} & 0 \\ 0 & \frac{1}{1-x} \end{bmatrix} \cdot \begin{bmatrix} \mathcal{A} & \mathcal{C} \\ \mathcal{B} & \mathcal{D} \end{bmatrix} = \begin{bmatrix} \frac{1}{x} \cdot \mathcal{A} & \frac{1}{x} \cdot \mathcal{C} \\ \frac{1}{1-x} \cdot \mathcal{B} & \frac{1}{1-x} \cdot \mathcal{D} \end{bmatrix}$$

now is a vector in the direction of the principal eigenvector (α_2, β_2) of $M^{\circlearrowleft A,d}$. We thus find that $\mathcal{A} = cst_2 \cdot \alpha_2 \cdot x$ and $\mathcal{B} = cst_2 \cdot \beta_2 \cdot (1-x)$, where $1/cst_2$ is a normalisation constant. We conclude that

$$\begin{aligned} l &= cst_2 \cdot (\alpha_1 \cdot \alpha_2 \cdot x \cdot \lambda_1 + \beta_1 \cdot \beta_2 \cdot (1-x) \cdot \lambda_2) \\ &= cst_2/cst_1 \cdot (\widetilde{\Pr}(a \mid d) \cdot \lambda_1 + \widetilde{\Pr}(\bar{a} \mid d) \cdot \lambda_2) \\ o &= cst_2 \cdot (\beta_1 \cdot \beta_2 \cdot (1-x) \cdot \lambda_1 + \alpha_1 \cdot \alpha_2 \cdot x \cdot \lambda_2) \\ &= cst_2/cst_1 \cdot (\widetilde{\Pr}(\bar{a} \mid d) \cdot \lambda_1 + \widetilde{\Pr}(a \mid d) \cdot \lambda_2) \end{aligned}$$

With $\Pr(a \mid d) = l/(l+o)$ and $\Pr(\bar{a} \mid d) = o/(l+o)$, we now find that

$$\Pr(a_i \mid d) = \widetilde{\Pr}(a_i \mid d) - \frac{\lambda_2}{\lambda_1 + \lambda_2} \cdot (2 \cdot \widetilde{\Pr}(a_i \mid d) - 1)$$

For $\bar{d}$, the derivation is analogous. For node B similar expressions are found. We would like to note that the transition matrices for nodes A and B have the same eigenvalues. The transition matrix for an entire cycle may, more or less, be viewed as the result of multiplication of the transition matrices between the neighbouring nodes in the cycle. Eventually, the clockwise matrices for nodes A and B result from the multiplication of the same two 2 by 2 matrices, yet in different order. The same applies to the counterclockwise matrices. As a consequence, the transition matrices per cycle for nodes A and B have the same eigenvalues.

Example We consider again the Bayesian network from Fig. 2. Upon performing loopy propagation on this network, a single clockwise cycle serves to include the information

$$M^{\circlearrowleft A,d} = \begin{bmatrix} 0.0540 & 0.6192 \\ 0.1240 & 0.5408 \end{bmatrix}$$

into the message vector of node A back to itself. The eigenvalues of this matrix are $\lambda_1 \approx 0.6662$ and $\lambda_2 \approx -0.0714$. Its normalised principal eigenvector is

$$\begin{bmatrix} \alpha_1 \\ \beta_1 \end{bmatrix} \approx \begin{bmatrix} 0.5028 \\ 0.4972 \end{bmatrix}$$

A single counterclockwise cycle serves to include the information

$$M^{\circlearrowleft A,d} = \begin{bmatrix} 0.0540 & 0.4960 \\ 0.1548 & 0.5408 \end{bmatrix}$$

into the message vector of node A back to itself. The eigenvalues of this matrix are again $\lambda_1 \approx 0.6662$ and $\lambda_2 \approx -0.0714$. Its normalised the principal eigenvector is

$$\begin{bmatrix} \alpha_2 \\ \beta_2 \end{bmatrix} \approx \begin{bmatrix} 0.4476 \\ 0.5524 \end{bmatrix}$$

The approximate probabilities found for node A upon loopy propagation given d now are equal to

$$cst_1 \cdot \begin{bmatrix} 0.5028 \cdot 0.4476 \cdot 0.2 \\ 0.4972 \cdot 0.5524 \cdot 0.8 \end{bmatrix} \approx \begin{bmatrix} 0.17 \\ 0.83 \end{bmatrix}$$

We thus find that $\widetilde{\Pr}(a \mid d) \approx 0.17$ and $\widetilde{\Pr}(\bar{a} \mid d) \approx 0.83$. We recall that the exact probability $\Pr(a \mid d)$ equals $l/(l+o)$ and can be read from the diagonal of the transition matrices to be $0.0540/0.5948 \approx 0.09$. For the relationship between $\Pr(a \mid d)$ and $\widetilde{\Pr}(a \mid d)$ we indeed observe that $\Pr(a \mid d) = \widetilde{\Pr}(a \mid d) - \lambda_2/(\lambda_1 + \lambda_2) \cdot (2 \cdot \widetilde{\Pr}(a \mid d) - 1)$ holds as $0.09 \approx 0.17 + 0.0714/0.5948 \cdot (2 \cdot 0.17 - 1)$. Note that the error in $\widetilde{\Pr}(a \mid d)$ computed from the network equals $0.09 - 0.17 = -0.06$ and hence is small. Moreover, given a threshold value of 0.0001, the algorithm had converged within just five cycles. □

5 The Decisiveness of the Approximations

In the previous section we studied, for a Bayesian network with a simple loop, the cycling error that may arise in the probabilities computed upon loopy propagation for the inner loop nodes. More specifically, we derived an expression that relates the exact probabilities for the inner loop nodes to the computed approximate probabilities. We now build upon this analysis to state some properties of the approximations in terms of the specification of the network. We say that an approximation is *overconfident* if it is closer to one of the extremes, that is to 0 or 1, than the exact probability; the approximation is *underconfident* if it is closer to 0.5. We use the term *decisiveness* to refer to the over- or underconfidence of an approximation. As an example, Fig. 5 depicts, for the network from Fig. 2, the exact and approximate probabilities $\Pr(a \mid d)$ and $\widetilde{\Pr}(a \mid d)$ as a function of $\Pr(a)$; Fig. 6 depicts $\Pr(a \mid \bar{d})$ and $\widetilde{\Pr}(a \mid \bar{d})$. We observe that the evidence d results in underconfident approximations, while the evidence $\bar{d}$ gives overconfident approximations for all possible values of $\Pr(a)$. We argue that the approximations for the inner nodes of a loop are either all pushed towards overconfidence or all pushed towards underconfidence as soon as the convergence node of the loop has an observed descendant. We further show that the decisiveness of the approximations depends on the sign of the qualitative influence between the parents of the convergence node and the sign of the intercausal influence that is induced between these parents by the entered evidence.

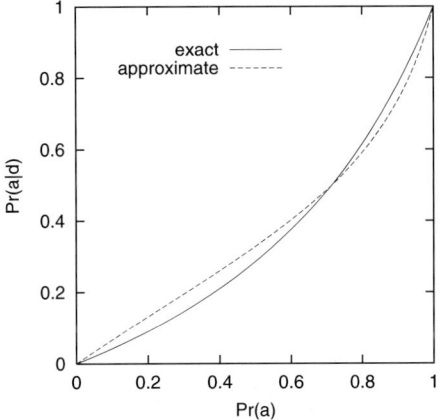

Fig. 5. $\Pr(a \mid d)$ and $\widetilde{\Pr}(a \mid d)$ as a function of $\Pr(a)$ for the network from Fig. 2.

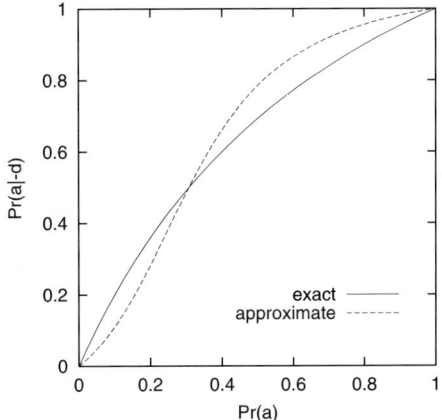

Fig. 6. $\Pr(a \mid \bar{d})$ and $\widetilde{\Pr}(a \mid \bar{d})$ as a function of $\Pr(a)$ for the network from Fig. 2.

We begin by introducing the two types of influence that we will exploit in the sequel. A *qualitative influence* [12] between two neighbouring nodes expresses how the values of the one node influence the probabilities of the values of the other node along their common arc. A positive qualitative influence of a parent B on its child C is found if

$$\Pr(c \mid bx) - \Pr(c \mid \bar{b}x) \geq 0$$

for any combination of values x for the set $p(C) \setminus \{B\}$ of parents of C other than B. In Fig. 1, for example, a positive influence of B on C is found if both $r - s \geq 0$ and $t - u \geq 0$. A negative and a zero qualitative influence are defined

Table 1. The ⊗- operator for combining signs.

⊗	+	−	0	?
+	+	−	0	?
−	−	+	0	?
0	0	0	0	0
?	?	?	0	?

analogously. When no consistent sign can be found for the influence given the different combinations of values x, we say that the influence is ambiguous. Positive, negative, zero and ambiguous qualitative influences are indicated by the signs $+$, $-$, 0 and $?$, respectively. Qualitative influences are symmetric, that is, a positive qualitative influence of B on C implies a positive influence of C on B. Qualitative influences further adhere to the property of transitivity, that is, influences along a trail with at most one incoming arc for each variable combine into a net influence whose sign is defined by the ⊗-operator from Table 1.

Intercausal influences [13] are dynamic in nature and can only arise after evidence has been entered into the network. In the prior state of a network, that is, when no evidence has been entered as yet, the parents of a node are d-separated from one another along the trail that includes their common child. As soon as evidence is entered for this child or for one of its descendants, however, the two parents may become dependent along this trail. The influence that is thus induced, is termed an intercausal influence. For example, for a node C with the independent parents A and B we find a positive intercausal influence of node B on node A with respect to c, if

$$\Pr(a \mid bc) - \Pr(a \mid \bar{b}c) \geq 0$$

Intercausal influences, like qualitative influences, adhere to the properties of symmetry and transitivity. We informally review the effect of the intercausal influence. For a node C with the parents A and B, entering evidence will influence the probability distribution for node A along the trail $A \rightarrow C$ and the probability distribution for node B along the trail $B \rightarrow C$. The influence of node C on the one parent now typically may change with a subsequent change of the probability distribution for the other parent. A positive intercausal influence between nodes A and B with respect to c implies that, given the evidence c, an increase in the probability of b will result in an increase, in terms of positivity, of the influence of node C on node A along the trail $A \rightarrow C$. 'An increase in positivity' will say that a negative influence becomes weaker and a positive influence becomes stronger. Given a positive intercausal influence, moreover, an increase of a, will result in an increase in positivity of the influence of node C on node B along the trail $B \rightarrow C$. Analogously, a negative intercausal influence implies that an increase in the probability of b

will result in a decrease of positivity of the influence of C on A along the trail $A \to C$ and that an increase in the probability of a will result in a decrease in positivity of the influence of C on B along the trail $B \to C$.

For the network from Fig. 1 we now derive an expression that captures the intercausal influence that is induced between the nodes A and B by the evidence d. We assume that the exact probability distributions for nodes A and B are non-degenerate; note that for degenerate distributions in fact no loop is present. To separate the intercausal influence from the direct influence, we suppose that A and B are independent in the prior network. The intercausal influence then is captured by

$$\Pr(a \mid bd) - \Pr(a \mid \bar{b}d) = \frac{\Pr(abd)}{\Pr(bd)} - \frac{\Pr(a\bar{b}d)}{\Pr(\bar{b}d)}$$

We find that

$$\frac{\Pr(abd)}{\Pr(bd)} = \frac{x \cdot e}{x \cdot e + (1-x) \cdot g}$$

and that

$$\frac{\Pr(a\bar{b}d)}{\Pr(\bar{b}d)} = \frac{x \cdot f}{x \cdot f + (1-x) \cdot h}$$

and hence that

$$\Pr(a \mid bd) - \Pr(a \mid \bar{b}d) = \frac{(x - x^2) \cdot (e \cdot h - f \cdot g)}{(x \cdot e + (1-x) \cdot g) \cdot (x \cdot f + (1-x) \cdot h)}$$

where

$$e = r \cdot y + (1-r) \cdot z$$
$$f = s \cdot y + (1-s) \cdot z$$
$$g = t \cdot y + (1-t) \cdot z$$
$$h = u \cdot y + (1-u) \cdot z$$

Because the denominator in the expression for $\Pr(a \mid bd) - \Pr(a \mid \bar{b}d)$ is positive, we have that the sign of the intercausal influence that is induced between nodes A and B, is equal to the sign of the numerator $(x - x^2) \cdot (e \cdot h - f \cdot g)$. Because $x = \Pr(a) \in (0, 1)$, moreover, the sign of the intercausal influence equals the sign of $e \cdot h - f \cdot g$, that is, the sign of

$$(y - z)^2 \cdot (r \cdot u - s \cdot t) + z \cdot (y - z) \cdot (r + u - s - t)$$

Similarly, we find that the sign of the intercausal influence induced by the evidence $\bar{d}$ equals the sign of

$$(z - y)^2 \cdot (r \cdot u - s \cdot t) + (1 - z) \cdot (z - y) \cdot (r + u - s - t)$$

We will use $II(A, B \mid d_j)$ to denote these expressions for $d_j \in \{d, \bar{d}\}$.

We now relate the two qualitative features reviewed above to the over- or underconfidence of the approximations computed for the inner loop nodes by the loopy-propagation algorithm. We recall that, since all entries of the transition matrix $M^{\circlearrowleft A,d}$ are positive, we have that its eigenvalues λ_1 and λ_2 are real numbers and both λ_1 and $\lambda_1 + \lambda_2$ are positive. From the relationship

$$\Pr(a_i \mid d_j) = \widetilde{\Pr}(a_i \mid d_j) - \frac{\lambda_2}{\lambda_1 + \lambda_2} \cdot (2 \cdot \widetilde{\Pr}(a_i \mid d_j) - 1)$$

established in the previous section, we therefore find that overconfident approximations will be computed for the probability $\Pr(a_i \mid d_j)$ whenever $\lambda_2 \geq 0$; underconfident approximations are found if $\lambda_2 \leq 0$. With $\lambda_2 = 0$, the exact probabilities are computed. From $\lambda = \frac{1}{2} \cdot [(l+o) \pm \sqrt{(l+o)^2 - 4 \cdot (l \cdot o - m \cdot n)}]$ we observe that the sign of the eigenvalue λ_2 equals the sign of the expression $l \cdot o - m \cdot n$. By simple manipulation of the terms involved, we find that, for $\Pr(a) \in (0,1)$, the sign of this expression is equal to the sign of

$$(p - q) \cdot II(A, B \mid d_j)$$

We thus find that the approximate probabilities given the evidence d_j, established for node A, are overconfident if the sign of $p - q$ is equal to the sign of $II(A, B \mid d_j)$, that is, if the sign of the qualitative influence between nodes A and B is equal to the sign of the intercausal influence that is induced between A and B by the evidence d_j; the approximations are underconfident otherwise. We recall that the transition matrices of A and B have the same eigenvalues. These nodes therefore will have the same decisiveness.

Example We consider once again the network from Fig. 2. The sign of the qualitative influence between the nodes A and B equals the sign of $1.0 - 0.2 = 0.8$ and hence is positive. The sign of the intercausal influence between nodes A and B that is induced by the evidence d, equals the sign of $(0.2-0.9)^2 \cdot (0.9 \cdot 0.3 - 0.0 \cdot 0.4) + 0.9 \cdot (0.2-0.9) \cdot (0.9+0.3-0.0-0.4) \approx -0.37$ and thus is negative. The qualitative and intercausal influences between nodes A and B therefore have opposite signs and the approximations established for the inner loop nodes will be underconfident. Indeed, we find the underconfident approximations $\widetilde{\Pr}(a \mid d) \approx 0.17$ and $\widetilde{\Pr}(b \mid d) \approx 0.30$ for the exact probabilities $\Pr(a \mid d) \approx 0.09$ and $\Pr(b \mid d) \approx 0.26$ upon loopy propagation. Given the evidence $\bar{d}$, on the other hand, the sign of the intercausal influence between A and B equals the sign of $(0.9 - 0.2)^2 \cdot (0.9 \cdot 0.3 - 0.0 \cdot 0.4) + (1.0 - 0.9) \cdot (0.9 - 0.2) \cdot (0.9 + 0.3 - 0.0 - 0.4) \approx 0.19$ and hence is positive. The qualitative and intercausal influences now have equal signs and overconfident approximations will result. Upon loopy propagation, we indeed find the overconfident approximations $\widetilde{\Pr}(a \mid \bar{d}) \approx 0.28$ and $\widetilde{\Pr}(b \mid \bar{d}) \approx 0.52$ for the exact probabilities $\Pr(a \mid \bar{d}) \approx 0.36$ and $\Pr(b \mid \bar{d}) \approx 0.51$. □

We found that loopy propagation will result in exact probabilities whenever $\lambda_2 = 0$, that is, whenever $(p - q) \cdot II(A, B \mid d_j) = 0$. If the factor $p - q$ equals zero, then the nodes A and B are a priori independent and in fact no loop is present. If the factor $II(A, B \mid d_j)$ equals zero, then the messages that node C sends to node B are independent of the probabilities for A and the messages that C sends to A are independent of the probabilities for B. Nodes A and B therefore receive the correct messages from node C. These messages, moreover, do not change in a next cycle of the algorithm. The algorithm will therefore converge and yield exact probabilities in just a single cycle. We would further like to note that for the probabilities $\Pr(a_i \mid d_j) = 0.5$, loopy propagation will result in exact probabilities, irrespective of the eigenvalues of the transition matrices. This observation can easily be verified by calculating $\widetilde{\Pr}(a \mid d_j)$ from the previously established relationship $\Pr(a_i \mid d_j) = \widetilde{\Pr}(a_i \mid d_j) - \lambda_2/(\lambda_1 + \lambda_2) \cdot (2 \cdot \widetilde{\Pr}(a_i \mid d_j) - 1)$ for $\Pr(a \mid d_j) = 0.5$.

We are now also able to explain the different approximation behaviour of the loopy-propagation algorithm demonstrated in Figs. 3 and 4. We recall that Fig. 3 pertains to the algorithm's approximation behaviour for node A given the evidence d in the network from Fig. 2. The qualitative influence between A and B is positive; the evidence d, moreover, serves to induce a negative intercausal influence between the two nodes. We now first consider the algorithm's behaviour during a clockwise cycle. If, in a particular cycle, the influence of entering the evidence d on the probability of a along the trail $D \leftarrow C \leftarrow A$ is positive, then, because of the positive qualitative influence between A and B, the influence on the probability of b along the trail $D \leftarrow C \leftarrow A \rightarrow B$ will be positive as well. Because of the negative intercausal influence between nodes A and B, this positive influence on b will result in a decrease in the strength of the positive influence of C on A. In the next cycle, therefore, the increase in the probabilities of a and b induced by the clockwise process, will be less than in the previous cycle. In the subsequent cycle, the positive influence between of C on A, induced by the clockwise process again will be stronger than in the previous cycle, and so on. Such an oscillating behaviour would also be found given a negative influence of D on A along the trail $D \leftarrow C \leftarrow A$. In the counterclockwise cycles of the algorithm a similar behaviour is observed. Both processes add up to the oscillating behaviour that we observed in Fig. 3 for our example network. Alternatively, if the qualitative influence and the intercausal influence have equal signs, the approximations will go steadily towards their final values. If, for example, both influences between the nodes A and B are positive, the increase in the probability of b induced in the clockwise cycle, will result in an even stronger positive influence of C on A. In the next cycle, the increase of the probability of b induced by the clockwise process will further increase, and so on.

So far, we addressed the situation where evidence is entered for node D in the network from Fig. 1. If the convergence node C itself is observed, a similar analysis holds. The expressions for the intercausal influence between the nodes A and B then reduce to $r \cdot u - s \cdot t$ and $r \cdot u - s \cdot t - (r + u - s - t)$ after the observation of c and $\bar{c}$, respectively. Although we analysed the influence of the cycling error for the simple network from Fig. 1 only, the essence of our analysis extends to networks of more complex topology. Our analysis, for example, extends directly to networks with simple loops with more than two inner nodes, to networks with multiple simple loops, and to networks with simple loops in which the inner nodes have additional neighbours outside the loop. The essence of our analysis also extends to networks with simple loops having more than one convergence node. In such a network, a cycling error occurs only if for each convergence node of a loop, either the node itself or one of its descendants is observed. An intercausal influence is then found between the parent nodes of each convergence node. We address an arbitrary convergence node C with parents A and B on the loop. Because we consider simple loops only, there will be exactly two trails between A and B in the loop. By exploiting the property of transitivity, we can derive the sign of the indirect influence between A and B along the trail not containing C. We consider this influence to be 'the' qualitative influence between A and B. The intercausal influence between the nodes A and B that is induced by evidence for C or for one of its descendants now can be looked upon as 'the' intercausal influence between A and B. The decisiveness of the approximations established for the inner loop nodes then can be derived, as before, from the signs the two influences mentioned above. Because, effectively this procedure comes down to $\otimes$-combining the signs of all influences along the loop, the convergence node C, used for establishing the decisiveness, indeed can be chosen arbitrarily.

Example We consider the example network with a loop with multiple convergence nodes from Fig. 7. Given the evidence d and f, the loopy-propagation algorithm will compute approximate probabilities for the inner loop nodes A and B. To establish the decisiveness of these approximations, we compute the signs of the two influences between A and B. We choose C to be 'the' convergence node of the loop. The sign of 'the' intercausal influence between

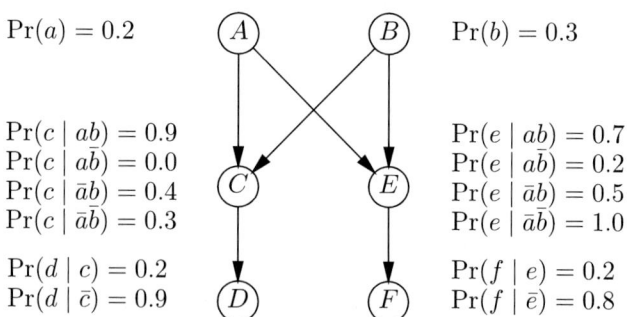

Fig. 7. An example Bayesian network, containing a loop with two convergence nodes.

A and B given d, then equals the sign of $(0.2 - 0.9)^2 \cdot (0.9 \cdot 0.3 - 0.0 \cdot 0.4) + 0.9 \cdot (0.2 - 0.9) \cdot (0.9 + 0.3 - 0.0 - 0.4) \approx -0.37$ and hence is negative. The sign of 'the' qualitative influence between A and B now equals the sign of the intercausal influence between A and B given f. This sign equals the sign of $(0.2-0.8)^2 \cdot (0.7 \cdot 1.0 - 0.2 \cdot 0.5) + 0.8 \cdot (0.2-0.8) \cdot (0.7+1.0-0.2-0.5) \approx -0.26$ and hence also is negative. Since the two signs are equal, we derive that the approximations for the inner loop nodes A and B will be overconfident. We indeed find the overconfident approximations $\widetilde{\Pr}(a \mid d) \approx 0.36$ and $\widetilde{\Pr}(b \mid d) \approx 0.31$ for the probabilities $\Pr(a \mid d) \approx 0.38$ and $\Pr(b \mid d) \approx 0.40$. □

We would like to note that in networks of more complex topology, the probabilities that are needed to determine the signs of the qualitative and intercausal influences between the parents of a convergence node, are not necessarily part of the specification of the network. In some situations, however, these signs can be derived by qualitative reasoning with an abstraction of the network [14; 15].

6 Conclusions

When Pearl's propagation algorithm for singly connected networks is applied to networks with loops, the algorithm is no longer exact and approximate probabilities are yielded. In this paper, we identified the two types of error that may be introduced into the approximations: the convergence error and the cycling error. For binary Bayesian networks with simple loops, we identified the factors that determine the effect of the cycling error on the decisiveness of the approximations calculated for the inner loop nodes. We found that this effect depends on the sign of the qualitative influence between the parents of the convergence node of the loop and the sign of the intercausal influence that is induced between these parents; the approximations are overconfident if these signs are equal and underconfident otherwise. Knowledge of the specification of the network thus provides directly for establishing properties of the approximate probabilities computed for the inner loop nodes. So far, we studied the effect of the cycling error on the decisiveness of the approximations for the nodes of a loop in isolation. An overall analysis, involving multiple compound loops, unfortunately, is much more complicated. In a network with multiple loops, for example, approximate probabilities may enter a loop as a result of errors introduced in other parts of the network and will have their own effect on the resulting approximations. In future research, we will focus on gaining further insight into the general performance of the loopy-propagation algorithm by means of controlled experiments. The insights yielded by our study of simple loops will serve to set up such experiments.

Aknowledgment

This research was (partly) supported by the Netherlands Organisation for Scientific Research (NWO).

References

[1] J. Pearl. *Probabilistic Reasoning in Intelligent Systems: Networks of Plausible Inference.* Morgan Kaufmann Publishers, Palo Alto, 1988
[2] G.F. Cooper. The computational complexity of probabilistic inference using Bayesian belief networks. *Artificial Intelligence*, **42**, 393–405, 1990
[3] P. Dagum, M. Luby. Approximate inference in Bayesian networks is NP hard. *Artificial Intelligence*, **60**, 141–153, 1993
[4] K. Murphy, Y. Weiss, M. Jordan. Loopy belief propagation for approximate inference: an empirical study. In K.B. Laskey, H. Prade (editors) *Proceedings of the Fifteenth Conference on Uncertainty in Artificial Intelligence.* Morgan Kaufmann Publishers, San Francisco, 467–475, 1999
[5] B.J. Frey. *Bayesian Networks for Pattern Classification, Data Compression and Channel Coding.* MIT Press, 1998
[6] D. Mackay, R.M. Neal. Good codes based on very sparse matrices. In C. Boyd (editor) *Cryptography and Coding: 5th SIAM Conference.* Lecture Notes in Computer Science, Springer-Verlag, **1025**, 100–111, 1995
[7] Y. Weiss. Interpreting images by propagating Bayesian beliefs. In M.C. Mozer, M.I. Jordan, T. Petsche (editors) *Advances in Neural Information Processing Systems*, **9**, 908–915, 1997
[8] Y. Weiss. Correctness of local probability propagation in graphical models with loops. *Neural Computation*, **12**, 1–41, 2000
[9] Y. Weiss, W.T. Freeman. Correctness of belief propagation in Gaussian graphical models of arbitrary topology. *Neural Computation*, **13**, 2173–2200, 2001
[10] J.H. Bolt, L.C. van der Gaag. The convergence error in loopy propagation. Paper presented at the International Conference on Advances in Intelligent Systems: Theory and Applications, in cooperation with the IEEE computer society, Luxembourg, 2004
[11] G. Strang. *Linear Algebra and Its Applications.* Academic Press, 1980
[12] M.P. Wellman. Fundamental concepts of qualitative probabilistic networks. *Artificial Intelligence*, **44**, 257–303, 1990
[13] M.P. Wellman, M. Henrion. Qualitative intercausal relations, or explaining "explaining away". In *KR-91, Principles of Knowledge Representation and Reasoning: Proceedings of the Second International Conference.* Morgan Kaufmann Publishers, Menlo Park, California, 535–546, 1991
[14] M.J. Druzdzel. *Probabilistic Reasoning in Decision Support Systems: From Computationto Common Sense.* PhD thesis, Department of Engineering and Public Policy, Carnegie Mellon University, Pittsburgh, Pennsylvania, 1993

[15] M.J. Druzdzel, M. Henrion. Efficient reasoning in qualitative probabilistic networks. In *Proceedings of the Eleventh National Conference on Artificial Intelligence*. AAAI Press, Menlo Park, California, 548–553, 1993

Lazy Inference in Multiply Sectioned Bayesian Networks Using Linked Junction Forests

Yang Xiang and Xiaoyun Chen

University of Guelph, Canada

Summary. Lazy propagation reduces the space complexity from HUGIN inference. Multiply Sectioned Bayesian Networks extend Bayesian Networks a cooperative multiagent paradigm. To combine the benefits of the two, a framework was proposed earlier to apply lazy propagation to inference in MSBNs. We propose an alternative framework with a simpler compiled structure. The issues of lazy communication and observation entering in a multiagent setting are considered. We prove that the inference is exact.

1 Introduction

Multiply Sectioned Bayesian Networks (MSBNs) [8] extend BNs [4] to the multiagent paradigm. The inference method is an extension of HUGIN method for BNs using the junction tree (JT) representation. Lazy propagation [3] extends the applicability of HUGIN inference method to larger domains. It uses a factorized representation for belief, performs only the necessary multiplication and marginalization, and results in reduced space complexity.

A framework was proposed earlier [6] to apply lazy propagation to inference in MSBNs. The compiled runtime representation requires the maintenance of multiple local graphical structures for each subnet. In this work, we propose an alternative framework for multiagent systems where only a single local structure is needed. We propose a set of algorithms for local lazy inference at each agent, for lazy communication among agents, and for entering observations. We prove that the lazy inference is autonomous and exact.

The alternative framework has the following advantages: Its local structure is isomorphic to that for standard inference in MSBNs. Hence, the same set of structure compilation algorithms [8] are applicable and the same compilation software components (such as those in WEBWEAVR [1]) can be reused. It can also lead to space savings (one local structure versus several) and resultant simplified control. Experimental evidence is expected from our ongoing research.

We briefly overview the framework of MSBNs and lazy propagation in Sections 2 and 3. Our overview assumes the knowledge on HUGIN and Shafer-Shenoy inference methods in JT representations of BNs. Readers unfamiliar with these are directed to [2; 5]. Readers who desire in-depth understanding of MSBNs are directed to [8]. The remaining sections develop the lazy propagation based new inference scheme for MSBNs. MSBNs are intended for large and complex domains. However, many relevant concepts can and should be illustrated with simple examples. Readers are reminded of the discrepancy between the complexity of the examples in the paper and that of intended applications.

2 Overview of MSBNs

2.1 Multiply Sectioned Bayesian Networks

A BN [4] can be used to structure the knowledge of a single agent. What is its counterpart for a cooperative multiagent system? From a small set of assumptions, it has been shown [7] that the resultant representation of a cooperative multiagent system is an MSBN:

1. exact probability measure of belief,
2. communication by belief over small sets of shared variables,
3. a simpler organization of agents,
4. DAG domain structuring, and
5. joint belief admitting agents' beliefs on internal variables and combining their beliefs on shared variables.

Although an MSBN can be applied under the single agent paradigm, our presentation follows the multiagent paradigm.

An MSBN M is a collection of Bayesian subnets, one from each agent, that together defines a BN. M represents probabilistic dependence of a *total universe* partitioned into multiple *subdomains* each of which is represented by a subnet. Agents cooperate to reason about what is going on [8]. Without confusion, we refer to an agent, its subdomain, and its subnet interchangeably from time to time. To ensure correct, distributed inference, subnets are required to satisfy certain conditions [7] described below:

Let $G_i = (N_i, E_i)$ $(i = 0, 1)$ be two graphs (directed or undirected). G_0 and G_1 are said to be *graph-consistent* if the subgraphs of G_0 and G_1 spanned by $N_0 \cap N_1$ are identical. Given two graph-consistent graphs $G_i = (N_i, E_i)$ $(i = 0, 1)$, the graph $G = (N_0 \cup N_1, E_0 \cup E_1)$ is referred to as the *union* of G_0 and G_1, denoted by $G = G_0 \sqcup G_1$. Given a graph $G = (N, E)$, a partition of N into N_0 and N_1 such that $N_0 \cup N_1 = N$ and $N_0 \cap N_1 \neq \emptyset$, and subgraphs G_i of G spanned by N_i $(i = 0, 1)$, G is said to be *sectioned* into G_0 and G_1. Sectioning is useful in defining the dependence between variables shared by subdomains in a graphical model:

Definition 1 *Let $G = (N, E)$ be a connected graph sectioned into subgraphs $\{G_i = (N_i, E_i)\}$. Let the subgraphs be organized into an undirected tree Ψ where each node is uniquely labeled by a G_i and each link between G_k and G_m is labeled by the non-empty* **interface** *$N_k \cap N_m$ such that for each i and j, $N_i \cap N_j$ is contained in each subgraph on the path between G_i and G_j in Ψ. Then Ψ is a* **hypertree** *over G. Each G_i is a* **hypernode** *and each interface is a* **hyperlink**. *A pair of hypernodes connected by a hyperlink is said to be* **adjacent**.

Each hyperlink serves as the information channel between subnets connected and is referred to as an agent *interface*. Agents communicate by exchanging beliefs over their interfaces. An interface must be a *d-sepset*, as defined below:

Definition 2 *Let G be a directed graph such that a hypertree over G exists. A node x contained in more than one subgraph with its parents $\pi(x)$ in G is a* **d−sepnode** *if there exists at least one subgraph that contains $\pi(x)$. An interface I is a* **d−sepset** *if every $x \in I$ is a d-sepnode.*

The overall structure of an MSBN is a hypertree MSDAG:

Definition 3 *A* **hypertree MSDAG** *$\mathcal{G} = \bigsqcup_i G_i$, where each G_i is a DAG, is a connected DAG such that (1) there exists a hypertree ψ over $\mathcal{G}$, and (2) each hyperlink in ψ is a d-sepset.*

Graphically, a hyperlink separates the hypertree MSDAG into two subtrees. Semantically, this corresponds to conditional independence given the d-sepset. An MSBN is then defined as follows:

Definition 4 *An MSBN M is a triplet $M = (\mathcal{N}, \mathcal{G}, \mathcal{P})$. $\mathcal{N} = \bigcup_i N_i$ is the* **total universe** *where each N_i is a set of variables. $\mathcal{G} = \bigsqcup_i G_i$ (a hypertree MSDAG) is the* **structure** *where nodes of each DAG G_i are labeled by elements of N_i. Let x be a variable and $\pi(x)$ be all the parents of x in G. For each x, exactly one of its occurrences (in a G_i containing $\{x\} \cup \pi(x)$) is assigned $P(x|\pi(x))$, and each occurrence in other DAGs is assigned a constant table. $\mathcal{P} = \prod_i P_i(N_i)$ is the* **jpd**, *where each $P_i(N_i)$ is the product of probability tables associated with nodes in G_i. Each triplet $S_i = (N_i, G_i, P_i)$ is called a* **subnet** *of M. Two subnets S_i and S_j are said to be* **adjacent** *if G_i and G_j are adjacent on the hypertree MSDAG.*

An example MSBN is shown in Fig. 1.

2.2 Linked Junction Forest

Inference in an MSBN is performed based on message passing. *Local inference* within each agent passes intra-subnet messages which bring a subnet into consistency. *Communication* among agents passes inter-subnet messages

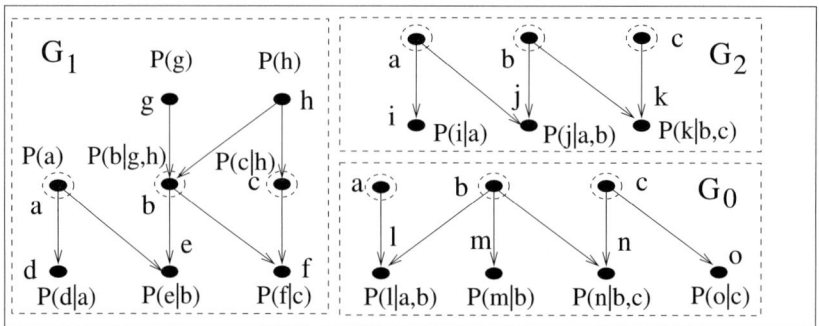

Fig. 1. A trivial MSBN where each d-sepnode is shown with a dashed circle. The hypertree has the structure $G_1 - G_0 - G_2$ and each d-sepset is $\{a, b, c\}$.

which brings the system into global consistency. These messages are marginal probability distributions. The key issue is to use messages over small subsets of variables so that inference is efficient.

To compute intra-subnet messages and propagate them effectively, each agent compiles its subnet into a junction tree (JT), where variables are grouped into *clusters* with intersection of adjacent clusters referred to as *separators*. Note that the hypertree of a MSDAG is a JT if each hypernode is labeled by the corresponding subdomain N_i. Without confusion, we simply refer to this JT as hypertree.

Similarly, to facilitate computation of inter-subnet messages, agents compile each d-sepset into a JT, called a *linkage tree*. With local JTs and linkage trees combined, the resultant representation is called a *linked junction forest* (LJF). For details on compilation, see [8]. See Fig. 2 for linkage trees L_1 between T_0 and T_1 and L_2 between T_0 and T_2. Each cluster in a linkage tree is called a `linkage`. Linkage $\{b, c\}$ is an information channel between cluster $\{b, c, f\}$ in T_1 and cluster $\{b, c, n\}$ in T_0. They are referred to as the `linkage hosts` of $\{b, c\}$.

Parallel to the structure compilation, probability tables in MSBN are converted to *potentials* (non-normalized probability distributions) associated with clusters, separators and linkages. From them, the *joint system potential* of LJF is defined that is equivalent to the jpd $\mathcal{P}$ of the MSBN. When observations are available, each agent performs local inference in its local JT using the HUGIN method. Communication among agents is performed by propagation on hypertree along hyperlinks (technically along linkages). After the communication, probabilistic queries posed to any agent can be answered exactly relative to observations entered in the entire LJF. We refer to the inference method as *HUGIN-like inference with LJFs*.

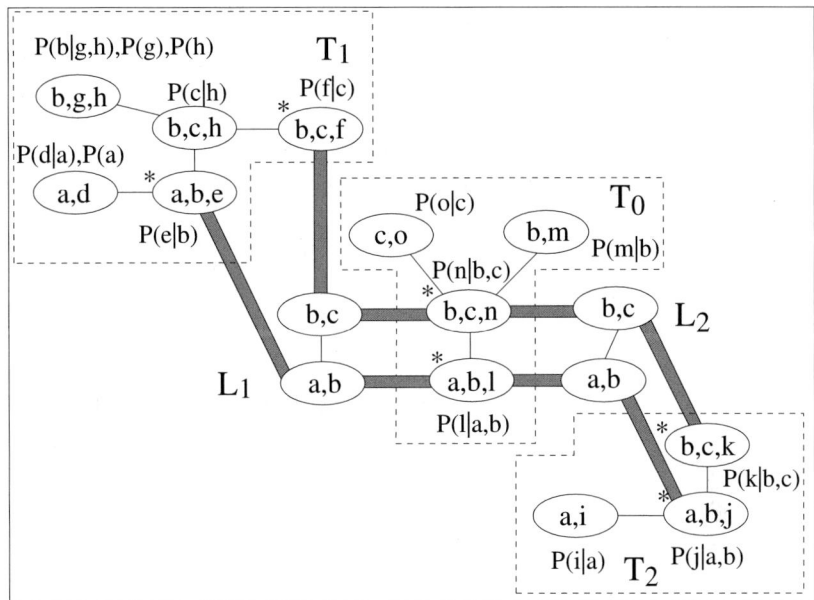

Fig. 2. JTs and linkage trees obtained from Fig. 1. Each linkage host is labeled by *. The thick links show the relation between each linkage and its hosts.

3 Overview of Lazy Propagation

Lazy propagation [3] is performed using the JT structure of a BN. Each cluster is associated with a set of potentials from the BN. We refer to the cluster of current focus by C and its set of potentials by β. When no potential is assigned to a cluster, $\beta = \emptyset$. The *joint system potential* of the JT is then the product of all potentials in all clusters, denoted as $B(N)$.

Each separator S between two adjacent clusters C and C' is associated with two buffers. One buffer is used to store the message from C to C' and the other from C' to C. We formalize lazy propagation below as pseudo-code algorithms so that we can refer to them in the new inference algorithms for MSBNs. Given a cluster C, for each separator S, we shall refer to the two buffers *locally* as the *in-buffer* and the *out-buffer* relative to C.

A cluster executes the following algorithm to compute and send a message to an adjacent cluster, where \ is the set difference operator.

Algorithm 1 (SendPotential) *Let C be a cluster with β. Let adjacent clusters be $C_1, ..., C_m$. Let β_i be the set of potentials in the in-buffer from C_i. When SendPotential relative to C_k is called in C, C does the following:*

(1) $\beta' = \beta \cup_{i \neq k} \beta_i$.

(2) Marginalize out variables $C \setminus C_k$ from β' . (To marginalize out variable x, multiply potentials with x in the domain and apply marginalization to the product.)

(3) Send the resultant set of potentials to the out-buffer to C_k.

In the following two algorithms, C is a cluster and `caller` is an adjacent cluster or the JT. The following algorithm is executed recursively by each cluster for inward message passing.

Algorithm 2 (CollectPotential) *When caller calls CollectPotential in cluster C, C does the following:*

(1) *If caller is the only adjacent cluster, perform SendPotential relative to caller.*

(2) *Otherwise, for each adjacent cluster Q except caller, call CollectPotential in Q. After all calls are completed, perform SendPotential relative to caller if it is an adjacent cluster.*

The following algorithm is executed recursively by each cluster for outward message passing.

Algorithm 3 (DistributePotential) *When caller calls DistributePotential in C, for each adjacent cluster Q except caller, C performs SendPotential relative to Q followed by a call of DistributePotential in Q.*

The following algorithm is executed by a JT for a full round of message passing.

Algorithm 4 (UnifyPotential) *Select a cluster C arbitrarily. Call CollectPotential in C. Call DistributePotential in C.*

The following proposition establishes the effect of UnifyPotential, where *const* denotes a constant:

Proposition 5 (Proposition 3.4 in [5]) *Let UnifyPotential be performed in a JT. For any cluster C with β and in-buffer messages β_i ($i = 1, ..., m$) from separators R_i with adjacent clusters, denote the product of potentials in β as $\beta(C)$ and the product of potentials in β_i as $\beta_i(R_i)$. Then*

$$\beta(C) \prod_{i=1}^{m} \beta_i(R_i) = const \sum_{N \setminus C} B(N).$$

When observations are available, for each cluster, update each potential whose domain contains observed variables and remove the observed variables from the domain. Store the observed values for subsequent queries. The following algorithm is used to enter the observation on a variable to the JT.

Algorithm 5 (EnterObservation) *When a variable x is observed at value x_0, for each cluster C (with β) containing x, do the following:*

(1) *Remove each potential $f(x)$ from β.*

(2) *For each potential $f(x, Y)$ in β, where $Y \neq \emptyset$, replace it by*

$$g(Y) = f(x = x_0, Y).$$

The effect of EnterObservation is such that the new joint system potential corresponds to the posterior distribution given the observation. After EnterObservation is performed for each observed variable, followed by an UnifyPotential, the posterior probabilities for each variable can be obtained from any cluster that contains it.

4 Lazy Inference With LJFs

We apply lazy propagation to inference in MSBNs. The on-line message computation will be guided by LJFs, but factorized beliefs and messages will be used as in lazy propagation. Each agent A_i is associated with the subnet S_i and local JT T_i.

4.1 Potential Assignment

Conditional probability tables (CPTs) in an MSBN are assigned to clusters in its LJF as potentials: For each node x in each subnet S_i, if it is assigned with a non-constant CPT (see Def 4), then assign the CPT to a cluster in local JT T_i that contains x and its parents in S_i. The potential associated with a local JT T_i is then

$$B_{T_i}(N_i) = \prod_j \prod_k \beta_{i,j,k},$$

where j indexes clusters, $\beta_{i,j}$ denotes the set of potentials assigned to the jth cluster, and $\beta_{i,j,k}$ is the kth potential in the set. The *joint system potential* of the LJF is

$$B_F(\mathcal{N}) = \prod_i B_{T_i}(N_i).$$

$B_F(\mathcal{N})$ is identical to jpd of the MSBN.

4.2 Lazy Inference: An Example

Lazy inference consists of lazy communication among agents followed by local lazy propagation. During lazy communication, inter-subnet messages are sent through linkage trees. Messages are passed through a linkage tree in both directions. Hence, a linkage between subnets S and R is associated with two message buffers, one for each direction.

Fig. 3 illustrates inward propagation with root agent A_0. First, UnifyPotential is performed by A_1 and A_2. At T_1, it causes message

$$B(b,c) = \sum_h P(c|h) B(b,h)$$

to be sent from cluster $\{b,c,h\}$ to $\{b,c,f\}$, where

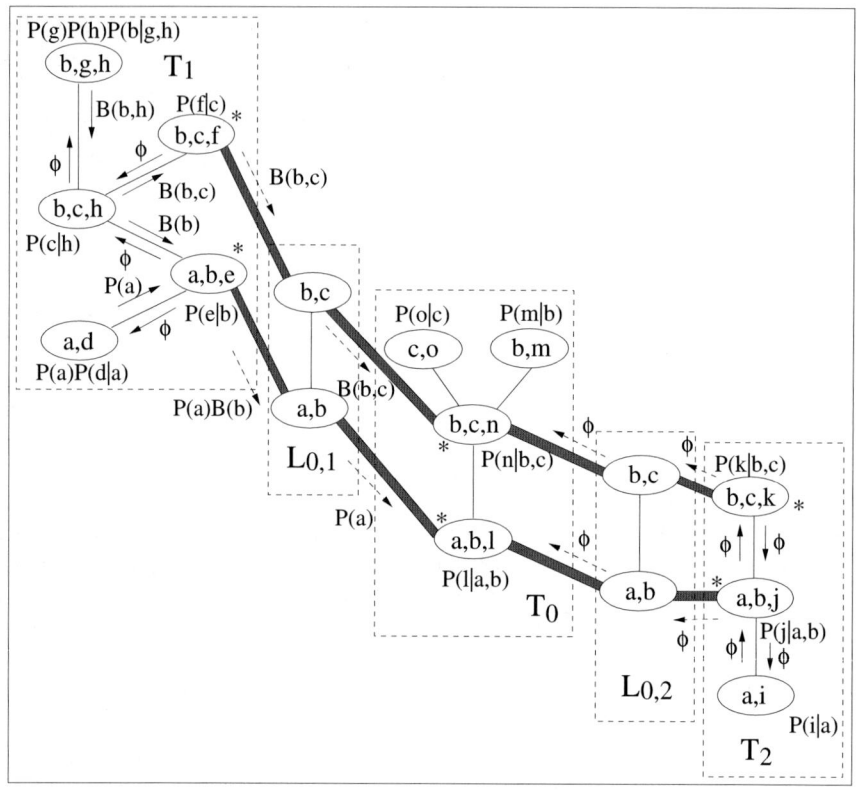

Fig. 3. Inward propagation in LJF.

$$B(b,h) = P(h) \sum_g P(b|g,h) P(g).$$

Similarly, messages $P(a)$ and $B(b) = \sum_h B(b,h)$ are sent from clusters $\{a,d\}$ and $\{b,c,h\}$ to cluster $\{a,b,e\}$, respectively. At linkage host $\{b,c,f\}$, message to linkage $\{b,c\}$ is computed based on local potentials plus the message from cluster $\{b,c,h\}$. The resultant message is $B(b,c)$. At linkage host $\{a,b,e\}$, message $P(a)B(b)$ to linkage $\{a,b\}$ is computed. As a consequence, both linkages in $L_{0,1}$ contain information on variable b: a duplication. To remove the duplication, A_1 examines potentials at linkage $\{a,b\}$ and identify $B(b)$ as the duplicated information on b. After $B(b)$ is deleted, messages from $L_{0,1}$ to T_0 become $B(b,c)$ through linkage $\{b,c\}$ and $P(a)$ through linkage $\{a,b\}$.

At A_2, UnifyPotential generates only empty messages among clusters. Messages from linkage hosts $\{b,c,k\}$ and $\{a,b,j\}$ to linkages are also empty. This concludes inward propagation.

Outward propagation follows, during which A_0 sends messages to A_1 and A_2. To calculate messages to A_1, A_0 performs UnifyPotential using linkage

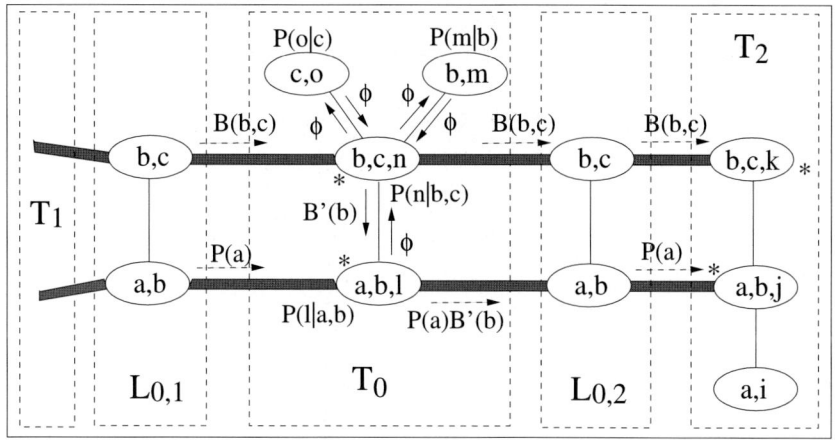

Fig. 4. Outward propagation from T_0 to T_2.

messages (empty) from A_2 but not those from A_1. All messages (intra as well as inter-subnet) are empty in this case.

Figure 4 shows outward propagation from A_0 to A_2. A_0 performs UnifyPotential using linkage messages from A_1 but not those from A_2. Message from cluster $\{b,c,n\}$ to $\{a,b,l\}$ is $B'(b) = \sum_c B(b,c)$ and all other intra-subnet messages are empty. Message from linkage host $\{b,c,n\}$ through linkage $\{b,c\}$ to A_2 is $B(b,c)$. The message through linkage $\{a,b\}$ to A_2 is $P(a)B'(b)$. Again, information on variable b is duplicated in the two linkage messages. After duplication $B'(b)$ is deleted from the message to linkage $\{a,b\}$, the resultant messages from A_0 to A_2 are $B(b,c)$ through linkage $\{b,c\}$ and $P(a)$ through $\{a,b\}$. Lazy communication is now complete.

After communication, each agent performs inference in its JT, which allows the prior probability of each variable x to be obtained from any cluster containing x in any subnet. The local inference extends UnifyPotential by including messages from linkages. For instance, to perform UnifyPotential in T_2, cluster $\{b,c,k\}$ includes linkage message $B(b,c)$ in computing the message to cluster $\{a,b,j\}$. To answer a query on $P(b)$, A_2 picks a cluster that contains b, say, $\{b,c,k\}$, and marginalizes the product of local potential $P(k|b,c)$, message from cluster $\{a,b,j\}$ (empty in this case) and linkage message $B(b,c)$. Below, we present inference algorithms of which the above example is a trace.

4.3 Local Lazy Propagation

The most primitive operation is SendPotential. To take into account message passing over linkages, we extend SendPotential (Algorithm 1) by extending the notion of *adjacency*: Two clusters are *adjacent* if

(1) they are directly connected in a JT, or

(2) they are hosts of a linkage between two JTs.

We refer to the extended Algorithm 1 as SendPotential*.

We redefine CollectPotential (Algorithm 2) and DistributePotential (Algorithm 3) to process messages over linkages. They use extended adjacency. In the algorithms, C is a cluster in a JT and `caller` is the local agent or an adjacent cluster not connected through a linkage.

Algorithm 6 (CollectPotential*) *When caller calls CollectPotential* in cluster C, C does the following:*

(1) If caller is the only adjacent cluster, perform SendPotential relative to caller.*

(2) Otherwise, for each adjacent cluster Q not connected through a linkage except caller, call CollectPotential in Q. After all calls are completed, perform SendPotential* relative to caller if it is an adjacent cluster.*

Note that CollectPotential* only receives messages from linkage in-buffers and does not send to linkage out-buffers because calling CollectPotential* across linkages is disallowed. Under the multiagent paradigm, CollectPotential* is a local operation of an agent, while sending messages across linkages involves a remote agent. CollectPotential* can be executed autonomously to answer local queries, while message passing across linkages requires coordination and incurs communication cost. Next, we redefine DistributePotential.

Algorithm 7 (DistributePotential*) *When the caller calls operation DistributePotential* in cluster C, for each adjacent cluster Q not connected through a linkage except caller, C performs SendPotential* relative to Q followed by a call of DistributePotential* in Q.*

Local lazy propagation uses Algorithm 4, with CollectPotential* and DistributePotential*, which we refer to as UnifyPotential*.

4.4 Lazy Communication

During communication, messages are sent from one agent with JT T to an adjacent agent with JT T' through their linkage tree. The messages are originated from linkage hosts in T. To ensure that each linkage host has the necessary information, UnifyPotential* must be performed before these messages are computed. This renders T locally consistent. As a result, for every two linkages adjacent in the linkage tree, the same information on their shared variables will be sent by their hosts. If such messages are directly passed to T', the new belief in T' will be incorrect due to information duplication. We consider below how to compute cross-linkage messages without information duplication.

To compute messages going from a source JT T to a destination JT T', the linkage tree L can be directed. For each linkage Q in L, the following message buffers are then allocated.

in-buffer$_1$ in-buffer from the host cluster in T.
in-buffer$_2$ in-buffer from the parent linkage in L. If Q has no parent linkage, its in-buffer$_2$ is null.
out-buffer$_1$ out-buffer to the host cluster in T'.
out-buffer$_2$, out-buffer$_3$, ... out-buffers to child linkages in L.

The message from Q to T' is computed as follows:

Algorithm 8 (SendLinkageMsg)
For each linkage Q, Q requests its linkage host to fill in-buffer$_1$ by SendPotential relative to Q. After both in-buffers are filled, Q does the following:*

(1) For each child linkage Q', marginalize out variables $Q \setminus Q'$ from potentials in in-buffer$_1$, and send resultant potentials to the out-buffer to Q'.

(2) Divide the set α of potentials in in-buffer$_1$ by the set γ of potentials in in-buffer$_2$ as follows and sends the resultant α to out-buffer$_1$:

(2.1) If a potential appears in both α and γ, delete it from both.

(2.2) For each potential f in γ, delete f from γ, multiply the set θ of potentials in α whose domains overlap with that of f, and divide the product by f. Replace θ in α by the result of the division.

Note that sending to out-buffer$_1$ involves inter-agent message transmission. Using SendLinkageMsg, algorithms below perform lazy communication in LJFs. In the algorithms, A is an agent and `caller` is the MSBN or an adjacent agent of A. CollectBeliefLLJF defines inward lazy communication along hypertree.

Algorithm 9 (CollectBeliefLLJF) *When caller calls CollectBeliefLLJF in agent A, A does the following:*

(1) If caller is not the only adjacent agent, call CollectBeliefLLJF in each adjacent agent except caller. After all calls are completed, receive linkage messages from each adjacent agent except caller.

(2) If caller is an adjacent agent, do UnifyPotential using linkage messages from each adjacent agent except caller, followed by SendLinkageMsg relative to caller.*

The inward propagation described in Section 4.2 is a trace of a call of CollectBeliefLLJF in A_0. A_0 then calls in A_1 and A_2. DistributeBeliefLLJF below defines outward lazy communication along hypertree.

Algorithm 10 (DistributeBeliefLLJF) *When caller calls DistributeBeliefLLJF in A, for each adjacent agent A' except caller, A does UnifyPotential* using linkage messages from each adjacent agent except A', followed by SendLinkageMsg relative to A' and a call of DistributeBeliefLLJF in A'.*

The outward propagation described in Section 4.2 is a trace of a call of DistributeBeliefLLJF in A_0. A_0 then calls it in A_1 and A_2. Since A_1 and A_2

have no adjacent agents except A_0, recursive calls terminate. Communicate-BeliefLLJF below combines above algorithms to accomplish lazy inference in a LJF.

Algorithm 11 (CommunicateBeliefLLJF) *Select an agent A arbitrarily. Call CollectBeliefLLJF in A. Call DistributeBeliefLLJF in A. Each agent performs UnifyPotential* using linkage messages from all adjacent agents.*

An agent A calls UnifyPotential* before sending messages to each adjacent agent. If A has k adjacent agents, then one call is made during CollectBeliefLLJF, $k-1$ calls are made during DistributeBeliefLLJF, and a final call is made at the end of CommunicateBeliefLLJF. Hence, a total of $k+1$ rounds of local lazy propagations are needed to complete CommunicateBeliefLLJF.

5 Soundness

In the following, we use *const* to denote a positive constant. Proposition 6 says that messages sent over a linkage tree define the marginal potential over the d-sepset.

Proposition 6 *Let T over N be a local JT, T' be a local JT adjacent of T, I be their d-sepset, and L be the linkage tree over I. Let UnifyPotential* be performed in T followed by SendLinkageMsg relative to T'. Let $B(N)$ be the potential*

$$B(N) = \prod_{C \in T} \beta(C) \prod_{Q' \notin L} \beta(Q'),$$

where $\beta(C)$ is the product of potentials assigned to a cluster C, $\beta(Q')$ is the product of potentials received from a linkage Q', and only linkages other than those in L are included. For each linkage $Q \in L$, let $\alpha(Q)$ be the product of potentials that Q sends to T' by SendLinkageMsg. Then

$$\prod_{Q \in L} \alpha(Q) = const \sum_{N \setminus I} B(N).$$

Proof:
First, we consider the effect of UnifyPotential* by applying Proposition 5. To do so, for each cluster C in T, we define the equivalent cluster potential of C as

$$\beta'(C) = \beta(C) \prod_{Q' \to C} \beta(Q'),$$

where $Q' \to C$ means that Q' is a linkage that feeds a message to C. We can then disregard each Q' in the remaining proof and Proposition 5 is now directly applicable.

Next, for any linkage $Q \in L$, consider its linkage host X. From Proposition 5, after UnifyPotential*, the set of potentials (including those from its in-buffers) associated with X defines the marginal of $B(N)$ onto X. This set, marginalized onto Q, is sent to in-buffer$_1$ of Q. Denote the product of potentials in in-buffer$_1$ by $\alpha'(Q)$ and the product of potentials in in-buffer$_2$ by $\theta'(Z)$, where Z is the separator between Q and its parent linkage. By Proposition 7.5 of reference [8], a linkage tree is a JT. Hence,

$$\prod_{Q \in L} \alpha'(Q) / \prod_{Q \in L} \theta'(Z) = const \sum_{N \setminus I} B(N).$$

The proposition follows since the message that Q sends to out-buffer$_1$ is

$$\alpha(Q) = \alpha'(Q)/\theta'(Z).$$

□

The following theorem says that the local potential of an agent and linkage tree messages it receives define the marginal of the joint system potential:

Theorem 7 *Let F over $\mathcal{N}$ be the LJF of an MSBN with the joint system potential $B_F(\mathcal{N})$ and let CommunicateBeliefLLJF be performed in F. Let T be any local JT over N and $B(N)$ be the potential*

$$B(N) = \prod_{C \in T} \beta(C) \prod_{Q \to T} \beta(Q),$$

where $\beta(C)$ is the product of potentials assigned to a cluster C, $\beta(Q)$ is the product of potentials received from a linkage Q into T (denoted by $Q \to T$). Then,

$$B(N) = const \sum_{\mathcal{N} \setminus N} B_F(\mathcal{N}).$$

Proof:
Denote the agent in charge of T as A. Given T, F can be viewed as a directed hypertree with A at the root. During CommunicateBeliefLLJF, only inter-agent messages directed towards A has an impact on $B(N)$. These messages are sent in semi-parallel order from leaves to the root. We analyze the impact of these messages by letting agents send one by one starting from any leaf agent A'.

Since A' (with subdomain N') is a leaf, it is adjacent to only one agent A'' (with subdomain N''). By Proposition 6, messages A' sent to A'' define the marginal of $B(N')$ onto their d-sepset. Since these messages are the only impact that A' has on $B(N)$ and they are received by A'', agent A' is effectively removed from the system. The new joint system potential defined by the local potentials in the remaining agents and the messages A'' received is

$$const \sum_{N' \setminus N''} B_F(\mathcal{N}).$$

By applying the above argument recursively to each leaf agent, eventually, all other agents in F will be removed except A. The result follows. □

The following corollary states that the local potentials in a cluster and its in-buffer messages define the marginal of the joint system potential. In the corollary, in-buffers include both those from adjacent clusters in the same local JT and those from linkages.

Corollary 8 *Let F over $\mathcal{N}$ be the LJF of an MSBN with the joint system potential $B_F(\mathcal{N})$ and let CommunicateBeliefLLJF be performed in F. Let C be any cluster in any local JT and $B(C)$ be the potential*

$$B(C) = \beta(C) \prod_{R \to C} \beta(R) \prod_{Q \to C} \beta(Q),$$

where $\beta(C)$ is the product of potentials assigned to C, $\beta(R)$ is the product of potentials received from the in-buffer associated with a separator R with an adjacent cluster of C, and $\beta(Q)$ is the product of potentials received from a linkage Q into C. Then,

$$B(C) = const \sum_{\mathcal{N} \setminus C} B_F(\mathcal{N}).$$

Proof:
It follows from Theorem 7 and Proposition 5. Theorem 7 ensures the marginal of $B_F(\mathcal{N})$ onto the subdomain of the local JT and Proposition 5 ensures further marginalization onto C. □

6 Enter Observations

When observation is obtained on a private variable, it can be entered using EnterObservation (Algorithm 5). The effect is that the new joint system potential corresponds to the posterior distribution given the observation. Consider a variable x with its value x_0 observed. If x is a root variable, EnterObservation does two things: (a) It removes $P(x)$. (b) For each child variable y of x, it replaces $P(y|x, \pi(y) \setminus \{x\})$ by $P(y|x = x_0, \pi(y) \setminus \{x\})$. Hence, the new joint system potential corresponds to $P(\mathcal{N} \setminus \{x\}|x = x_0)$.

If x is not a root variable, (a) is not applicable. EnterObservation will, in addition to (b), replace $P(x|\pi(x))$ by $P(x = x_0|\pi(x))$. This is equivalent to an operation

$$\sum_x P(\mathcal{N} \setminus \{x\}, x = x_0) = P(\mathcal{N} \setminus \{x\}|x = x_0).$$

When observation is obtained by an agent A on a public variable, however, the above is not sufficient. By performing EnterObservation in A, the local belief of A is updated. However, x may have parents or children in other agents.[1] Unless they take corresponding actions, the joint system potential has not been updated correctly. We do not require other agents to do so immediately following A's observation as agent communication is costly. Instead, it's desirable that the coordinated observation entering is delayed until the next communication. We therefore modify CollectBeliefLLJF into CollectBeliefLLJF* below.

- For (1), receive observations on d-sepnodes as well as linkage messages.
- For (2), EnterObservation before UnifyPotential*, and send observations on d-sepnodes to caller before SendLinkageMsg.

Similarly, DistributeBeliefLLJF is modified into DistributeBeliefLLJF* by performing EnterObservation before UnifyPotential*.

We refer to Algorithm 11, modified with CollectBeliefLLJF* and DistributeBeliefLLJF*, as CommunicateBeliefLLJF*. The following theorem establishes its effect whose proof is straightforward given Corollary 8 and the above discussion.

Theorem 9 *Let F over $\mathcal{N}$ be the LJF of an MSBN with jpd $P(\mathcal{N})$. Let Obs be the set of variables observed at value obs. Let CommunicateBeliefLLJF* be performed in F after observations on Obs have been entered by the corresponding agents through EnterObservation. Let C be any cluster in any local JT and B_C be the potential*

$$B_C = \beta(C) \prod_{R \to C} \beta(R) \prod_{Q \to C} \beta(Q),$$

where $\beta(C)$ is the product of potentials associated with C, $\beta(R)$ is the product of those received from the in-buffer associated with a separator R with an adjacent cluster of C, and $\beta(Q)$ is the product of potentials received from a linkage Q into C. Then,

$$B_C = const \sum_{C \cap Obs} \sum_{\mathcal{N} \setminus C} P(\mathcal{N}|obs).$$

Note that we have used word 'associated' instead of 'assigned' regarding potentials in C to emphasize the possible change of these potentials due to EnterObservation. We have also used notation B_C instead of $B(C)$ to emphasize that the product does not include observed variables in its domain. The inner summation above marginalizes $P(\mathcal{N}|obs)$ to variables in C and the outer summation marginalizes out any variable in C that has been observed.

The following theorem establishes inference autonomy for each agent. Its proof is trivial given Theorem 9 and Proposition 5.

---

[1] See [8] for reasons why x may not be observed by all relevant agents.

Theorem 10 Let observations $Obs' = obs'$ be obtained by agent A after global observations $Obs = obs$ followed by CommunicateBeliefLLJF*. Let A perform EnterObservation relative to obs' followed by UnifyPotential*. Then, for each cluster C in A's local JT,

$$B_C = const \sum_{C \cap Obs \cap Obs'} \sum_{\mathcal{N} \setminus C} P(\mathcal{N}|obs, obs').$$

7 Remarks

We presented an alternative exact method for multiagent inference in MSBNs with a simpler run-time structure than a previously proposed method. In the worst case, the complexity of lazy inference in MSBNs is upper-bounded by that of HUGIN-like inference. However, in average cases, it is expected to be much reduced due to factorized representation of cluster potentials. Further experimental investigation will provide empirical evidence on the actual complexity and comparison among the three inference methods: HUGIN-like inference, that of [6], and the method presented.

Acknowledgements

The funding support from Natural Sciences and Engineering Research Council (NSERC) of Canada to the first author is acknowledged.

References

[1] P. Haddawy. 1999. An overview of some recent developments in Bayesian problem-solving techniques. *AI Magazine*, 20(2):11–19.
[2] F.V. Jensen. 1996. *An Introduction To Bayesian Networks*. UCL Press.
[3] A.L. Madsen and F.V. Jensen. 1998. Lazy propagation in junction trees. In *Proc. 14th Conf. on Uncertainty in Artificial Intelligence*.
[4] J. Pearl. 1988. *Probabilistic Reasoning in Intelligent Systems: Networks of Plausible Inference*. Morgan Kaufmann.
[5] G. Shafer. 1996. *Probabilistic Expert Systems*. Society for Industrial and Applied Mathematics, Philadelphia.
[6] Y. Xiang and F.V. Jensen. 1999. Inference in multiply sectioned Bayesian networks with extended Shafer-Shenoy and lazy propagation. In *Proc. 15th Conf. on Uncertainty in Artificial Intelligence*, pages 680–687, Stockholm.
[7] Y. Xiang and V. Lesser. 2003. On the role of multiply sectioned Bayesian networks to cooperative multiagent systems. *IEEE Trans. Systems, Man, and Cybernetics-Part A*, 33(4):489–501.
[8] Y. Xiang. 2002. *Probabilistic Reasoning in Multi-Agent Systems: A Graphical Models Approach*. Cambridge University Press.

Part III

Learning

A Study on the Evolution of Bayesian Network Graph Structures

Jorge Muruzábal[1] and Carlos Cotta[2]

[1] Departamento de Estadística e Investigación Operativa, ESCET, Universidad Rey Juan Carlos, 28933 - Móstoles, Spain
[2] Departamento de Lenguajes y Ciencias de la Computación, ETSI Informática, Universidad de Málaga, Campus de Teatinos, 29071 - Málaga, Spain

Summary. Bayesian Networks (BN) are often sought as useful descriptive and predictive models for the available data. Learning algorithms trying to ascertain automatically the best BN model (graph structure) for some input data are of the greatest interest for practical reasons. In this paper we examine a number of evolutionary programming algorithms for this network induction problem. Our algorithms build on recent advances in the field and are based on selection and various kinds of mutation operators (working at both the directed acyclic and essential graph level). A review of related evolutionary work is also provided. We analyze and discuss the merit and computational toll of these EP algorithms in a couple of benchmark tasks. Some general conclusions about the most efficient algorithms, and the most appropriate search landscapes are presented.

1 Introduction

A Bayesian Network (BN) is a graphical model postulating a joint distribution for a target set of discrete random variables. Critical qualitative aspects relate to stochastic dependencies and are determined by the underlying graphical structure, a *Directed Acyclic Graph* (DAG). To deal with the problem of learning sensible BN models from data (a problem known to be NP-hard), a number of algorithms have been considered to search various target spaces, including most notably the space of DAG structures (b-space) and the space of equivalence classes of DAG structures (e-space), see e.g. [33; 2]. The field is very active and further representation schemes keep emerging in the literature, see e.g. Studený's algebraic approach [42]. For the most familiar search spaces, some key insights and guiding principles of interest have emerged over time [25; 11; 9; 36]. We adhere here to these principles as we try to evaluate their components in a rich evolutionary framework.

Evolutionary algorithms have been successful by now in many applications; in particular, they have been considered in this context as well [31]. This family of algorithms can be seen as an interesting class of population-based score-and-search methods, where the *fitness* measure is equated to some standard scoring metric like the marginal likelihood [25], and we can enjoy the benefits of our experience and theoretical results in the evolutionary computation field. We focus here on the *evolutionary programming* (EP) paradigm, see [18; 17] for general reference. The EP paradigm is based on the pressure exerted by selection and mutation alone, i.e., no recombination is used. Recombination is of course an important, often useful heuristic for mixing genetic material and indeed it has been often explored in b-space [31; 14; 43]. To the best of our knowledge, however, no previously proposed evolutionary algorithm (be it EP or otherwise) follows the aforementioned theoretical principles as our EP algorithms do.

The specific approaches we consider differ in either search space or type of *neighborhood*. The latter turns out to be a key concept for search algorithms: the neighborhood of a graph (in b- or e-space) equals the set of DAGs that can be reached from that DAG in a single mutation. This clearly depends on the battery of operators available. In this work, we consider on one hand the approach based on essential graphs (equivalence classes or e-space) suggested by the results of [11]. On the other hand, we consider two families of algorithms working directly in b-space inspired by the results of [9]. We denote these three algorithms as EPQ, EPNR and EPAR respectively. We are specifically interested in analyzing the relative performance of these approaches, and the computational tradeoffs involved in their application to the induction of BN structures.

2 Learning Bayesian Networks

A Bayesian Network $(\mathbf{G}, \theta)$ encompasses the Directed Acyclic Graph (DAG) $\mathbf{G}$ and a set of probability distributions attached to $\mathbf{G}$, say $\theta = \theta(\mathbf{G})$. The DAG is the set of links or *arcs* among variables or *nodes*. If we denote the whole set of discrete variables as $\mathbf{X} = \{X_1, X_2, ..., X_n\}$, each X_i has a set of *parents* denoted by $\Pi_i = \{X_j \in \mathbf{X} \mid (X_j \to X_i) \in \mathbf{G}\}$. Then, the DAG $\mathbf{G}$ represents the joint distribution

$$P(\mathbf{X}) = \prod_{i=1}^{n} P(X_i \mid \Pi_i)$$

with the parameterization $P(X_i = k \mid \Pi_i = j) = \theta_{ijk}$, $j = 1, ..., q_i$; $k = 1, ..., r_i$; r_i is the number of distinct values that X_i can assume, and q_i is the number of different configurations that Π_i can present.

Two DAGs are (*Markov*) *equivalent* if they encode the same set of independence and conditional independence statements. Each equivalence class, say $[\mathbf{G}]$, can be represented by the *essential* graph [2; 11], a unique *partially directed acyclic graph* or PDAG, say $\bar{\mathbf{G}}$. If an arc $X \to Y$ shows up in all

$\mathbf{H} \in [\mathbf{G}]$, then that arc is *compelled* in $[\mathbf{G}]$. If an arc is not compelled, then it is *reversible*, i.e., there exist $\mathbf{H}, \mathbf{K} \in [\mathbf{G}]$ such that $\mathbf{H}$ contains $X \to Y$ and $\mathbf{K}$ contains $Y \to X$. The unique PDAG $\bar{\mathbf{G}}$ representing $[\mathbf{G}]$ contains a directed arc for each compelled arc in $[\mathbf{G}]$ and an undirected arc for each reversible arc in $[\mathbf{G}]$. Our e-space refers precisely to the space of those PDAGs which represent some $[\mathbf{G}]$, see below.

There exist two different approaches to learning graphical structure from data, namely, those based on prior conditional independence testing and those usually referred to as score-and-search approaches. The first approach seeks to establish well-founded constraints on the graphical structure, thus simplifying the search space considerably, see [19; 23; 3]. Score-and-search methods omit this step and proceed directly to evaluate all tentative graph structures provided by some method via a suitable scoring metric [25; 24; 12; 7]. While there are also proposals that try to combine the best from each class of methods, here we shall be concerned with the latter class of methods almost exclusively. Besides the scoring metric itself, which is known not to make a big difference in practice for large sample size N, we need to specify the set of *traversal* operators that will be used to search for better solutions locally.

Given a BN $(\mathbf{G}, \theta)$ and a data matrix $\mathbf{D}$ with n columns and N (exchangeable) rows, there are several ways to measure the quality of fit to the data [24]. We focus here on the *marginal likelihood*:

$$P(\mathbf{D}|\mathbf{G}) = \int P(\mathbf{D}|\mathbf{G}, \theta)\pi(\theta|\mathbf{G})d\theta .$$

A closed-form expression is available for $P(\mathbf{D}|\mathbf{G})$ in the case of suitable Dirichlet-based priors $\pi(\theta|\mathbf{G})$ under certain assumptions [25]. Specifically, we take

$$\pi(\theta|\mathbf{G}) \propto \prod_{i,j}\prod_{k} \theta_{ijk}^{\alpha_{ijk}-1}$$

where $\alpha = \{\alpha_{ijk}\}$ is the *virtual count* hyperparameter ($\alpha_{ijk} > 0$). These α_{ijk} must be supplied by the user (just like the complete data set $\mathbf{D}$), but we denote our fitness or basic DAG scoring metric as $\Psi = \Psi(\mathbf{G}; \mathbf{D}) = \log P(\mathbf{D}|\mathbf{G})$ for simplicity.

A given measure Ψ is called *score-equivalent* if it is constant over each equivalence class $[\mathbf{G}]$. The present Ψ is score-equivalent if $\alpha_i = \sum_{j,k}\alpha_{ijk} \equiv \alpha$ for some $\alpha > 0$, the so-called BDe metric [25]. We consider below the BDeu(α) metric $\alpha_{ijk} = \alpha/r_i q_i$, see e.g. [8]. Another typical option is $\alpha_{ijk} = 1$, the well-known (but not score-equivalent) K2 metric [13]. Note that the score of a given PDAG $\bar{\mathbf{G}}$ is taken as the constant value assumed by members of the associated $[\mathbf{G}]$; genuine equivalence class metrics can be defined too [10].

2.1 Learning Equivalence Classes

Let $\bar{\mathbf{G}}$ denote the unique PDAG structure representing some equivalence class $[\mathbf{G}]$. Among other things, we know that $\bar{\mathbf{G}}$ and any $\mathbf{G} \in [\mathbf{G}]$ share the same skeleton or connectivity pattern (ignoring directionality) and the same *v-structures*. A v-structure is a substructure $X \to Z \leftarrow Y$ where X is not linked to Y directly. Note that not all PDAGs represent equivalence classes, only *completed* PDAGs (CPDAGs) do. A related class of PDAG models is discussed in [1]; however, this class does not exhibit the nice one-to-one correspondence that we have between equivalence classes and CPDAGs.

Chickering [11] presents six operators for introducing local variation in existing CPDAGs, namely, InsertU, DeleteU, InsertD, DeleteD, ReverseD and MakeV. The first five operators are rather self-explanatory. As to the sixth, it transforms a substructure $X - Z - Y$ (where X is not linked to Y directly) into the v-structure $X \to Z \leftarrow Y$. Note that each of these operators changes either the skeleton or the number of v-structures and thus guarantees that a new equivalence class is visited. An example of the application of each of these operators is provided in Figure 1.

The modified CPDAGs need not be evaluated from scratch: efficient score-updating formulae are provided for each operator [11]. The key idea behind this local scoring is that a *decomposable*, score-equivalent metric Ψ is typically used (for example, both K2 and BDeu(α) are decomposable). A metric Ψ is said decomposable if, for some function σ (and implicit data), $\Psi(\mathbf{G}) = \sum_{i=1}^{n} \sigma(X_i, \Pi_i)$, where calculation is restricted in each summand to a single node X_i and its parents Π_i. Thus, only those nodes whose Π_i is changed by the operator need to be updated. To illustrate, the change in score attributed to a particular (valid) move deleting $X \to Y$ in $\bar{\mathbf{G}}$ (and leading to some $\bar{\mathbf{H}}$) can be expressed as $\Psi(\bar{\mathbf{H}}) - \Psi(\bar{\mathbf{G}}) = \sigma(Y, \Lambda_1) - \sigma(Y, \Lambda_2)$, where $\Lambda_1 \subset \mathbf{X}$ is the set of nodes connected to Y (with either a directed or undirected arc), and $\Lambda_2 = \Lambda_1 \cup \{X\}$. Similar or slightly more complex expressions hold for the remaining operators.

While the six operators are all local in principle, there may be "cascading" implications in some moves. For example, as seen in Figure 1, DeleteD and ReverseD may make other directed arcs undirected. Or, after applying MakeV, many arcs may switch from undirected to directed. Hence, it is difficult to predict the behavior of the tentative graphs produced along the way, and indeed surprises may arise in some cases, see [15] and below. In practice, we find the outcome $\bar{\mathbf{H}} = \omega(\bar{\mathbf{G}})$ by applying two key algorithms in turn [11]. We first use the PDAG-to-DAG routine to extract a member DAG $\mathbf{H}$ from the raw result of the mutation, say $\bar{\mathbf{H}}_r$. If no such $\mathbf{H}$ can be found, the intended mutation is not valid (the PDAG can not be completed; a compact validity test is provided for each operator to prevent unnecessary computations). Otherwise we call the DAG-to-CPDAG routine (with input this $\mathbf{H}$) to determine the resulting (validated) $\bar{\mathbf{H}}$. As discussed below, these two routines can be used in reverse order to move within the same equivalence class in a random way (we have included a stochastic component in the PDAG-to-DAG routine, namely, the order in which the nodes will be traversed so as to assign directionality to undirected arcs).

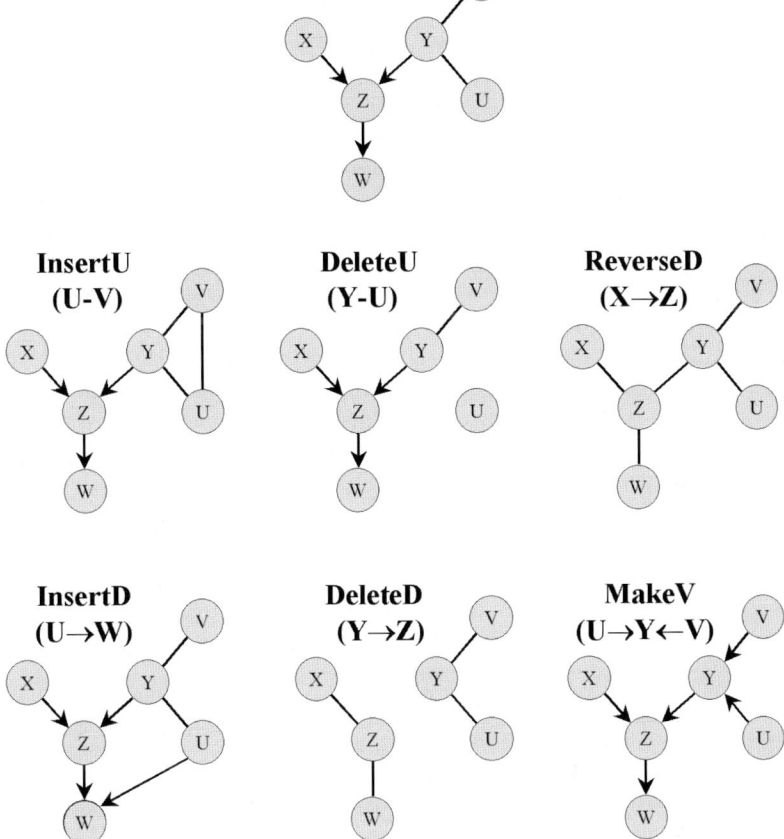

Fig. 1. Operators for traversing e-space.

2.2 Inclusion-Driven Learning

Castelo and Kočka [9] and others argue that traversal operators which respect the *inclusion boundary* (IB) condition or principle have appealing theoretical properties. Briefly, if the true distribution generating the data can be expressed as a BN model, and if certain reasonable assumptions concerning the score function are made, then, in the long run (for large sample size N) we are guaranteed to reach the target model when we use traversal operators that verify this principle. A traversal operator respects the IB condition if the neighborhood associated to a given **G** always contains its IB, say $\mathcal{IB}(\mathbf{G})$. The $\mathcal{IB}(\mathbf{G})$ collection of models contains all those "immediately next" to **G** in a precise distributional sense. The bottom line is that traversal operators should be designed so that they possibly visit any "sufficiently close neighbor" in this sense. On this matter, the standard NR and AR (No and All Reversals respectively) neighborhoods are a primary reference. In the case of NR, only the usual `InsertD` (Insert directed arc at a random valid location, that is, wherever the insertion does not form a cycle) and the similarly defined `DeleteD` operators are allowed, whereas AR incorporates also `ReverseD`. Natural questions are: How well do the basic NR and AR neighborhoods do with regard to the IB condition? Can we find a traversal operator whose neighborhood coincides exactly with $\mathcal{IB}(\mathbf{G})$? If so, how much should $\mathcal{IB}(\mathbf{G})$ be augmented (if at all)?

The answer to the second question is given by the particular neighborhood ENR (Equivalence class-based NR jump), see e.g. [9]. For a given $\mathbf{G}$, this consists of the union of all DAGs that belong to the standard NR neighborhood of $\mathbf{G}$ together with all DAGs that belong to the same neighborhood of all other $\mathbf{H} \in [\mathbf{G}]$. The idea in the implicit intra-class navigation is that certain areas of $[\mathbf{G}]$ may be closer to some intermediate equivalence classes of interest than others. A tentative improvement over ENR is provided by the ENCR neighborhood [9]. This is defined just like ENR, except now the `ReverseD` operator is allowed and restricted to non-covered arcs. A given arc $Y \to X$ is said to be *covered* in the DAG $\mathbf{G}$ if $\Pi_X = \{Y\} \cup \Pi_Y$ holds. In other words, if there exists another arc $Z \to X$ then $Z \to Y$ must also exist (and vice versa). Hence, a covered $Y \to X$ can not be part of any v-structure. It follows that the reversal of a covered arc does not change neither the skeleton nor the number of v-structures. Therefore, non-covered arc reversals are guaranteed to leave the current equivalence class.

Of course ENR encompasses a huge number of graphs and hence needs to be *simulated* by a random walk or otherwise. Given a DAG $\mathbf{G}$, we can move within $[\mathbf{G}]$ by iterated (random) covered arc reversal. Let r the number of calls to be made for each move. It is argued in [9] that r need not be very large because equivalence classes contain an average of less than four DAGs [21]. In practice, the algorithms will need to handle (DAGs from) equivalence classes close to the target. Hence, if the target equivalence class is believed to be large, then r may need to be larger. In any case, once the r stipulated random reversals have taken place, the resulting structure is modified according to the standard NR (implementing ENR) or whatever neighborhood is implied by the traversal operators (implementing also ENCR).

Note that there exist other learning algorithms which also respect the IB condition. These include the GES algorithm proposed by Chickering [12] (a fully greedy algorithm which begins with an empty graph) and the KES generalization considered by Nielsen et al. [36]. Furthermore, complying Markov Chain Monte Carlo (MCMC) algorithms can also be devised [9]. We believe that our current EP approach is likely to cooperate effectively along these alternative lines too, see the concluding section.

3 Evolutionary Approaches to Network Induction

Several evolutionary algorithms have been proposed for the present graphical model induction task. The seminal paper by Larrañaga et al. [31] presents the first genetic algorithm (GA) in b-space, see also [41]. The parallel paper by Larrañaga and coworkers [29] examines the role of the GA when restricted to explore the space of topological orderings (essentially permutations) of the variables. A popular heuristic algorithm like K2 [13] takes one such topological order as input and returns a fitted BN that respects that order.

More recent work in the area by this research team involves the so-called Estimation of Distribution Algorithms, see e.g. [7]. Evolutionary algorithms of this sort replace crossover with sampling from a model fitted to the best individuals in the current population [39; 30]. Model distributions here refer in principle to the space of DAGs; hence, they must remain relatively simple to be tractable. For example, it is not uncommon to model univariate (marginal) arc behavior or simple arc-to-arc dependencies. It follows that graphs sampled from these models may include cycles and thus need repairing [7].

Wong et al. [44] use an EP approach that aims to enjoy the advantages of both the prior testing and the score-and-search approaches mentioned earlier. They preprocess the data and use the standard Mutual Independence measure to compute a matrix that evaluates the strength of every possible arc in the emerging DAG. Mutation operators include the classical operators considered here as well as other, less frequent operators. These operators are sensitive to the MI information, in the sense that, for instance, the weakest arc is more likely to be deleted when the deletion operator is called and so on. This bias acts in the same way throughout the run. Note that the MI measure can only capture the value of a given arc *taken in isolation*. In actual learning runs, however, we should find that the value of a given arc $Y \to X$ depends more naturally on whether and which other arc(s) $Z \to X$ are concurrent. In fact, it has been pointed out that the offspring produced by this so-called MDLEP method are often worse than their parents [46]. We shall explicitly consider this improvement rate in our experiments below.

Harwood and Scheines [23] propose an *annealed* GA to search over the space of equivalence classes of DAGs or e-space. These authors provide a good discussion of the pros and cons of the various strategies available for inducing networks from data. They handle the slightly different problem in which variables are continuous and linear local regressions are computed at each node so as to provide arcs with a certain coefficient. These coefficients come along with the graph and play the role of the conditional distributions θ in the discrete setting. Harwood and Scheines suggest to improve the standard evolutionary process by adding an annealing scheme that slowly increases the penalty for complex models in the fitness function (so that the system works initially with relatively dense networks). They further prune the search space along the run by permanently banning adjacencies from future consideration if these adjacencies become extinct in the current population. Harwood and

Scheines also tackle the issue of dealing with relatively few data (compared to the number of available variables, $N \ll n$).

Cotta and Muruzábal [14] propose various crossover operators in b-space. Their approach is based on *preventing* the formation of cycles (to avoid the costly repairing, see [16]). It also involves a (surrogate of) the *conditional* mutual independence measure, which assesses the value of arcs in the context of their relevant parent sets Π in the parent DAGs. This measure is used to rank the goodness of the various arcs possibly transmitted from parents to offspring. The proposed crossover operators exploit this information in various ways making important arcs more likely to be transmitted from parent to offspring DAGs. Cotta and Muruzábal also suggest that *respectful* strategies (transmitting routinely all arcs shared by both parent DAGs) might prove advantageous to crossover operators in this setting.

Wong and Leung [46] present the hybrid evolutionary algorithm HEA. Their HEA is based on the so-called *merge* operator, a parent-set-based version of crossover. Given two parent DAG structures, merge chooses the parent set Π_i of variable i in the offspring from the corresponding Π_i in the parent DAGs, with the goal that offspring exhibit better overall scores than their progenitors. HEA is also based on cycle-prevention. The merge operator is given priority over the mutation operators used in [44] since it exhibits several desirable properties, most notably, that score information related to conserved Π_i can be reused economically.

Van Dijk and Thierens [43] recognize the potential gain provided by the PDAG-based non-redundant encoding and discuss a GA that allows searching in e-space. They point out that the standard DAG representation may jeopordize the fusion of useful building blocks and thus lead to a poor crossover operator in general. However, in their implementation PDAGs are instantiated as DAGs prior to crossover; then the DAG offspring are cleaned up (cycles are broken) and reinserted as PDAGs. Van Dijk and Thierens [43] dismiss the framework of Chickering [11] (adopted here) arguing that this "requires a more complicated implementation and is only of practical interest". While they also acknowledge the computational cost of the numerous DAG-to-PDAG and CPDAG-to-DAG calls in their proposal, the type of neighborhood implemented by the set of traversal (and/or crossover) operators is crucial for our score-and-search algorithms to be able to escape local optima.

Another type of parent-set-based evolutionary algorithm has been proposed by Wong et al. [45], namely the cooperative co-evolutionary GA or CCGA. This builds upon the two-phase approach discussed earlier [44], so that the search space is constrained by the "verified assertions" made in the prior testing phase. The idea is then to search (separately) over the space of parent configurations Π_i in each case. The maximum size of Π is kept limited throughout. The optimal parent sets for the different variables (species) are assembled together to form the final complete DAGs (again, some postprocessing may be needed for cycle removal). The fitness measure for the individual search processes does include a component that evaluates the degree of cooperation of tentative Π_i. This component is based on the scores of the "collaborative structures" assembled from good representatives of each species along the way. Similar ideas have been known for a while in the wider evolutionary context, see e.g. [34].

To summarize, evolutionary approaches have flourished in recent years for the graph induction problem. These algorithms can be tailored to deal with this problem in various ways, and a host of ideas frequently found useful are brought to play to our advantage. However, no conclusive assessment has been reached yet, so more detailed comparative work and benchmarking are in order.

4 Evolutionary Search Landscapes for Optimal Network Induction

In this section we review the details of our evolutionary algorithms. We first review EPQ in e-space, then continue with EPNR and EPAR in b-space. It is useful to begin by describing the common skeleton in these EP algorithms.

4.1 Standard EP Setting

The common steps in our algorithms are the following: (i) we begin with a population of P randomly initialized graphs and we evaluate them using the fitness or scoring metric Ψ. (ii) At each generation, P members of the current population are selected by means of binary tournament (two graphs are randomly drawn and the highest score wins). (iii) Each selected graph may be preprocessed. (iv) Either the original or the preprocessed graph is mutated once by selecting an operator ω from the available battery Ξ according to some distribution Ω, and applying it at a random (valid) entry point in the target graph. (v) All P mutated graphs are (locally) evaluated and stored. (vi) Finally, the best P out of the $2P$ available structures at this point are selected for the next generation, the remaining P are discarded and a new iteration takes place.

The probability distribution Ω over the battery Ξ may be fixed in evolutionary time, or it may be dynamic (in various ways). At the moment, we use a stationary, uniform Ω throughout the process and for all individuals.

Initialization of DAG structures can be pursued either in a purely random way or heuristically. In the first case, we have devised a simple randomization routine in which parameter $\delta \in [0,1]$ controls the arc density of the resulting graph. More sophisticated approaches to uniformly distributed DAG generation exist, see e.g. [27]. In the second case, the K2 heuristic is used, taking a random permutation of the variables as seed. The process is further controlled by $\pi_{\max}$, the maximum number of parents allowed per variable. Note that this limit is set only on the initial structures, it is not enforced along the run. Initial (valid) PDAGs are easily generated from random DAGs by using the DAG-to-CPDAG routine mentioned in Section 2.1.

4.2 Neighborhoods in E-Space and B-Space

No preprocessing is carried out in the case of EPQ. The particular neighborhood used here does not quite contain the target ENR (although it typically contains many other DAGs), so EPQ does not really enjoy the convergence property discussed in Section 2.2. Recall that the operators analyzed in [11] are `InsertU`, `DeleteU`, `InsertD`, `DeleteD`, `ReverseD` and `MakeV`. It was shown in [35] that all operators are needed for best performance, so we allow all in Ξ here. Recall also that all of them change the current equivalence class and all can be scored efficiently.

Note that some operators may not find a suitable entry point in the selected CPDAG and hence may become non-applicable (in which case a different operator should be selected etc.). If, on the other hand, one or more appropriate entry points can be found for the selected ω, then the operator is tentatively applied at a randomly selected point. If the mutated CPDAG $\bar{\mathbf{H}} = \omega(\bar{\mathbf{G}})$ passes the corresponding validity test, it is incorporated to the offspring population. We track the *success* ratio of each operator $\varepsilon = \varepsilon(\omega)$ during the replacement stage, i.e., the number of CPDAGs that ω produced and made it to the next population.

We now continue with our EP algorithms in b-space, EPNR and EPAR. Note that preprocessing of a selected DAG $\mathbf{G}$ refers to the *navigation* within the class $[\mathbf{G}]$ containing $\mathbf{G}$. As explained in Section 2.2, this navigation is achieved via a series of covered arc reversals.

Castelo and Kočka [9] discuss implementations of the ENR, ENCR and other neighborhoods. They refer to the version covering ENR as RCARNRr (for r Repeated Covered Arc Reversals followed by a NR jump); it goes hand by hand with our EPNR(r) algorithm. As noted above, ENCR is simulated similarly and the corresponding algorithm is denoted as RCARRr. We adopt below a simpler implementation of ENCR allowing all arc reversals, which we call EPAR(r). The case $r = 0$ (no navigation at all) transforms radically the associated neighborhoods, with the result that the theoretical support is lost [9]. However, we still consider EPNR(0) (equal, of course, to the standard NR) and EPAR(0) for the sake of reference. We also consider the case in which all navigation steps are collapsed into two as follows: firstly DAG-to-CPDAG is applied to $\mathbf{G}$ to obtain $\bar{\mathbf{G}}$; then, PDAG-to-DAG is used on $\bar{\mathbf{G}}$ to extract another DAG $\mathbf{H} \in [\mathbf{G}]$. This scheme is denoted as $r = \infty$. It had not been proposed previously, although we feel it is a natural competitor for the $r > 0$ alternatives [35].

5 Experiments and Results

The algorithms described above have been deployed on two conspicuous networks: ALARM –a 37-variable network for monitoring patients in intensive care [4]– and INSURANCE –a 27-variable network for evaluating car insurance risks [6]. The equivalence class [ALARM] is represented by a CPDAG with 4 undirected and 42 directed arcs. As to [INSURANCE], it is a larger and denser equivalence class, represented by a CPDAG with 18 undirected arcs and 34 directed arcs. In both cases, a training set of $N = 10,000$ examples was created once by random probabilistic sampling as customary. The BDeu($\alpha = 1$) metric $\Psi(\mathbf{G}|\mathbf{D}, \alpha) = \log P(\mathbf{D}|\mathbf{G})$ is the fitness function (to be maximized). Previous work [35] indicates that this setting $\alpha = 1$ provides the best results (fake dependencies abound for larger values of α, whereas for lower values some true dependencies are lost).

All experiments have been performed using a population size of $P = 100$ individuals. The termination criterion is reaching a number of 500 generations, i.e., 50,000 networks generated. Such a termination criterion follows the common practice in evolutionary computation, where fitness computation is the basic cost unit. Nevertheless, this particular application has a distinctive feature: the goodness of a generated structure is not calculated from scratch, but by means of local evaluations (recall the decomposability of our fitness function). Since the number of such local evaluations depends on the operator and on the value of r, we have monitored the accumulated number of local evaluations across the run, to obtain another –possibly more representative– figure of cost. Two different initialization settings have been considered: random initialization using density value $\delta = 0.05$, and K2 initialization with maximum number of parents per variable $\pi_{\max} = 2$.

The results are shown in Figure 2. Notice firstly the results of EPNR(0). These are remarkably inferior to those of any EPNR($r > 0$) for both networks. This confirms the limitations of the basic NR neighborhood. As soon as $r > 0$, there is a sharp performance improvement. This improvement clearly supports the usefulness (for this particular neighborhood) of intra-class navigation, for it increases the connectivity of the search space (and hence decreases the number of local optima). This is also true for EPAR, although the difference among the several values of r is not so large in this case. The effect of using the denser AR neighborhood is here dominant. Indeed, by taking $r > 0$ new paths in b-space are possible, although the enhanced inter-class navigation capability offered by ReverseD remains the prime feature (as it can be noted by comparing the behavior of EPAR(0) with that of EPNR(r)). Actually, the performance of EPAR(r) is always superior to that of its EPNR(r) counterpart. EPQ is also better than EPNR(0) and tends to perform similarly to EPAR. Since the connectivity in e-space is very rich, it is worth investigating which operators are most useful.

As can be seen in Figure 3, there exists naturally a general decreasing trend in success rates (improvements are less frequent in the latter stages of

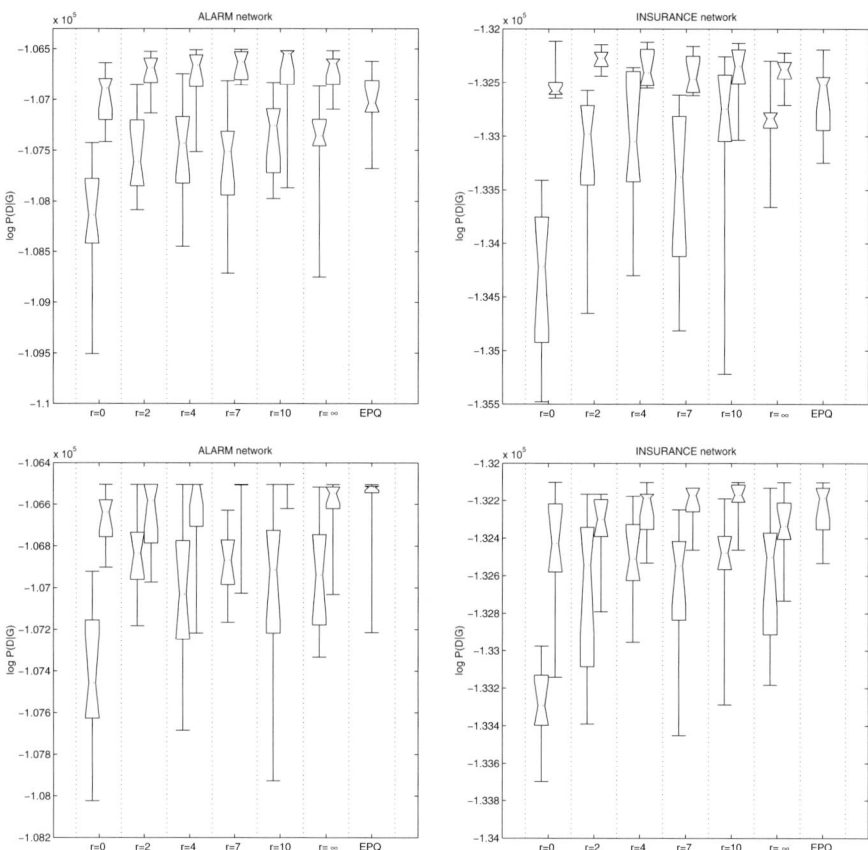

Fig. 2. Results by the proposed EP algorithms (each boxplot summarizes ten runs). (Top) Random initialization (bottom) Heuristic initialization. For each value of r, the left boxplot corresponds to EPNR, and the right one to EPAR. Notice the use of different scales in each plot.

the run). Also for this reason, lower success levels are obtained when using heuristic initialization (the algorithm performs its run at a higher fitness level). The different decreasing rates shed some light on the relative contribution of operators. In particular, note that the undirected-arc-based, "brick and mortar" operators `InsertU` and `MakeV` tend to maintain the highest success ratios by the end of the run. The injection of v-structures appears thus crucial for balancing the adequate proportion of directed and undirected arcs. Indeed, if `MakeV` were removed from the set of available operators, directed arcs would begin to vanish very quickly [35]. On the other hand, `DeleteD`, `ReverseD` and `DeleteU` are the ultimately least useful.

This overall success picture may suggest the following EPQ dynamics. It appears that the typical behavior of this algorithm is to first trim massively

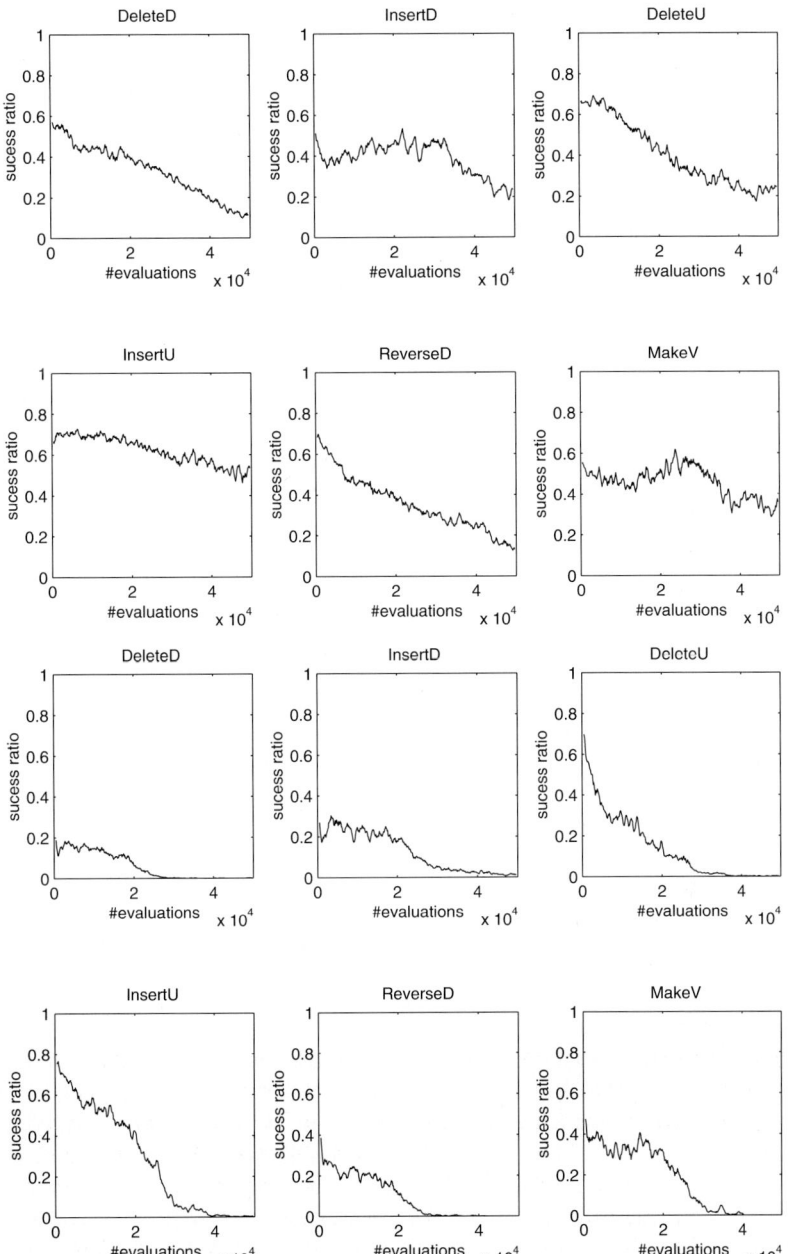

Fig. 3. Mean success ratio (averaged for ten runs) of e-space operators when using random initialization (two upper rows) and heuristic initialization (two lower rows). Results correspond to the ALARM network.

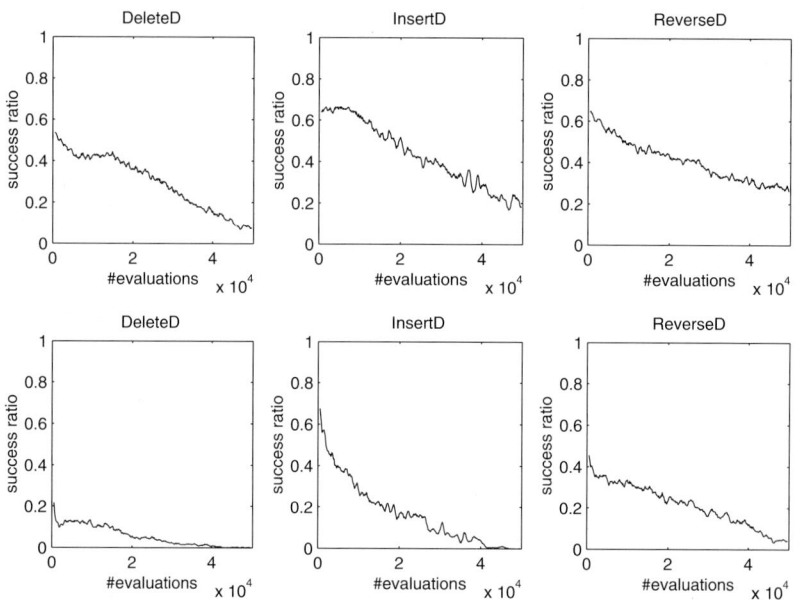

Fig. 4. Mean success ratio (averaged for ten runs) of b-space operators when using random initialization (upper row) and heuristic initialization (lower row). Results correspond to the ALARM network.

all irrelevant arcs, ending up with a relatively small structure with mostly undirected arcs. As the structure grows from this basis, some useful set of both compelled and reversible arcs is secured. Tentative directionality is assigned by MakeV and a higher number of sensible directed arcs and v-structures emerge over time.

It must also be noted that ReverseD plays an important role in helping the networks size down appropriately, as indicated by the denser networks obtained when this operator is removed [35]. This important role is also clear in the case of traversing b-space –see Figure 4– where its removal results in handicapped search capabilities as indicated by the poor results of EPNR with respect to EPAR.

Tables 1 and 2 show the structural properties of the networks evolved using heuristic initialization. Two facts must be highlighted: firstly, the number of recovered arcs (in the true equivalence class) is always bigger for EPAR; also, networks tend there to be smaller (like in the case of EPQ). The best run for ALARM recovers all but one of the arcs. For INSURANCE, the best network has a Hamming distance of 11 with respect to the original one. From an absolute point of view, the quality of these results is high, and comparable to the state-of-the-art.

A final comment must be done regarding the somewhat hidden cost of performing intra-class navigation, namely the fact that local score-updates

Table 1. Structure of the networks generated by EPNR and EPAR using heuristic initialization (averaged for ten runs). From left to right in each case: number of recovered undirected arcs, number of recovered directed arcs, and total number of arcs.

	EPNR						EPAR					
	ALARM			INSURANCE			ALARM			INSURANCE		
r	S_U	S_D	n_{arcs}	S_U	S_D	n_{arcs}	S_U	S_D	n_{arcs}	S_U	S_D	n_{arcs}
0	2.4	17.8	66.0	4.9	15.4	53.3	3.8	32.8	50.2	6.4	24.5	47.5
2	3.4	34.1	50.9	7.3	23.6	48.5	3.7	38.5	48.5	7.4	25.9	46.7
4	3.4	35.3	51.9	6.0	25.6	47.4	3.9	38.0	48.0	9.9	27.2	46.6
7	3.2	34.2	52.7	5.9	23.8	48.0	4.0	39.4	46.7	9.4	27.9	46.2
10	3.0	34.6	54.0	8.3	24.5	47.3	4.0	40.5	46.3	10.0	28.2	46.2
∞	3.0	32.4	53.6	6.9	23.7	48.1	3.8	35.0	48.7	7.9	25.9	46.7

Table 2. Structure of the networks generated by EPQ using heuristic initialization (averaged for ten runs). Interpretation is as before.

ALARM			INSURANCE		
S_U	S_D	n_{arcs}	S_U	S_D	n_{arcs}
4.0	36.4	48.0	10.6	27.0	46.4

are required whenever a covered arc is reversed. Recall that, while the sum of local scores $\sum \sigma(X_i, \Pi_i)$ would remain unchanged, the inner terms would change. That is, when reversing the covered arc $X_i \to X_j$ we would have

$$\sigma(X_i, \Pi_i) + \sigma(X_j, \Pi_j) = \sigma(X_i, \Pi_i \cup \{X_j\}) + \sigma(X_j, \Pi_j \setminus \{X_i\}), \quad (1)$$

but each summand would be different in general (and hence they need being computed in order to keep the consistency of the overall score, see [40] for a related argument). Figure 5 shows the evolution of fitness for the first 70,000 such local evaluations. It turns out that EPAR(0) provides the best tradeoff between computational cost and quality achieved. The difference is remarkable for ALARM; in the case of INSURANCE, EPAR(∞) manages to catch up with EPAR(0) at around 60,000 local evaluations. The remaining settings of r result in slower convergence. Slightly better networks may be attained at the end of the run, but each new network generated required a larger computational effort.

6 Conclusions, Discussion, and Future Work

We have considered a variety of EP algorithms for learning Bayesian Network graph structures from data. Our primary aim has been to investigate the role of intra-class navigation in the enhanced AR and NR neighborhoods,

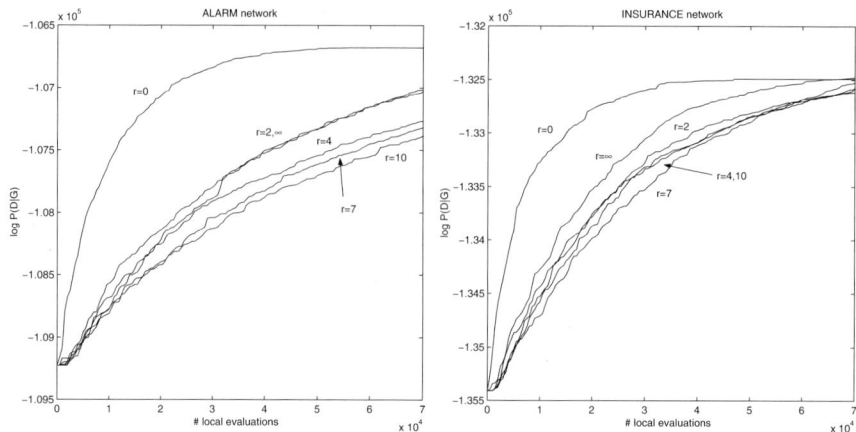

Fig. 5. Evolution of fitness in EPAR as a function of the number of local evaluations. (Right) ALARM (left) INSURANCE.

and the adequacy of the corresponding fitness landscapes for evolutionary exploration. We have reproduced and confirmed in this new context previously reported phenomena such as the poor performance of EPNR(0) and the usefulness of intra-class navigation in this case. Our assessment of the behavior of EPAR indicates that the inter-class navigation capability featured by ReverseD dominates the situation though. Furthermore, the extra cost due to intra-class navigation needs to be taken into account. As a result, it might be profitable to disable this latter feature when using the enhanced AR neighborhood, at least during the initial stages of evolution. After all, it is only in the latter stages of the run when the algorithm is more likely to be in a local optimum (or in the basin of attraction thereof), and the increased connectivity provided by covered arc reversals may be more useful. In earlier stages, the benefit would be probably overcome by its associated computational cost. A related computational concern refers to r, the number of covered arc reversals made at each step. Precisely because there appears to be some uncertainty about the best values for this parameter, we believe that the use of an adaptive (or even *self-adaptive* [5]) scheme for varying r across the run may be highly interesting. This is a line for future developments.

There is also a huge potential for exploiting phenotypic information in this context. Our current operators are essentially genotypic (all decisions are fully randomized), and hence blind to quality. The usage of such information can have a positive effect in the convergence properties of the learning algorithms [14]. Confirming the results obtained in this work for these phenotypic operators, and indeed inquiring about their limits with respect to the IB condition, is another appealing line of work.

Another interesting approach mimics the chain irreducibility requirement in MCMC algorithms [33; 20]. MCMC algorithms constitute a major reference for EP and other evolutionary algorithms in the graphical model induction arena [32; 22]. The reason is perhaps best seen by noticing that, in the multiple-chain case, valid jump proposal distributions can be advantageously based on information from several individuals (chains), see e.g. [26]. The idea is then close to the recombination aim of GA-based algorithms. It may be suspected that insights provided in either area can transfer profitably to the other. Specifically, to encompass the basic ENR neighborhood in our simulation, we can use algorithms that behave like EPNR with probability $p < 1$ only, jumping to e-space and following (selected operators in) EPQ with probability $1 - p$. Perhaps the most useful operators `InsertU` and `MakeV` should be given priority here, at least for the various initialization methods that we have implemented. Again, parameter p can vary along the evolutionary run.

Switching to more practical matters, it is often acknowledged that, for the purpose of data mining, not one but several good candidate graphs will have to be produced. The idea is to first train a *diverse* set of BN models, then isolate recurring features in these models. The significance of these features (not only single edges but also more complex constructs like *Markov blankets*, see below) can be assessed reliably, see e.g. [38]. This problem seems also particularly prone to benefit from an evolutionary approach, since many techniques have been proposed to maintain diversity in the population [17]. In [28], for example, *speciation* based on the so-called fitness-sharing technique is enforced in b-space. The basic similarity measure between DAGs is based on the number of recovered, reversed and missing arcs. Several representative models are extracted from the final population, and these models are combined leading to more robust predictions.

Finally, it has been argued sometimes that, when the main goal is feature extraction and classification, Markov blanket learning may be more appropriate (perhaps the only hope for scalability in some cases) than full BN learning. Given a DAG **G**, the Markov blanket (or boundary, MB) of a given variable X (with respect to **G**), say $MB(X)$, is defined as the union of all X's parents, all X's children, and all parents of X's children. It is always the case that X is conditionally independent of all other variables given $MB(X)$. Thus, if explaining and predicting the behavior of X is the primary concern, then the graphical substructure depicting the dependencies between X and variables in its MB is all we care about. This observation simplifies the problem considerably. For examples of techniques capable of searching for MB directly and critical reviews of the literature on this subject, see the recent contributions [3] and [37]. The authors of the latter paper also argue that, conversely, full BN learning can be greatly facilitated by learning first each MB separately. In a similar vein, Riggelsen [40] has recently suggested that a MB-based MCMC approach improves upon the more usual, single-arc-based variants. These ideas appear indeed likely to continue to play an important role in future developments in the area.

Acknowledgement

The authors would like to express their gratitude to Robert Castelo, David Chickering, Paolo Giudici, Juan Antonio Lozano and David Ríos. Financial support by grants from Spanish and European agencies is greatly appreciated.

References

[1] S. Acid and L.M. de Campos. Searching for Bayesian Network structures in the space of restricted acyclic partially directed graphs. *Journal of Artificial Intelligence Research*, 18:445–490, 2003.

[2] S.A. Andersson, D. Madigan, and M.D. Perlman. A characterization of Markov equivalence classes for acyclic digraphs. *Annals of Statistics*, 25:505–541, 1997.

[3] Xue Bai, Clark Glymour, Rema Padman, Joseph Ramsey, Peter Spirtes, and Frank Wimberly. PCX: Markov blanket classification for large data sets with few cases. Technical Report CMU-CALD-04-102, Center for Automated Learning and Discovery, School of Computer Science, Carnegie Mellon University, 2004.

[4] I.A. Beinlich, H.J. Suermondt, R.M. Chavez, and G.F. Cooper. The ALARM monitoring system: A case study with two probabilistic inference techniques for belief networks. In J. Hunter, J. Cookson, and J. Wyatt, editors, *Proceedings of the Second European Conference on Artificial Intelligence and Medicine*, volume 38 of *Lecture Notes in Medical Informatics*, pages 247–256, Berlin, 1989. Springer-Verlag.

[5] H.-G. Beyer. Toward a theory of evolution strategies: Self adaptation. *Evolutionary Computation*, 3(3):311–347, 1996.

[6] J. Binder, D. Koller, S. Russell, and K. Kanazawa. Adaptive probabilistic networks with hidden variables. *Machine Learning*, 29:213–244, 1997.

[7] R. Blanco, I. Inza, and P. Larrañaga. Learning Bayesian Networks in the space of structures by Estimation of Distribution Algorithms. *International Journal of Intelligent Systems*, 18:205–220, 2003.

[8] W. Buntine. Theory refinement in Bayesian Networks. In P. Smets, B. D'Ambrosio, and P.P. Bonissone, editors, *Proceedings of the Conference on Uncertainty in Artificial Intelligence*, pages 52–60. Morgan Kaufmann, 1991.

[9] R. Castelo and T. Kočka. On inclusion-driven learning of Bayesian Networks. *Journal of Machine Learning Research*, 4:527–574, 2003.

[10] R. Castelo and M.D. Perlman. Learning essential graph Markov models from data. In J.A. Gámez and A. Salmerón, editors, *First European Workshop on Probabilistic Graphical Models*, pages 17–24, 2003.

[11] D.M. Chickering. Learning equivalence classes of Bayesian-network structures. *Journal of Machine Learning Research*, 2:445–498, 2002.

[12] D.M. Chickering. Optimal structure identification with greedy search. *Journal of Machine Learning Research*, 3:507–554, 2002.
[13] G. Cooper and E. Herskovits. A Bayesian method for the induction of probabilistic networks from data. *Machine Learning*, 9:309–347, 1992.
[14] C. Cotta and J. Muruzábal. Towards a more efficient evolutionary induction of Bayesian Networks. In J. J. Merelo et al., editors, *Parallel Problem Solving From Nature VII*, volume 2439 of *Lecture Notes in Computer Science*, pages 730–739. Springer, Berlin Heidelberg, 2002.
[15] C. Cotta and J. Muruzábal. On the learning of Bayesian Network graph structures via Evolutionary Programming. In P. Lucas, editor, *Second European Workshop on Probabilistic Graphical Models*, pages 65–72, 2004.
[16] C. Cotta and J.M. Troya. Analyzing Directed Acyclic Graph recombination. In B. Reusch, editor, *Computational Intelligence: Theory and Applications*, volume 2206 of *Lecture Notes in Computer Science*, pages 739–748. Springer-Verlag, Berlin Heidelberg, 2001.
[17] A.E. Eiben and J.E. Smith. *Introduction to Evolutionary Computing*. Springer, Berlin Heidelberg, 2003.
[18] L.J. Fogel, A.J. Owens, and M.J. Walsh. *Artificial Intelligence Through Simulated Evolution*. John Wiley, New York NY, 1966.
[19] N. Friedman, I. Nachman, and D. Pe'er. Learning Bayesian Network structures from massive datasets: The sparse candidate algorithm. In H. Dubios and K. Laskey, editors, *Proceedings of the Fifteenth Conference on Uncertainty in Artificial Intelligence*, pages 206–215, San Francisco CA, 1999. Morgan Kaufmann.
[20] W.R. Gilks, S. Richardson, and D.J. Spiegelhalter. *Markov Chain Monte Carlo in Practice*. Chapman and Hall, 1996.
[21] S.B. Gillespie and M.D. Perlman. Enumerating Markov equivalence classes of acyclic digraph models. In M. Goldszmidt, J. Breese, and D. Koller, editors, *Proceedings of the Seventh Conference on Uncertainty in Artificial Intelligence*, pages 171–177, Seatle WA, 2001. Morgan Kaufmann.
[22] P. Giudici and R. Castelo. Improving Markov chain Monte Carlo model search for data mining. *Machine Learning*, 50(1–2):127–158, 2003.
[23] S. Harwood and R. Scheines. Genetic Algorithm search over causal models. Technical Report CMU-PHIL-131, Department of Philosophy, Carnegie Mellon University, 2002.
[24] D. Heckerman. A tutorial on learning with Bayesian Networks. In M.I. Jordan, editor, *Learning in Graphical Models*, pages 301–354. Kluwer Academic, Dordrecht, 1998.
[25] D. Heckerman, D. Geiger, and D.M. Chickering. Learning Bayesian Networks: the combination of knowledge and statistical data. *Machine Learning*, 20(3):197–243, 1995.

[26] C.C. Holmes and B.K. Mallick. Parallel Markov chain Monte Carlo sampling: an evolutionary based approach. Technical report, Mathematics Department (Statistics Section), Imperial College, London, 1998.
[27] J. S. Ide, F. G. Cozman, and F. T. Ramos. Generating random Bayesian Networks with constraints on induced width. In R. López de Mántaras and L. Saitta, editors, *Proceedings of the 16th European Conference on Artificial Intelligence*, pages 323–327, 2004.
[28] Kyung-Joong Kim, Ji-Oh Yoo, and Sung-Bae Cho. Robust inference of Bayesian Networks using speciated evolution and ensemble. In Mohand-Said Hacid et al., editors, *Foundations of Intelligent Systems: 15th International Symposium*, volume 3488 of *Lecture Notes in Computer Science*, pages 92–101, Berlin Heidelberg, 2005. Springer-Verlag.
[29] P. Larrañaga, C.M.H. Kuijpers, R.H. Murga, and Y. Yurramendi. Learning Bayesian Network structures by searching for the best ordering with Genetic Algorithms. *IEEE Transactions on Systems, Man and Cybernetics*, 26(4):487–493, 1996.
[30] P. Larrañaga and J. A. Lozano. *Estimation of Distribution Algorithms. A New Tool for Evolutionary Computation*. Kluwer Academic, 2002.
[31] P. Larrañaga, M. Poza, Y. Yurramendi, R.H. Murga, and C.M. H. Kuijpers. Structure learning of Bayesian Networks by Genetic Algorithms: A performance analysis of control parameters. *IEEE Transactions on Pattern Analysis and Machine Intelligence*, 10(9):912–926, 1996.
[32] K.B. Laskey and J. Myers. Population Markov chain Monte Carlo. *Machine Learning*, 50(1–2):175–196, 2003.
[33] D. Madigan, S.A. Andersson, M.D. Perlman, and C.T. Volinsky. Bayesian model averaging and model selection for Markov equivalence classes of acyclic digraphs. *Communications in Statistics - Theory and Methods*, 25:2493–2520, 1996.
[34] D. E. Moriarty and R. Miikkulainen. Forming neural networks through efficient and adaptive coevolution. *Evolutionary Computation*, 5(4):373–399, 1997.
[35] J. Muruzábal and C. Cotta. A primer on the evolution of equivalence classes of Bayesian-Network structures. In X. Yao et al., editors, *Parallel Problem Solving from Nature VIII*, volume 3242 of *Lecture Notes in Computer Science*, pages 612–621. Springer-Verlag, Berlin Heidelberg, 2004.
[36] J. D. Nielsen, T. Kočka, and J. M. Peña. On local optima in learning Bayesian Networks. In C. Meek and U. Kjaerulff, editors, *Proceedings of the Nineteenth Conference on Uncertainty in Artificial Intelligence*, pages 435–442, 2003.
[37] J. M. Peña, J. Björkegren, and J. Tegnér. Scalable, efficient and correct learning of Markov boundaries under the faithfulness assumption. In *Proceedings of the Eighth European Conference on Symbolic and Quantitative Approaches to Reasoning under Uncertainty*, volume 3571 of *Lecture*

Notes in Artificial Intelligence, pages 136–147, Berlin Heidelberg, 2005. Springer-Verlag.
[38] J. M. Peña, T. Kočka, and J. D. Nielsen. Featuring multiple local optima to assist the user in the interpretation of induced Bayesian Network models. In *Proceedings of the Tenth International Conference on Information Processing and Management of Uncertainty in Knowledge-Based Systems*, pages 1683–1690, 2004.
[39] M. Pelikan, D.E. Goldberg, and E. Cantú-Paz. Linkage problem, distribution estimation, and Bayesian Networks. *Evolutionary Computation*, 8(3):311–340, 2000.
[40] Carsten Riggelsen. MCMC learning of Bayesian Network models by Markov blanket decomposition. In J. Gama et al., editors, *Proceedings of the 16th European Conference on Machine Learning*, volume 3720 of *Lecture Notes in Computer Science*, pages 329–340, Berlin Heidelberg, 2005. Springer-Verlag.
[41] B. Sierra and P. Larrañaga. Predicting survival in malignant skin melanoma using Bayesian Networks automatically induced by Genetic Algorithms. an empirical comparison between different approaches. *Artificial Intelligence in Medicine*, 142:215–230, 1998.
[42] M. Studený. *Probabilistic Conditional Independence Structures*. Springer, 2005.
[43] S. van Dijk and D. Thierens. On the use of a non-redundant encoding for learning Bayesian Networks from data with a GA. In X. Yao et al., editors, *Parallel Problem Solving from Nature VIII*, volume 3242 of *Lecture Notes in Computer Science*, pages 141–150, Berlin Heidelberg, 2004. Springer-Verlag.
[44] M.L. Wong, W. Lam, and K.S. Leung. Using Evolutionary Programming and Minimum Description Length principle for data mining of Bayesian Networks. *IEEE Transactions on Pattern Analysis and Machine Intelligence*, 21(2):174–178, 1999.
[45] M.L. Wong, S.Y. Lee, and K.S. Leung. Data mining of Bayesian networks using cooperative coevolution. *Decision Support Systems*, 38:451–472, 2004.
[46] M.L. Wong and K.S. Leung. An efficient data mining method for learning Bayesian Networks using an evolutionary algorithm-based hybrid approach. *IEEE Transactions on Evolutionary Computation*, 8(4):378–404, 2004.

Learning Bayesian Networks with an Approximated MDL Score

Josep Roure Alcobé

Escola Universitària Politècnica de Mataró
Av. Puig i Cadafalch 101-111, 08303 Mataró. Spain

Summary. In this paper we present an approximation to the mutual information between a single variable and a set of variables. The main aim of our approach is to reduce the amount of sufficient statistics (i.e. frequency counts) required to calculate the mutual information. To do so, we use the chain rule and assume different independence statements between variables. We will use our approximation to calculate the MDL of a given Bayesian network. We will show that our approximated approach to the MDL measure is score equivalent and we will use it in order to learn Bayesian networks from data. We will experimentally see that learning algorithms that use our approach obtain high quality Bayesian networks. We also note that our approach can be used in any information based measures.

1 Introduction

We will present, in this work, an approximation to the mutual information, $I(\mathbf{X}; Y) = \sum_{\mathbf{X}Y} P(\mathbf{X}Y) \log P(\mathbf{X}Y)/P(\mathbf{X})P(Y)$, between a single variable Y and a set of variables $\mathbf{X}$ that avoids estimating, from data, joint probabilities of large set of variables. We will use our approach to calculate the MDL score and learn Bayesian networks.

It is widely reported in the *statistical pattern recognition* literature, see [1], that the performance of a classifier depends on the interrelationship between sample sizes, number of features, and classifier complexity. It has been often observed in practice that adding variables to a classifier may actually degrade its performance if the number of data instances that are used to learn the classifier is small relative to the number of variables. This is known as the *peaking phenomenon* which is a consequence of the *curse of dimensionality* [1], usually stated as follows: in order to estimate a joint probability, the number of required data instances grows exponentially with the number of variables. This is due to the fact that the required number of parameters in order to estimate a joint probability distribution grows exponentially with the number of variables, i.e the number of counts of a contingency table. This is also illustrated by Hastie et al. [2], when they state that the sampling density is proportional to $N^{1/p}$, where p is the number of variables and N is the sample size. Thus, if $N_1 = 100$ represents a dense sample for a single input problem, then $N_{10} = 100^{10}$ is the sample size required for the same sampling density with 10 inputs. Thus in high dimensions all feasible training samples sparsely populate the input space.

The main aim of our approach is to reduce the memory required to store *sufficient statistics*. This is very important in some learning environments. In *incremental* environments, where new data are processed as long as they are available without re-processing the previously learned ones, it is required to store all the *sufficient statistics* [3]. In these sort of environments, in order to scale up learning algorithms, it is very important to reduce the amount of *sufficient statistics* required.

Also algorithms that learn from very large data streams, see [4], can benefit from our approach. In this environment, learning algorithms attempt to minimize the number of data instances used to produce the model. For doing so, algorithms iteratively learn models increasing the number of data instances used and stop when they observe that the model quality does not grow when additional data is used. Some works, like Meek et al. [5], observe that to gather the number of data to be used it suffices to use an approximated algorithm that is cheap in computing time, and to learn the final model with the *full* learning algorithm using the appropriate number of data instances.

In this paper, we will use the chain rule to obtain another expression for the mutual information that has two desirable properties. Firstly, the expression is incremental in the number of variables, that is, we will be able to express the mutual information when a new variable Z is added to the set of variables $\mathbf{X}$ as the mutual information between Y and $\mathbf{X}$ plus the information due to the variable Z. Secondly, the obtained expression will allow us to drastically reduce the number of variables considered at each term but still take into account all the relationships between pairs of variables in $\mathbf{X}$ and variable Y. The first property is useful for Bayesian network learning since algorithms build networks from arc-less structures by incrementally adding variables to the sets of parents.

The rest of the paper is organized as follows. In the rest of this section we introduce Bayesian networks and a well-known learning algorithm. We also introduce the MDL based quality measure for Bayesian networks that uses the mutual information between a variable X_i and the set of its parents $\mathbf{Pa}_i$. In Section 2, we will use the chain rule to obtain the above mentioned expression for the mutual information. In Section 3, we will introduce the approximated mutual information and compare it to the exact one. In Section 4, we will use our approach to obtain an approximated MDL measure and we will show that it is *score equivalent*. In Section 5 we will study the error introduced by our approximation and how it may affect learning Bayesian networks. Finally, we give some experimental results.

1.1 Learning Bayesian Networks

A *Bayesian network* is an annotated directed acyclic graph that encodes a joint probability distribution of a set of random variables $\mathbf{X} = \{X_1, \ldots, X_n\}$ each of which has a domain of possible values. Formally, a Bayesian network for $\mathbf{X}$ is a pair $BN = (B_S, B_P)$ where the first component, B_S, is a directed acyclic graph (DAG) whose vertexes correspond to the random variables $X_1, \ldots, X_n$, and whose edges represent directed dependencies between variables. The parents of X_i, denoted as $\mathbf{P}a_i$, is the set of variables with an arc to X_i in the graph. The model structure yields to a factorization of the joint probability distribution for $\mathbf{X}$, $P(\mathbf{X}) = \prod_{i=1}^{n} P(X_i|\mathbf{P}a_i)$. The structure, B_S, encodes a set of independence statements $I(B_S)$ that can be *read* using the notion of *d-separation*. We will say that, in a network structure B_S, the sets of variables $\mathbf{X}$ and $\mathbf{Y}$ are independent given the set $\mathbf{Z}$, $(\mathbf{X} \perp \mathbf{Y}|\mathbf{Z}) \in I(B_S)$, if $\mathbf{X}$ and $\mathbf{Y}$ are *d-separated* given $\mathbf{Z}$. In order to define *d-separation* we need to classify the connections between different sets of variables into three categories:

- **Serial connection** $X \to Y \to Z$, or Markov chain, where a variable X has an influence to a variable Y which in turn has an influence to a variable Z
- **Converging connection** $X \to Y \leftarrow Z$, or V-structure, where variables X and Z have and influence to Y
- **Diverging connection** $X \leftarrow Y \to Z$, where Y has an influence to both X and Z.

Definition 1. *Let $\mathbf{X}$, $\mathbf{Y}$ and $\mathbf{Z}$ be three sets of variables in B_S. We say that $\mathbf{X}$ and $\mathbf{Y}$ are d-separated given $\mathbf{Z}$ if for all paths between any node $X \in \mathbf{X}$ and $Y \in \mathbf{Y}$ there is an intermediate variable $Z \in \mathbf{Z}$ such that either the connection is serial or diverging and the state of Z is known; or the connection is converging and neither Z nor any of Z's descendants have received evidence.*

The second component of a Bayesian network, B_P, represents the parameters that quantifies the network. It has a parameter $\theta_{ijk} = P(X_i = x_i^k|\mathbf{P}a_i = \mathbf{p}a_i^j)$ for each possible state x_i^k of X_i and for each configuration $\mathbf{p}a_i^j$ of $\mathbf{P}a_i$.

Most of the learning algorithms found in the literature are hill-climbing searchers that begin with the arc-less network and perform the operator that most increases the score of the resulting structure and does not introduce a cycle into the network. Algorithms stop when the use of a single operator cannot increase the network's score. The most common operators are add, delete and reverse a single arc. The difference between the algorithms is the domain of models and the operators they use. For our experiments we will use algorithm B [6] that yields full DAG structures and the neighborhood of a given network structure, B_S, is the set of all networks that can be obtained from B_S by adding a single arc $X_i \to X_j$ to B_S such that does not introduce a cycle,

$$\mathcal{N}_H(B_S) = \{(\mathbf{X}, E')|E' = E \cup \{(X_i, X_j)\} \wedge (X_i, X_j) \notin E \wedge B'_S \text{ is a DAG}\}$$

In order to measure the quality of the alternative structures we will use the MDL quality measure that we explain in the following subsection.

1.2 MDL Scoring Function

MDL approach to scoring functions for Bayesian networks is based on the idea that the best model of a database is the model that minimizes the sum of the encoding length of the model plus the encoding length of the data given the model, $MDL(B|D) = K(B) + DL(D|B)$. Friedman et al. [7] used $-\log P_B(\mathbf{u})$ the encoding length of each instance $\mathbf{u}$, and obtained the following expression for the encoding length, $DL(D|B)$, of the whole dataset D given the Bayesian network B,

$$DL(D|B) = -\sum_{j=1}^{N} \log P_B(\mathbf{u}_j) \qquad (1)$$

where N is the number of data instances in D. Now, if we sum over the different possible instances u and we denote $N_D(u)$ as the number of instances equal to u in the dataset D, we obtain

$$DL(D|B) = -\sum_{u} N(u) \log \prod_{i=1} P(x_i|\mathbf{pa}_i)$$
$$-\sum_{i}\sum_{x_i,\mathbf{pa}_i} N_D(x_i, \mathbf{pa}_i) \log P(x_i|\mathbf{pa}_i) \qquad (2)$$

Now, using standard arguments, we note that the parameters values that minimize this expression are $\theta_{x_i|\mathbf{pa}_i} = \hat{P}(x_i|\mathbf{pa}_i)$, that is, we need to compute the appropriate fractions from data. We can re-write this expression in a more convenient way in terms of conditional entropy: $NH(X_i|\mathbf{Pa}_i)$, where $H(X|Y) = -\sum_{x,y} P(x,y) log P(x|y)$ is the conditional entropy of X given Y. So, following again [7], we can re-write the expression above (Equation 2) as

$$DL(D|B) = N\sum_{i=1}^{n} H(X_i|\mathbf{Pa}_i) = N[\sum_{i}^{n} H(X_i) - I(X_i; \mathbf{Pa}_i)] \qquad (3)$$

where n is the number of variables, $H(X) = -P(X) \log P(X)$ is the entropy of variable X and $I(X;Y)$ is the mutual information between variables X and Y. Note, that term $H(X_i)$ in Equation 3 does not depend on the network structure and therefore to minimize $DL(D|B)$ we only need to maximize $I(X_i; \mathbf{Pa}_i)$. Lam and Bacchus [8] took the Kullback-Leibler divergence as a measure of the encoding length of the data given the Bayesian network. They observed that the Kullback-Leibler divergence is a monotonically decreasing function of $\sum_{i}^{n} I(X_i; \mathbf{Pa}_i)$ arriving to a similar expression.

The encoding length of the network structure is usually expressed as the total number of parameters that we need to store for the network. Note that for each variable X_i we need to store $|\mathbf{P}a_i|(|X_i|-1)$ parameters. The number of bits used for each of these parameters is usually taken to be $1/2 \log N$, and so, the encoding length, $K(B)$, for the whole Bayesian network structure is

$$K(B) = \frac{1}{2} \log N \sum_{i=1}^{n} |\mathbf{P}a_i|(|X_i|-1)$$

Note that the MDL measure is factored, or equivalently, the sum property holds. That is, the encoding length of the whole Bayesian network is expressed as the sum of the encoding length of each variable and its parent set.

$$MDL(B|D) = \sum_{i=1}^{n} MDL(X_i, \mathbf{P}a_i)$$

This property if very important for learning algorithms since it localizes the score effect of an addition (or removal) of an arc to the families affected (i.e. a variable and its parent set). This property also holds for the Bayesian approach to quality measures [9].

Another property that is usually required for scoring measures is that they give the same quality score to Bayesian network structures that are equivalent, that is, structures that define the same probability distribution, or in other words, network structures that state the same set of (in)dependencies between variables. When this property holds for a given quality measure it is said to be *score equivalent*. Note that the MDL approach to quality functions is *score equivalent* [10].

2 Incremental Mutual Information

The mutual information $I(\mathbf{X}; Y)$ and the joint entropy $H(\mathbf{X}, Y)$ can be incrementally expressed using the chain rule [11]. Let $\mathbf{X}^{(n)} = \{X_1, X_2, \ldots, X_n\}$,

$$I(\mathbf{X}^{(n)}; Y) = I(X_1; Y) + I(X_2; Y|X_1) + \cdots + I(X_n; Y|\mathbf{X}^{(n-1)})$$

which can be expressed as

$$I(\mathbf{X}^{(n)}; Y) = I(X_1; Y) + \sum_{i=2}^{n} I(X_i; Y|\mathbf{X}^{(i-1)}) \qquad (4)$$

Note that this expression is incremental in the number of variables, that is, when a new variable, Z, is added to a set of variables, $\mathbf{X}$, we can express the mutual information as

$$I(\mathbf{X}^{(n)}Z; Y) = I(\mathbf{X}^{(n)}; Y) + I(Z; Y|\mathbf{X}^{(n)}) \qquad (5)$$

Now we will further decompose the mutual information using the following identity $I(X;Y|Z) = I(X;Y) + I(X;Z|Y) - I(X;Z)$ and reordering terms, from Equation (4) we obtain

$$I(\mathbf{X}^{(n)}; Y) = \sum_{i=1}^{n} I(X_i; Y)$$

$$+ \sum_{i=2}^{n-1} I(\mathbf{X}^{(i-1)}; X_i|Y) + I(\mathbf{X}^{(n-1)}; X_n|Y) \quad (6)$$

$$+ \sum_{i=2}^{n-1} -I(\mathbf{X}^{(i-1)}; X_i) - I(\mathbf{X}^{(n-1)}; X_n) \quad (7)$$

Using the chain rule and the identity stated above to the last two terms (6 and 7) of the former equation, we obtain

$$I(\mathbf{X}^{(n)}; Y) = \sum_{i=1}^{n} I(X_i; Y) + \sum_{i=1}^{n-1} [I(X_i; X_n|Y) - I(X_i; X_n)]$$

$$+ \sum_{i=2}^{n-2} I(\mathbf{X}^{(i-1)}; X_i|YX_n) + I(\mathbf{X}^{(n-2)}; X_{n-1}|YX_n)$$

$$+ \sum_{i=2}^{n-2} -I(\mathbf{X}^{(i-1)}; X_i) - I(\mathbf{X}^{(n-2)}; X_{n-1})$$

Performing repeatedly the same substitutions, we obtain

$$I(\mathbf{X}^{(n)}; Y) = \sum_{i=1}^{n} I(X_i; Y)$$

$$+ \sum_{i=2}^{n} \sum_{j=1}^{i-1} I(X_i; X_j|YX_n \ldots X_{i+1})$$

$$- \sum_{i=2}^{n} \sum_{j=1}^{i-1} I(X_i; X_j|X_n \ldots X_{i+1})$$

and where the incremental expression is

$$I(\mathbf{X}^{(n)}Z;Y) = I(\mathbf{X}^{(n)};Y)$$
$$+ I(Z;Y) \qquad (8)$$
$$+ \sum_{i=1}^{n} I(Z;X_i|YX_n\ldots X_{i+1}) \qquad (9)$$
$$- \sum_{i=1}^{n} I(Z;X_i|X_n\ldots X_{i+1}) \qquad (10)$$

Let us take a look at this last expression. We can see that the contribution to the mutual information between a set of variables $\mathbf{X}^{(n)}$ and a variable Y when a new variable, Z, is added to $\mathbf{X}^{(n)}$ can be divided into three parts, two of them positive and one negative. The first one, term (8), measures the contribution of the new variable to the mutual information with Y. The second one, term (9), measures the relationship between Z, $\mathbf{X}^{(n)}$ and Y. The third one, term (10), measures the relationship between Z and $\mathbf{X}^{(n)}$ when no additional information is known. This last term is negative, and roughly speaking, it measures the amount of information Z provides with that is already provided by the set of variables $\mathbf{X}^{(n)}$.

To take a closer look let us consider the terms (9) and (10) when $i = n$, $I(X_n;Z|Y) - I(X_n;Z)$. We note that this term may be positive or negative [12]: when the joint probability distribution of variables forms a **Markov chain** (see Figure 1 (a)), then $I(X_n;Z|Y) \leq I(X_n;Z)$. Further note that, in this case, $I(X_n;Z|Y) = 0$ and that $I(Y;X_n) \geq I(Z;X_n)$. On the contrary, when the joint probability distribution forms a **V-structure** (see Figure 1 (b)), then $I(X_n;Z|Y) \geq I(X_n;Z)$, and also that $I(X_n;Z|Y)$ is maximum. A typical example of this joint probability structure is *The Burglar Alarm* problem [13].

So, from the considerations stated above, we can see that the contribution of a new variable Z to the mutual information $I(\mathbf{X}^{(n)}Z;Y)$:

- The higher the mutual information between Y and the new variable Z is, the higher its contribution to the whole mutual information, $I(\mathbf{X}^{(n)}Z;Y)$, is. That is, the more coupled is Z to Y the more it contributes to the mutual information.
- The more similar the joint probability distribution of $\mathbf{X}^{(n)}$, Z, and Y is to a V-structure the higher the contribution of Z to the whole mutual

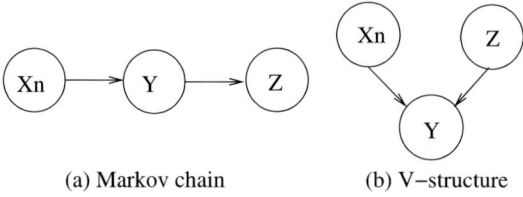

(a) Markov chain   (b) V-structure

Fig. 1. Joint probability distribution structures

information is. That is, the more coupled is Z to $\mathbf{X}^{(n)}$ given Y the more Z contributes to the mutual information.
- The higher the mutual information between Z and $\mathbf{X}^{(n)}$ is, the lower its contribution to the whole mutual information is. That is the more coupled Z is to $\mathbf{X}^{(n)}$ the lower is its contribution to the mutual information since the fewer *new* information it contributes with.

We also want to stress that the contribution of a variable to the mutual information is never negative, $I(\mathbf{X}^{(n)}; Y) \leq I(\mathbf{X}^{(n)} Z; Y)$ [11]. This can easily be seen in Equation (4) where each variable contributes with a non-negative amount since the mutual information is always non-negative. See also that when the underlying joint probability of Z, $\mathbf{X}^{(n)}$ and Y form a Markov chain that the contribution of the variable Z to $I(\mathbf{X}^{(n)} Z; Y)$ is zero. It is easily seen form Equation 5 since, in this case, $P(Z, Y|\mathbf{X}^{(n)}) = P(Z|\mathbf{X}^{(n)})P(Y|\mathbf{X}^{(n)})$, i.e. Z and Y are independent given $\mathbf{X}^{(n)}$.

For notational simplicity, the sum of terms (8), (9) and (10) will be noted by $\Delta(Z)$.

3 Approximated Mutual Information

In this section we present an approximation to the mutual information. The aim of our approximation is to drastically reduce the number of variables involved in the joint probabilities in order to reduce the memory space used to store the *sufficient statistics*.

To avoid using joint probabilities of large set of variables in the mutual information $I(\mathbf{X}^{(n)}; Y)$, we will discard conditioning the mutual information with variables in $\mathbf{X}^{(n)}$,

$$I(\mathbf{X}^{(n)}; Y) = \sum_{i=1}^{n} I(X_i; Y)$$
$$+ \sum_{i=1}^{n} \sum_{j=i+1}^{i-1} I(X_i; X_j|Y)$$
$$- \sum_{i=1}^{n} \sum_{j=i+1}^{i-1} I(X_i; X_j)$$

and where the incremental expression is

$$I(\mathbf{X}^{(n)} Z; Y) = I(\mathbf{X}^{(n)}; Y)$$
$$+ I(Z; Y) \tag{11}$$
$$+ \sum_{i=1}^{n} I(Z; X_i|Y) \tag{12}$$
$$- \sum_{i=1}^{n} I(Z; X_i) \tag{13}$$

Note that in our approximation we are calculating joint probabilities of sets of three variables at most. This reduces the amount of *sufficient statistics* (e.g. counts) used from k^{n+2}, in the *full* version of the mutual information, to k^3, where k is the arity of the variables, we assume that all variables have the same arity without loss of generality, and $n+2$ is the number of variables involved: $\mathbf{X}^{(n)}, Z, Y$.

Thus, we are assuming in each term i that the variable X_i is independent of the set of variables $X_n, \ldots, X_{i+1}$. Obviously terms (12) and (13) will, in general, give different results than the corresponding terms, (9) and (10), of the *full* mutual information. Observe though, that this new expression keeps the three contributions of a new variable Z to the mutual information that measures different aspects of the dependencies between the variables, or sets of variables, involved.

We will note as $\Delta_a(Z)$ the sum of terms (11), (12) and (13). See that while $0 \leq \Delta(Z)$, it does not hold for $\Delta_a(Z)$. Furthermore, observe that the approximated measure may be either equal, greater or lower than the exact measure. Let us analyze the error introduced in our approximated measure. Using Equation 5 and the equality above (Equations 11, 12 and 13) we obtain

$$\Delta(Z) - \Delta_a(Z) = I(Z;Y|\mathbf{X}^{(n)}) + (n-1)I(Z;Y) - \sum_{i=1}^{n} I(Z;Y|X_i)$$

To show that this error can be either equal, higher or lower than zero, consider again the case where the underlying joint probability of Z, $\mathbf{X}^{(n)}$ and Y forms a Markov chain, then $\Delta(Z) = 0$ and $\Delta_a(Z) = I(Z;Y) + \sum_{i=1}^{n}[I(Z;Y|X_i) - I(Z;Y)]$ and since in this case $I(Z;Y|X_i) << I(Z;Y)$, $\Delta_a(Z) =\leq 0$. Here we used the equality $I(Z;X|Y) - I(Z;X) = I(Z;Y|X) - I(Z;Y)$.

Let us further illustrate this fact with some examples. Take two variables, Y and Z, with uniform probabilities over binary words of length n. Note that since both distributions are uniform we have that $H(Y) = H(Z) = \log_2(2^n) = n$, in other words, we need n bits to represent 2^n equally probable symbols. We will now consider different cases that will rise different values of the error. The first two examples show two cases in which both measures are equal.

Example 1 ($\boldsymbol{\Delta}(Z) = \Delta_a(Z) = 0$). Let $Y = Z$ and let each variable $X_i \in \mathbf{X}^{(n)}$ be a copy of the i-th bit of variable Y. So, on the one hand we have that $\Delta(Z) = I(Z;Y|\mathbf{X}^{(n)}) = 0$ since Z and Y are independent given $\mathbf{X}^{(n)}$. And, on the other, we have that $I(Z;Y) = H(Y) = H(Z) = n$, $\Delta_a(Z) = (n-1)I(Z;Y) - \sum_{i=1}^{n} I(Z;Y|X_i) = (n-1)n - n\log_2(2^{n-1}) = 0$. Note that the mutual information between Z and Y given X_i is reduced by the single bit provided by X_i, $I(Z;Y|X_i) = n-1$. Thus, we have that $\Delta(Z) = \Delta_a(Z) = 0$. Note that Z does not provide with any new information once $\mathbf{X}^{(n)}$ is known so $\Delta(Z) = 0$. In this case, our approximation also captures this fact and yields the same result. We want to stress that all variables in $\mathbf{X}^{(n)}$ are independent of each other.

Example 2 ($\mathbf{\Delta}(Z) = \Delta_a(Z) \neq 0$). Let Y and Z be independent variables as before and let each variable $X_i \in \mathbf{X}^{(n)}$ represent the logic *or* operator of the i-th bits of variables Z and Y. In this case we have that $\Delta(Z) = \Delta_a(Z)$ since $I(Z;Y|\mathbf{X}^{(n)}) = \sum_{i=1}^{n} I(Z;Y|X_i)$. This last equality holds because $H(\mathbf{X}^{(n)}) = \sum_{i=1}^{n} H(X_i)$ when variables in $\mathbf{X}^{(n)}$ are independent of each other.

The following two cases are similar to the ones above but now variables in $\mathbf{X}^{(n)}$ will not be independent anymore. In this two new examples our approximated measure introduces an error.

Example 3 ($\mathbf{\Delta}(Z) = 0 \leq \Delta_a(Z)$). As in the first example let $Y = Z$, and so $I(Z;Y) = n$. Now let $X_i \in \mathbf{X}^{(n)}$ be a variable the values of which are a copy of all bits of variable Y but the i-th one, and so the values of X_i are represented by $n-1$ bits. As in the first example we have that $\Delta(Z) = I(Z;Y|\mathbf{X}^{(n)}) = 0$, but now $\sum_{i=1}^{n} I(Z;Y|X_i) = n \cdot 1$ since each X_i already provides with information about $n-1$ bits of variable Y. Thereof $0 \leq \Delta_a(Z) = (n-1)n - n = n^2 - 2n$ and the inequality holds when $n \geq 2$.

Example 4 ($\mathbf{\Delta}(Z) \neq 0 \leq \Delta_a(Z)$). As in the second example let Y and Z be independent, frow where $I(Z;Y) = 0$. Now let each variable $X_i \in \mathbf{X}^{(n)}$ be a variable the values of which represent the logical *or* operator of all but the i-th bit of variables Z and Y. So the values of X_i are binary words of length $n-1$. We know that when $\mathbf{X}^{(n)}$ is a function of Z and Y then $I(Z;Y) < I(Z;Y|\mathbf{X}^{(n)})$ (see Example 6). We also can see that X_i and $\mathbf{X}^{(n)}$ share $n-1$ bits and that $\forall i,j \in [1,n], i \neq j, X_i$ and X_j share $n-2$ bits. So, we have that $I(Z;Y|X_i)$ is only slightly lower than $I(Z;Y|\mathbf{X}^{(n)})$ and that $\forall i,j \in [1,n], I(Z;Y|X_i) = I(Z;Y|X_j)$. Thereof, we can see that $I(Z;Y|\mathbf{X}^{(n)}) < \sum_{i=1}^{n} I(Z;Y|X_i)$, and thus $\Delta(Z) < \Delta_a(Z)$

Note that in these two last examples variables $X_i \in \mathbf{X}^{(n)}$ are not independent and also that $\forall i,j \in [1,n] : i < j, I(X_i;X_j) = n-2$ since any pair of variables will share $n-2$ bits.

Unfortunately, we have not been able to give an analytical bound for the error introduced by our approximated mutual information yet. However, since our approximated measure assumes that variables in $\mathbf{X}^{(n)}$ are independent, the less independent they are the bigger the error will be. In Section 5 we will study the effect of the error introduced in learning Bayesian networks.

4 Approximated MDL

In this section we will use our approximated mutual information in order to obtain an approximated MDL measure. First we express the mutual information in an incremental way. Given a variable, X_i, and its parent set, $\mathbf{Pa}_i = \{Pa_{i1}, \ldots, Pa_{in}\}$, when a new parent, Z, is added the MDL can be expressed as

$$MDL(X_i, \mathbf{Pa}_i Z) =$$
$$K(X_i, \mathbf{Pa}_i) + DL(D|X_i, \mathbf{Pa}_i) \qquad (14)$$
$$+ (\frac{1}{2}\log N)|\mathbf{Pa}_i|(|X_i| - 1)(|Z| - 1) \qquad (15)$$
$$- NI(Z; X_i) \qquad (16)$$
$$- N \sum_{j=1}^{n} I(Z; Pa_{ij}|X_i Pa_{in} \ldots Pa_{ij+1}) \qquad (17)$$
$$+ N \sum_{j=1}^{n} I(Z; Pa_{ij}|Pa_n \ldots Pa_{ij+1}) \qquad (18)$$

Note that in this equation we state the MDL for a given variable X_i and its parent set $\{\mathbf{Pa}_i\} \cup \{Z\}$. The MDL for the whole Bayesian network structure can be easily obtained summing the MDL over all variables and their parent sets, since the MDL is factored. See that the term (14) corresponds to $MDL(X_i, \mathbf{Pa}_i)$, term (15) corresponds to the new encoding length when a parent is added and the last three terms, (16), (17) and (18), correspond to $-N\Delta(Z)$. Note that the MDL measure is a trade-off between $Delta(Z)$, which needs to be maximized, and the number of new parameters introduced.

Now we state our approximated version of the MDL using the notation from the former section,

$$MDL_a(X_i, \mathbf{Pa}_i Z) =$$
$$K(X_i, \mathbf{Pa}_i) + DL_a(D|X_i, \mathbf{Pa}_i) \qquad (19)$$
$$+ (\frac{1}{2}\log N)|\mathbf{Pa}_i|(|X_i| - 1)(|Z| - 1) \qquad (20)$$
$$- NI(Z; X_i) \qquad (21)$$
$$- N \sum_{j=1}^{n} I(Z; Pa_{ij}|X_i) \qquad (22)$$
$$+ N \sum_{j=1}^{n} I(Z; Pa_{ij}) \qquad (23)$$

where $DL_a(D|X_i, \mathbf{Pa}_i)$ corresponds to the encoding length of data given the network structure using our approximation to mutual information. Note again that the last three term of the equation – (21), (22) and (23) – correspond to $-N\Delta_a(Z)$.

Now we will show that our approach is *score equivalent*. First, we need to define the notion of *covered* arc.

Definition 2. *An arc, $Y \to X$, in a network structure is covered if $\mathbf{Pa}_x = \mathbf{Pa}_y \cup \{Y\}$ (see Figure 2).*

Chickering et al. [10] showed that between any pair of equivalent Bayesian networks, there exists a sequence of distinct covered arc reversals that make

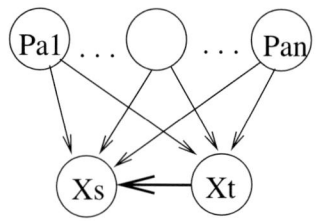

Fig. 2. $X_s \leftarrow X_t$ is a covered arc

both structures identical. Thus, to show that a scoring function is *score equivalent*, it suffices to see that it gives the same score to two structures that differ in a single covered arc reversal.

Lemma 1. *Let U be a set of variables, and let D be a database over U. Let B_1 and B_2 be two network structures over U. Furthermore, let X_s and X_t be two nodes in B_1 and B_2, where $\mathbf{Pa}_{X_s} = \{Pa_1, \ldots, Pa_n\}$ and $\mathbf{Pa}_{X_t} = \{Pa_1, \ldots, Pa_n\} \cup \{X_s\}$ in B_1 and $\mathbf{Pa}_{X_s} = \{Pa_1, \ldots, Pa_n\} \cup \{X_t\}$ and $\mathbf{Pa}_{X_t} = \{Pa_1, \ldots, Pa_n\}$ in B_2, and the parent sets for the rest of variables in U are the same in both structures. Then,*

$$MDL_a(B_1, D) = MDL_a(B_2, D)$$

Proof:

Since our approach measures in the same way the number of parameters of Bayesian networks than the *full* approach, we only need to show that the scoring length of the data given both structures, B_1 and B_2, is the same, $DL(B_1, D) = DL(B_2, D)$.

$$DL_a(B_1, D) = DL_a(B_2, D)$$
$$\Leftrightarrow$$
$$ML_a(X_s, \{Pa_1, \ldots, Pa_n\}) + ML_a(X_t, \{Pa_1, \ldots, Pa_n\}) + \Delta_a(X_s))$$
$$=$$
$$ML_a(X_s, \{Pa_1, \ldots, Pa_n\}) + \Delta_a(X_t) + ML_a(X_t, \{Pa_1, \ldots, Pa_n\})$$
$$\Leftrightarrow$$
$$\Delta_a(X_s) = \Delta_a(X_t)$$
$$\Leftrightarrow$$
$$I(X_t; X_s) + \sum_{i=1}^{n} I(X_s; Pa_i | X_t) - I(X_s; Pa_i)$$
$$=$$

$$I(X_s; X_t) + \sum_{i=1}^{n} I(X_t; Pa_i | X_s) - I(X_t; Pa_i)$$
$$\Leftrightarrow$$
$$\forall i \in [1, \ldots, n] \; I(X_s; Pa_i | X_t) - I(X_s; Pa_i) = I(X_t; Pa_i | X_s) - I(X_t; Pa_i)$$
$$\Leftrightarrow$$
$$\sum_{X_s X_t Pa_i} P(X_s X_t Pa_i) \log \frac{P(X_s X_t Pa_i) P(Pa_i) P(X_s) P(X_t)}{P(Pa_i X_t) P(X_s X_t) P(Pa_i X_s)}$$
$$=$$
$$\sum_{X_s X_t Pa_i} P(X_s X_t Pa_i) \log \frac{P(X_s X_t Pa_i) P(Pa_i) P(X_s) P(X_t)}{P(Pa_i X_t) P(X_s X_t) P(Pa_i X_s)}$$

□

Once that we know that our approximated measure behaves well in the sense that it equally scores networks that encode the same independence statements, we will compare its scores with the ones of the exact MDL measure when measures are used to learn Bayesian networks.

5 Comparing MDL and MDL$_a$ in learning Bayesian networks

In Section 3 we have seen that the error introduced by our approximated measure depends on the dependency degree of variables in $\mathbf{X}^{(n)}$. Now, we will analyze the effects of the error introduced by our approximated MDL in learning Bayesian networks from data.

The fact that the MDL measure is factorized allows us to restrict our study to a given family of variables, that is, to a variable Y and its parent set $\mathbf{Pa}_Y$. Observe also that our measure is exact when considering sets of tree variables, i.e. a variable and two parents. In our experiments, see Section 6 we use algorithm B [6] that performs a greedy search. It introduces to the structure the arc that most reduces the MDL score. Since the MDL is factored and algorithm B tries one arc at a time we can, without lost of generality, restrict our study to the introduction of the third parent to a given variable.

We will compare the value yield by the MDL and the MDLa when given a variable Y with two parents, $\mathbf{Pa}_Y = \{X_1, X_2\}$, the learning algorithm is looking for the best variable, Z, to be introduced as the third parent. We assume that X_1 is the first parent introduced by algorithm B, that is, $MDL(X_1 \rightarrow Y) \le MDL(X_2 \rightarrow Y)$. Note that the exact MDL measure and the approximated one equally encode the length of the model and so the difference between them is due to the measure of the encoding length of the data given the model. Thereof, we need to compare the contribution of a new parent in the encoding length of data given the model. Recall that the bigger $\Delta(Z)$ or $\Delta_a(Z)$ the more minimized MDL and MDL_a respectively are and the better Z is supposed to be as parent of Y when $\mathbf{Pa}_Y = \{X_1, X_2\}$. In the next cases we are going to compare both exact and approximated measures assuming different *true* underlying probabilities over variables. What we would like to see is that our approximated measure never over scores, compared to the exact measure, a variable Z that is considered to be a *bad* parent of Y and that it never under scores a variable that is considered to be a *good* one.

Case 1:

The network structure shown in Figure 3 (Case 1) encodes the following set of independence statements, $\{(X_1 \perp X_2|Z), (Z \perp Y|X_1, X_2)\}$. See that $\Delta(Z) = I(Z; Y|X_1, X_2) = 0$ since $(Z \perp Y|X_1, X_2)$. So now we need to analyze $\Delta_a(Z) = I(Z; Y) - I(Z; Y|X_1) - I(Z; Y|X_2)$. In a general case this quantity could be either equal, lower or bigger than zero. If we take a closer look we see that $\Delta_a(Z) > 0$ would imply that $I(Z; Y|X_1)$ and $I(Z; Y|X_2)$ are both zero or very close to zero, which in turn would imply that $(Z \perp Y|X_1)$ and $(Z \perp Y|X_2)$. In Lemma 2 we sate that $(Z \perp Y|X_1)$ and $(Z \perp Y|X_2)$ implies that $(X2 \perp Y|X_1)$. See that if this had happened, variable X_2 would not have been introduced as the second parent since $\Delta(X_2) = I(X_2; Y|X_1) \approx 0$ and thus the MDL score would grow, instead of decrease, considering the encoding length of the network. Thereof, in this case where Z is a *bad* parent of Y we have observed that $\Delta_a(Z) \leq \Delta(Z)$, that is, our approximated MDL_a under scores $Y \to Z$.

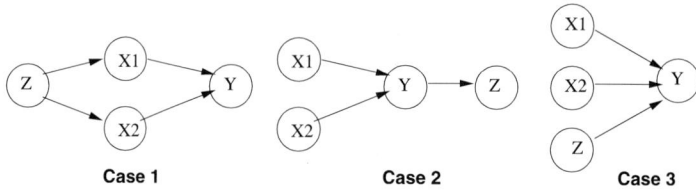

Fig. 3. Underlying *true* structures

Lemma 2. *Let B_S be the structure of a Bayesian network, where $MDL(B_S \cup \{X_1 \to Y\}) < MDL(B_S \cup \{X_2 \to Y\})$ and let $\{(Z \perp Y|X_1), (Z \perp Y|X_2)\} \in I(B_s)$ where $I(B_s)$ stands for the set of independence statements encoded by the network, then $(X_2 \perp Y|X_1) \in I(B_s)$*

Proof: $(Z \perp Y|X_2)$ means that there is a serial connection between Z and Y in which X_2 is in the middle. Suppose $\neg(X_2 \perp Y|X_1)$ and since $(Z \perp Y|X_2)$, then there is a serial connection between X_2 and Y in which X_1 is not in the middle. Thereof, X_1 is either in the serial connection between Z and X_2 or in a different one. Note that none of the two serial connections are possible. The first would mean that $MDL(B_S \cup \{X_1 \to Y\}) < MDL(B_S \cup \{X_2 \to Y\})$, and the second would mean that Z and Y were not *d-separated* given X_1 or given X_2.

Case 2:

The structure shown in Figure 3 (Case 2), encodes $\{(X_1 \perp X_2), (X_1 \perp Z|Y), (X_2 \perp Z|Y)\}$ as independence statements. Using this statements we obtain the following difference between measures,

$$\Delta(Z) - \Delta_a(Z) = I(Z;Y|X_1,X_2) - I(Z;X_1|X_2) - I(Z;X_1|Y) + I(Z;X_1)$$
$$= -I(X_1;X_2|Z) \leq 0$$

See that $I(Z;Y|X_1,X_2) = I(Z;X_1,X_2) = 0$ where we obtained the first equality using $(X_2 \perp Z|Y)$ and the second equality using $\{(X_1 \perp Z|Y), (X_2 \perp Z|Y)\}$. See also that $I(Z;X_1|Y) = 0$ since $(X_1 \perp Z|Y)$, and finally we obtained $-I(Z;X_1|X_2) + I(Z;X_1) = -I(X_1;X_2|Z)$ using $(X_1 \perp X_2)$. Thus this is a case in which MDL_a over scores a variable Z that is not a *good* parent of Y. However, note that when Z has no parents then the approximated measure will score Z as a better parent of Y than Y parent of Z since $-I(Z;Y|X_1) - I(Z;Y|X_2) \leq 0$. When Z and Y have the same set of parents $\{X_1, X_2\}$ then the arc $Z \to Y$ is covered and the measure will correctly give the same score to both directions. And finally, when Z has a set of parents $\{T_1, T_2\}$ where neither T_1 nor T_2 belongs to $\{X_1, X_2\}$ then the direction of the arc $Z \to Y$ will depend on how coupled is Z to $\{X_1, X_2\}$ and Y to $\{T_1, T_2\}$.

Case 3:

The structure shown in Figure 3 (Case 3) encodes $\{(X_1 \perp X_2), (X_1 \perp Z), (X_2 \perp Z)\}$. Because we want to explore different situations we will only consider $\{(X_1 \perp X_2)\}$, at this moment, in order to obtain a more general expression of the difference between both measures. If we only use that $(X_1 \perp X_2)$ we obtain, $\Delta(Z) - \Delta_a(Z) = I(X_1;X_2|ZY) - I(X_1;X_2|Y) - I(X_1;X_2|Z)$. Let us now consider three different cases:

- When $\{(X_1 \perp Z), (X_2 \perp Z)\}$ we see that $I(X_1;X_2|Z) = I(X_1;X_2)$ and now using that X_1 and X_2 are independent we obtain that $I(X_1;X_2|Z) = 0$. Finally we obtain: $\Delta(Z) - \Delta_a(Z) = I(X_1;X_2|ZY) - I(X_1;X_2|Y)$
- Let us take a second case where $(X_1 \perp X_2|Z)$, that is, we add the arcs $Z \to X_1$ and $Z \to X_2$ to the network structure in Figure 3 (Case 3). With these two independence statements, we obtain again the former expression since $I(X_1;X_2|Z) = 0$ follows from the independence statement.
- Now we add the arcs $X_2 \to Z$ and $X_2 \to Z$ to the network structure of Figure 3 (Case 3). In this case we cannot further reduce the expression for the difference and so we keep $\Delta(Z) - \Delta_a(Z) = I(X_1;X_2|Z,Y) - I(X_1;X_2|Y) - I(X_1;X_2|Z)$

Note that the two expressions for the difference between measures we just obtained can again be either equal, lower or greater than zero. We claim that when Z is independent of X_1 and X_2 the difference will be positive. In other words, when Z is independent of the other parents and thus is a *good* parent of Z since it provides with *new* information, the approximated measure will under score its contribution. On the contrary, when Z is not independent of the other parents the difference between the measures will be negative. That is, the approximated measure over scores a parent that is *not so good*. We can illustrate this fact with two examples.

Example 5. [Z is independent of X_1 and X_2] Let X_1, X_2 and Z be independent random binary variables, and let variable Y be the *or* logic operator of the former ones. Table 1 summarizes all the possible values of such a configuration of variables. Since all parents of Y are independent of each other we have that $\Delta(Z) - \Delta_a(Z) = I(X_1; X_2 | Z, Y) - I(X_1; X_2 | Y)$. From the configuration table we can calculate $I(X_1; X_2 | Z, Y) = 2/8 \log \frac{1/3}{2/3 \cdot 1/3} + 1/8 \log \frac{1/3}{2/3 \cdot 2/3} = 0.02840$ and $I(X_1; X_2 | Y) = 1/8 \log \frac{1/7}{3/7 \cdot 3/7} + 4/8 \log \frac{2/7}{4/7 \cdot 3/7} + 2/8 \log \frac{2/7}{4/7 \cdot 4/7} = 0.00533$. Thus, we obtained $\Delta(Z) - \Delta_a(Z) > 0$

Table 1. Configurations of variables: $Y = X_1 \vee X_2 \vee Z$

X_1	X_2	Z	Y
0	0	0	0
0	0	1	1
0	1	0	1
0	1	1	1
1	0	0	1
1	0	1	1
1	1	0	1
1	1	1	1

Example 6. [Z is dependent of X_1 and X_2] Let X_1 and X_2 be two independent binary random variables, let variable Z be a copy of X_1, and finally $Y = X_1 \vee X_2 \vee Z$ as before. So, we have that the difference between measures is $\Delta(Z) - \Delta_a(Z) = I(X_1; X_2 | Z, Y) - I(X_1; X_2 | Y) = H(X_1 | Z, Y) - H(X_1 | Z, Y, X_2) - I(X_1; X_2 | Y)$, where $H(X_1 | Z, Y) = H(X_1 | Z, Y, X_2) = 0$ since variable X_1 is completely determined given Z. Thereof, $\Delta(Z) - \Delta_a(Z) = -I(X_1; X_2 | Y) \leq 0$.

Even the analytical results of this section showed that our approximated measure does not always favor *good* parents, in the next section we will experimentally see that the Bayesian network structures obtained with it are reasonably similar to those obtained with the exact MDL measure.

6 Experimental Results

In this section we compare the performance of algorithm B (see Section 1.1) when it uses the *full MDL* and the approximated MDL_a. We used the datasets Adult(48.842 instances and 13 variables), Car (1.728 inst. and 7 var.), DNA (3.190 inst. and 61 var.), Letter Recognition (20.000 inst. and 16 var.) Mushroom (8.124 inst. and 23 var.) and Nursery (12.960 inst. and 9 var.) from the UCI machine learning repository [14], the Alarm dataset (20.000 inst. and 37 var.) [9] that is a standard benchmark in the Bayesian network literature and a synthetic dataset that were kindly donated by Robert Castelo [15], that we will call Synthetic (100.000 inst. and 25 var.). We used the discretization utility of MLC++ [16] in order to discretize the attributes of the Adult and Letter Recognition data sets.

In Table 2, we show the results obtained with both approaches. On the left side, there are the results for the structures learned using the *full* MDL as the scoring function, while on the right side we show the results learned using the approximated MDL. We measure all the structures with both the MDL and the MDL_a scores in order to compare their quality. We also show the number of arcs, columns #a, in order to see the complexity of the yielded networks.

Table 2. Comparing MDL vs. MDL_a

	Learned with MDL			Learned with MDL_a		
	MDL	MDL_a	#a	MDL	MDL_a	#a
Adult	617584	617816	21	617609	617609	20
Alarm	217760	216785	48	223069	205982	63
Car	13710	13710	5	13710	13710	5
DNA	258291	258291	60	258291	258291	60
Letters	472021	472021	20	472021	472021	20
Mushroom	97350	84379	52	115722	59014	74
Nursery	126345	126345	9	126345	126345	9
Synthetic	1.3574e+6	1.3475e+6	78	1.34998e+6	1.31531e+6	103

At a first glance, we can see that for half of the datasets (Alarm, Mushroom, and Synthetic), the algorithms learn different structures. The ones learned with the MDL_a are significantly more complex the the ones learned with the MDL. This is due to the fact the the MDL_a measure over scores *good* parents and thus tends to introduce into the network structures more arcs than the exact MDL measure. We observed that the variables of these datasets are related to each other (i.e $P(XY) \neq P(X)P(Y)$). Note that the rest of the network structures are very sparse, that is, there are almost the same number of arcs than number of variables. This indicates that the variables of these datasets are very independent (i.e. $P(XY) \approx P(X)P(Y)$) and thus the MDL and the MDL_a are very similar, as shown in Table 2.

We also compared the structures of the networks and observed that the arcs of the structures learned with the MDL are a subset of the arcs of the structures obtained with MDLa. That is, the structures learned with the MDL_a are (almost) the same than the ones learned with MDL with additional arcs. In the experiments we observed that the structure learned for the Alarm dataset with the $MDLa$ score (Alarm-MDLa) has 15 additional arcs. Mushroom-MDLa has 24 additional arcs, and Mushroom-MDL has 9 arcs not found in Mushroom-MDLa. Finally, Synthetic-MDLa has 25 additional arcs, and Synthetic-MDL has 10 arcs not found in the former structure.

We also compared the time spent by both algorithms. See that our approximated measure is cheaper, in computing time, than the exact one since the number of iterations performed to compute the mutual information is reduced from k^{n-i+2} to k^3 for the $i-th$ term of the MDL expression, see Equations (9) and (12). Table 3 shows the number of clock ticks spent by the algorithm with the exact MDL (row E) and by the algorithm with the approximated MDL (row A). We can see that the exact MDL spends more time than the MDL_a when the complexity of the structure is the same, but that it is faster than the $MDLa$ when the structure complexity of the later is higher.

Table 3. Clock ticks

	Adult	Alarm	Car	DNA	Letter	Mushroom	Nursery	Synthetic
Exact	937	7468	15	4562	609	984	79	15531
Appr.	906	8531	1	4734	781	1750	62	18422

7 Conclusions and Future Work

In this work we have introduced an incremental expression for the mutual information between a variable, Y, and a set of variables, $\mathbf{X}^{(n)}$, when a new variable, Z, is added to the set. We also decomposed the contribution of the new variable into three terms each considering different aspects of its contribution to the mutual information. From this decomposition we obtained an approximation that maintained the three aspects of the contributed information.

We analyzed the error introduced by our approximated measure and saw that there are cases in which *good* parents are over scored and other cases where they are under scored. Even though, we experimentally showed that in general *good* parents are over scored since the structures learned with our approximated MDL are usually the same than those obtained with the exact measure with some added arcs.

We think that there is still a lot of work to be done in order to further understand the error introduced with our approximation. We will quantitatively study the error introduced by our approach and to study in which situations, underlying probability distributions of data, this error is highest and lowest. We would also like to use our approximation to other non hill climbing strategies to learn Bayesian networks [17] and to relate our approach to others like [18].

Acknowledgments

This work has been supported by the Fulbright Program and the Generalitat de Catalunya. Also by FEDER and CICYT, TIC-2003-08382-C05-02.

References

[1] Jain, A.K., Duin, R.P.W., Mao, J.: Statistical pattern recognition: A review. IEEE Transactions on Pattern Analysis and Machine Intelligence **22** (2000) 4–37

[2] Hastie, T., Tibshirani, R., Friedman, J.: The elements of statistical learning. Datamining, inference and prediction. Springer Series in Statistics. Springer (2001)

[3] Roure, J.: Incremental augmented naive bayes classifiers. In: Proceedings of the sixteenth European Conference of Artificial Intelligence (ECAI 2004), IOS Press (2004)

[4] Hulten, G., Domingos, P.: Mining complex models from arbitrarily large databases in constant time. In: Proceedings of the Eighth ACM SIGKDD International Conference on Knowledge Discovery and Data Mining, ACM Press (2002) 525–531

[5] Meek, C., Thiesson, B., Heckerman, D.: The learning-curve sampling method applied to model-based clustering. Journal of Machine Learning Research (2002)

[6] Buntine, W.: Theory refinement on Bayesian networks. In D'Ambrosio, B., Smets, P., Bonisone, P., eds.: Proceedings of the 7th UAI. (1991)

[7] Friedman, N., Goldszmidt, M.: Learning Bayesian networks with local structure. In: Proceedings of the Twelveth Conference on Uncertainty in Artificial Intelligence. (1996)

[8] Lam, W., Bacchus, F.: Learning Bayesian belief networks. an approach based on the MDL principle. Computational Intelligence **10** (1994) 269–293

[9] Cooper, G., Herskovits, E.: A Bayesian method for the induction of probabilistic networks form data. Machine Learning **9** (1992) 309–347

[10] Chickering, D.M.: A transformational characterization of equivalent Bayesian networks. In Besnard, P., Hanks, S., eds.: Conference on Uncertainty in Artificial Intelligence, Morgan-Kaufman (1995) 87–98

[11] Cover, T.M., Thomas, J.A.: Elements of Information Theory. Wiley series in telecommunications. John Wiley & Sons, Inc. (1991)

12. McKay, D.J.: Information theory, inference and learning algorithms. http://wol.ra.phy.cam.ac.uk/mackay/itprnn/book.ps.gz (1999)
13. Pearl, J.: Probabilistic Reasoning in Intelligent Systems. Morgan Kaufmann (1988)
14. Murphy, P., Aha, D.: UCI repository of Machine Learning databases. http://www.ics.uci.edu/ mlearn/MLRepository.html. (1994) Irvine, CA: University of California, Department of Information and Computer Science.
15. Castelo, R., Kočka, T.: On inclusion-driven learning of Bayesian networks. Journal of Machine Learning Research (2003)
16. Kohavi, R., John, G., Long, R., Manley, D., Pfleger, K.: MLC++: A Machine Learning library in c++. In: Proceedings of the Sixth International Conference on Tools with Artificial Intelligence, IEEE Computer Society Press (1994) 740–743
17. Tian, J.: A branch-and-bound algorithm for MDL learning bayesian networks. In: Proceedings of Uncertainty in Artificial Intelligence, Morgan Kauffman (2000) 580–588
18. Friedman, N., Getoor, L.: Efficient learning using constrained sufficient statistics. In: International Workshop on Artificial Intelligence. (1999)

Learning of Latent Class Models by Splitting and Merging Components

Gytis Karčiauskas

Institute of Fundamental Technological Research, Polish Academy of Sciences
gkarc@ippt.gov.pl

Summary. A problem in learning latent class models (also known as naive Bayes models with a hidden class variable) is that local maximum parameters are often found. The standard solution of having many random starting points for the EM algorithm is often too expensive computationally. We propose to obtain better starting points for EM by splitting and merging components in models with already estimated parameters. This way we extend our previous work, where only a component splitting was used and the need for a component merging was noticed. We discuss theoretical properties of a component merging. We propose an algorithm that learns latent class models by performing component splitting and merging. In the experiments with real-world data sets, our algorithm in a majority of cases performs better than the standard algorithm. A promising extension would be to apply our method for learning cardinalities and parameters of hidden variables in Bayesian networks.

1 Introduction

Latent class analysis [9; 5] is a method for finding classes of similar cases from multivariate categorical data. The data is assumed to be generated by a *latent class* (LC) model, which has a structure shown in Fig. 1. An LC model consists

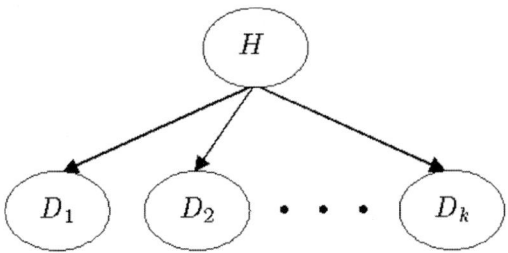

Fig. 1. Latent class model.

of a hidden *class variable* (H) and observed *manifest variables* ($D_1, \ldots, D_k$). Each state of the class variable corresponds to a different component (class). Manifest variables are assumed to be conditionally independent given the class variable. An LC model is also known as a *naive Bayes* model with a hidden class variable. *Parameters* of an LC model consist of a marginal probability distribution for the class variable and conditional probability distributions for each manifest variable given the class variable.

The goal in latent class analysis is for a given data over manifest variables to determine the optimal number of components ($|H|$) and model parameters. This is done by learning parameters for different $|H|$ and then selecting $|H|$ that gave the best model according to some criteria. Usually model parameters are learned by using the EM algorithm [3]. The well-known problem of EM is that local rather than global maximum parameters can be found [5; 10]. The standard way of dealing with this problem is to run EM many times from random starting points (starting parameterizations). The more starting points are used, the closer to the global maximum the parameters should be. However, often a high number of starting points is required, thus making the algorithm computationally very expensive. A more feasible computationally is a *multiple restart EM* approach [2], where many different starting points are used, but repeatedly, after a specified number of EM iterations, only parameterizations giving the highest likelihood are retained. Even though this algorithm is faster, often it still requires much time.

One could try to find better parameters in a much shorter time by using an information about parameters of an LC model with different $|H|$. The idea of splitting and merging components to obtain starting points for EM has already been applied in learning various models [11; 4; 12].

In this paper, we propose a method for learning LC models by repeatedly splitting and merging components. Starting points for EM are obtained only by performing a component splitting or merging in an LC model. This way, we extend a work of Karčiauskas et al. [7], where only a component splitting was used and the need for a component merging was noticed. After introducing a component merging, we discuss its theoretical properties. By combining a component splitting and merging, we introduce an operation for adjusting model parameters. In addition, we improve the implementation of a component splitting of Karčiauskas et al. [7] by introducing a partial EM. We use 20 real-world data sets to compare our algorithm with those that use standard starting points for EM.

In Sect. 2, we describe the operations on components and the algorithm based on them. In Sect. 3, we report the experiments performed. And in Sect. 4, we discuss the results of the experiments, relations between our and other similar work, and a possible future work.

2 Component Splitting and Merging

2.1 Notation and Definitions

A variable is denoted by an upper-case letter (for example, D_i), a state of a variable by a lower-case letter (for example, d_i). A vector of states for a set of variables is denoted by a bold lower-case letter (for example, $\mathbf{d}$). Training data over discrete variables $D_1, \ldots, D_k$ is $\mathbf{D} = (\langle \mathbf{d}_1, n_1 \rangle, \ldots, \langle \mathbf{d}_N, n_N \rangle)$, where $\mathbf{d}_i$ is a k-dimensional vector, n_i is a non-negative count, and $\mathbf{d}_i \neq \mathbf{d}_j$ for $i \neq j$. L is a latent class model with components $h_1, \ldots, h_m$. The probability of $\mathbf{d} = (d_1, \ldots, d_k)$ given L is

$$P_L(\mathbf{d}) = \sum_{i=1}^{m} P_L(h_i) P_L(\mathbf{d}|h_i) = \sum_{i=1}^{m} P_L(h_i) \prod_{j=1}^{k} P_L(d_j|h_i), \quad (1)$$

where P_L is specified by parameters of L. The log-likelihood of data $\mathbf{D}$ given model L is denoted as

$$LL(\mathbf{D}|L) = \sum_{i=1}^{N} n_i \ln P_L(\mathbf{d}_i). \quad (2)$$

$\mathbf{D}_{h_s}$ denotes the part of $\mathbf{D}$ that probabilistically belongs to component h_s. Formally, $\mathbf{D}_{h_s} = (\langle \mathbf{d}_1, n_{l1} \rangle, \ldots, \langle \mathbf{d}_N, n_{lN} \rangle)$, where $n_{li} = n_i \, P_L(h_s|\mathbf{d}_i)$ (so, $\sum_{l=1}^{m} n_{li} = n_i, \forall i = 1, \ldots, N$). Here $P_L(h_s|\mathbf{d}_i) = P_L(h_s) P_L(\mathbf{d}_i|h_s)/P_L(\mathbf{d}_i)$.

2.2 Component Splitting

Here we overview a component splitting. We say that model L^* is obtained from model L by *splitting* a component h_s if L^* instead of h_s contains components h_s^1 and h_s^2 that are both similar to h_s, and all the other components are identical in both models. More formally, if L contains components $h_1, \ldots, h_m$, L^* contains components $h_1, \ldots, h_{s-1}, h_s^1, h_s^2, h_{s+1}, \ldots, h_m$ and the following is true:

- $P_{L^*}(h_i) = P_L(h_i)$, $P_{L^*}(D_j|h_i) = P_L(D_j|h_i)$, $i = 1, \ldots, m$, $i \neq s$, $j = 1, \ldots, k$,
- $P_{L^*}(h_s^1) = P_{L^*}(h_s^2) = \frac{1}{2} P_L(h_s)$,
- $||P_{L^*}(D_j|h_s^1) - P_L(D_j|h_s)|| < \varepsilon$, $||P_{L^*}(D_j|h_s^2) - P_L(D_j|h_s)|| < \varepsilon$, $j = 1, \ldots, k$, where $||\cdot||$ is an L^2-norm of a vector and $\varepsilon \in \mathbb{R}$ is chosen in advance and is close to 0.

Component splitting, defined in this way, has the following good property, proved in our previous work [7; 6]. If for $\mathbf{D}$ large enough, a penalized likelihood score[1] of an LC model is not maximal, then with probability 1 it is possible to increase the model score by splitting a component.

---

[1] A penalized likelihood score of model L consists of two terms, one of which is the log-likelihood $LL(\mathbf{D}|L)$, and another – a penalty for the complexity of L.

2.3 Component Merging

Now we introduce a component merging. We say that model L^* is obtained from model L by *merging* components h_s and h_t if h_s in L^* is a weighted average of h_s and h_t in L, there is no component h_t in L^*, and all the other components are identical in both models. More formally, if L contains components $h_1, \ldots, h_m$, L^* contains components $h_1, \ldots, h_{t-1}, h_{t+1}, \ldots, h_m$ and the following is true:

- $P_{L^*}(h_i) = P_L(h_i)$, $P_{L^*}(D_j|h_i) = P_L(D_j|h_i)$, $i = 1, \ldots, m$, $i \neq s$, $i \neq t$, $j = 1, \ldots, k$,
- $P_{L^*}(h_s) = P_L(h_s) + P_L(h_t)$,
- $P_{L^*}(D_j|h_s) = P_L(D_j|h_s)P_L(h_s)/P_{L^*}(h_s) + P_L(D_j|h_t)P_L(h_t)/P_{L^*}(h_s)$, $j = 1, \ldots, k$.

Next we discuss theoretical properties of the component merging. First, let us see what properties we would like the component merging operation to have. We would use this operation when a model contains too many components. That is, when merging components, we expect that training data **D** can be described equally well (or almost equally well) by a model with one component less. If **D** is described perfectly by our current m–component model and it is possible to describe **D** perfectly by an m-1–component model as well, then we would obviously like the component merging operation to be able to obtain this m-1–component model from our current model.

We performed some experiments to check the properties of component merging. We parameterized randomly an m–component model L and generated perfectly described by L data **D** (that is, the distributions $P(D_1, \ldots, D_k)$ defined by L and by **D** are identical). Then we estimated the maximum likelihood (ML) parameters of an m+1–component model L' given **D**. That is, **D** is described perfectly by L' as well. For estimating the ML parameters, we have run the standard EM algorithm from a random starting point (for some L' having many parameters, this required running EM more than once, until the ML parameters with some precision have been found). We did this for small models L, because when L gets larger it becomes computationally difficult to find the ML parameters for L'.[2] Each time, we checked if it is possible to obtain model L from model L' by merging two components of L'. In some cases it was possible, and in some not. The results of these experiments are summarized in Table 1. Here a:kx2 indicates the structure with k binary manifest variables and a components, a-b:kx2 indicates all the structures with k binary manifest variables where the number of components ranges from a to b, $dim(L')$ indicates the standard dimension of model L' (i.e., the number

---

[2] Even for small models, we had to use iterative methods, such as the EM algorithm, for finding ML parameters with some precision. We have tried to find exact ML parameters with Maple^{TM}, but already for an LC model with two components and four binary manifest variables the number of equations and unknowns becomes too high.

Table 1. Results of experiments about properties of component merging.

Structure of L	Structure of L'	$dim(L') \leq dc$
2:4x2	3:4x2	yes
2-4:5x2	3-5:5x2	yes
2-8:6x2	3-9:6x2	yes
2:3x2	3:3x2	no
3:4x2	4:4x2	no
5:5x2	6:5x2	no
9:6x2	10:6x2	no

of independent parameters in L'), and dc indicates the complete dimension of $D_1 \times \ldots \times D_k$ (i.e., $dc = |D_1| \cdot \ldots \cdot |D_k| - 1$). For the models above the middle line, it is possible to obtain model L from model L' by merging some two components of L'. For the models below the middle line, it is not possible. Here we did not include the trivial case when L has only one component, because then merging the two components of L' always produces L. Also, we did not test models with 7 or more manifest variables, because then it becomes computationally too difficult to find the ML parameters for L'.

Based on these results, we make the following conjecture.

Conjecture 1. Assume that data **D** is described perfectly by an LC model L with binary manifest variables and that model L is unique (i.e., there is no other LC model having the same number of components as L and describing **D** perfectly)[3]. Assume that for an LC model L', having one component more than L, we have $dim(L') \leq dc$. Then, if **D** is described perfectly by L', model L can be obtained by merging two components of L'.

We have also explored the case where L contains manifest variables with more than two states. In Karčiauskas et al. [8], we have made a similar conjecture about this more general case. There, we have used the concept of model identifiability [5]. However, for this more general conjecture we have later found a counter-example [6].

In general, it seems that the component merging works for models where the number of manifest variables k is high enough or, alternatively, the number of components m is low enough. Our conjectures are the attempts of specifying conditions for m and k under which the component merging works.

So far, we have assumed that the current model contains one component too much. In practice, it can of course differ from the true model by more than one component. Can we expect that under some conditions it is possible to arrive to the true model by repeatedly performing a pairwise component merging and always keeping the maximum log-likelihood? We have made the

---

[3] It is assumed that models, which differ only in the ordering of components, are the same.

following test. The same way as in the experiments above, for a randomly parameterized model L with two components and five binary manifest variables, we estimated the ML parameters of model L' having four components. We repeated this several times for different parameterizations of L. In some cases, it was possible to arrive to model L by performing a pairwise component merging in L' and thus obtaining a 3–component model L'', which had the ML parameters, and then performing a pairwise component merging in L''. However, in other cases the parameters of L'' were not those of maximum likelihood, and model L could be obtained only by merging three components at once in L' (when more than two components are merged, they are combined into one new component the same way as h_s and h_t are combined in the definition of the component merging above). So, it seems that when the number of components is too high by more than one, a pairwise merging is not enough. In our algorithm, we use a pairwise merging because it has a lower complexity and because we believe that it does not make a big difference for practical data.

Finally, we should remind that all these results about the component merging are based on estimation of the ML parameters with some precision (and the larger the model, the lower the precision). Our judgments on whether the parameters are those of ML and whether merging some particular components leads to some particular model are based on those estimates of parameters.

In our algorithm, models obtained by the component splitting and merging will be used as starting points for EM.

2.4 Parameter Adjustment

Now we introduce an operation that tries to improve model parameters for a fixed $|H|$. First, we give a motivation for such an operation. It is possible to learn an LC model by repeatedly incrementing or decrementing $|H|$, where starting points for EM are obtained only by splitting or merging components, and it is required that any single increment or decrement of $|H|$ increases the model score. The algorithm would stop when neither incrementing nor decrementing $|H|$ increases the model score. However, the following situation can occur. The best model G has m components, and by incrementing and decrementing $|H|$ we have arrived at model L with m components that has however a lower score than G because of not optimal parameters. It can be that neither incrementing nor decrementing $|H|$ increases the score of L, and so L would be the final model. However, it can be that by *both* incrementing and decrementing $|H|$ we obtain model L' (with m components) that has a higher score than L.

So, we introduce an operation that we call *parameter adjustment*. It consists of the successive component splitting and merging (or component merging and splitting) with no requirement that a single increment or decrement of $|H|$ would increase the model score. More formally, we say that L' is obtained from L by a parameter adjustment if there exists model L^* such that:

- L^* is obtained from L by the component splitting and then running EM, and L' is obtained from L^* by the component merging and then running EM,

or

- L^* is obtained from L by the component merging and then running EM, and L' is obtained from L^* by the component splitting and then running EM.

For an m-component model L, there are $O(m^3)$ possible ways to perform the parameter adjustment, because for each of m ways to split there are $\binom{m+1}{2}$ ways to merge, or alternatively for each of $\binom{m}{2}$ ways to merge there are $m-1$ ways to split. However, in our implementation we will consider only $O(m^2)$ ways to perform the parameter adjustment, as described in the next section.

Even if all the $O(m^3)$ ways to perform the parameter adjustment are tried, there are no guarantees that such an operation will improve the parameters of L when they are not optimal. Also, the only reason for choosing a single splitting and a single merging rather than changing $|H|$ by more than 1 is that this approach is the fastest. However, in the experiments this operation often helped to find higher scoring models.

2.5 Algorithm

Here we provide our algorithm for learning LC models. Procedure *LearnLC* takes training data **D** as an input and returns an LC model. It tries to find an LC model that is the best according to function *score*, which can be any penalized likelihood scoring function. The algorithm starts with a one component model L, for which the maximum likelihood parameters are deterministically computed from **D**. The algorithm repeatedly tries to increase the score of L by either incrementing $|H|$ (procedure *Split*), decrementing $|H|$ (procedure *Merge*), or adjusting parameters (procedure *Adjust*). In phase 1, $|H|$ is repeatedly incremented until the score of L does not increase. Similarly, in phase 2, $|H|$ is repeatedly decremented until the score of L does not increase. Phases 1 and 2 are performed repeatedly one after another until none of them increases the score of L. Then one attempt to adjust parameters of L is made. If adjusting parameters increases the score of L, the algorithm goes back to running phases 1 and 2. Otherwise, model L is returned.

Procedure *Split* increments $|H|$ by splitting a component in L. For each component h_s, a prespecified number ($|\mathcal{L}_0|$) of independent random splits is performed. Component h_s is split randomly by making $P_{L^*}(D_j|h_s^1) - P_L(D_j|h_s) = \mathbf{r}_j$ and $P_{L^*}(D_j|h_s^2) - P_L(D_j|h_s) = -\mathbf{r}_j$. Each element of vector $\mathbf{r}_j$ is a random number from $[-p;p]$ (and at least one of them is exactly p or $-p$), where $p \in \mathbb{R}$ is a small positive parameter.[4] Also, $\forall j : \sum_{i=1}^{|D_j|} r_{j,i} = 0$.

---

[4] If $\mathbf{r}_j$ causes any parameter from $P_{L^*}(D_j|h_s^1)$ or $P_{L^*}(D_j|h_s^2)$ to be outside $[0.000001; 0.999999]$, $\mathbf{r}_j$ is scaled down so that all the parameters are inside this interval. If already $P_L(D_j|h_s)$ contains parameters outside this interval, $\mathbf{r}_j$ is set to **0**.

Procedure 1 *LearnLC*(**D**)

Let L be an LC model with one component.
$doPhase1 \leftarrow true$, $doPhase2 \leftarrow true$.
loop
  if $doPhase1$ **then**
    $L_0 \leftarrow L$.
    repeat
      $L' \leftarrow Split(L, \mathbf{D})$.
      if $score(L') > score(L)$ **then**
        $L \leftarrow L'$.
      end if
    until $L \neq L'$.
    if $score(L) > score(L_0)$ **then**
      $doPhase2 \leftarrow true$.
    end if
    $doPhase1 \leftarrow false$.
  end if
  if $doPhase2$ **then**
    $L_0 \leftarrow L$.
    repeat
      $L' \leftarrow Merge(L, \mathbf{D})$.
      if $score(L') > score(L)$ **then**
        $L \leftarrow L'$.
      end if
    until $L \neq L'$.
    if $score(L) > score(L_0)$ **then**
      $doPhase1 \leftarrow true$.
    end if
    $doPhase2 \leftarrow false$.
  end if
  if not $doPhase1$ and not $doPhase2$ **then**
    $L' \leftarrow Adjust(L, \mathbf{D})$.
    if $score(L') > score(L)$ **then**
      $L \leftarrow L'$, $doPhase1 \leftarrow true$, $doPhase2 \leftarrow true$.
    else
      Return L.
    end if
  end if
end loop

After splitting h_s, a partial EM is run – that is, only parameters for the two new components are updated. This can be understood as running a normal EM for a model containing only two new components and after that substituting h_s in L with those two components.[5] A model that has the highest score after a partial EM is selected and parameters of that model are updated by running a normal EM.

Procedure 2 $Split(L, \mathbf{D})$

    for all component h_s of L **do**
        Produce set $\mathcal{L}_0$ of two-component models by performing $|\mathcal{L}_0|$ random independent splits of h_s and for each split producing a model that contains only new components h_s^1 and h_s^2 from L^* (see Sect. 2.2).
        Obtain model L_0 by running a multiple restart EM with $\mathcal{L}_0$ as starting points and data $\mathbf{D}_{h_s}$.
        Obtain model L_s from L by substituting h_s with the two components from L_0.
    end for
    Let $L' = \arg\max_{L_s} score(L_s)$.
    Optimize parameters of L' by running EM with data $\mathbf{D}$.
    Return L'.

Procedure *Merge* decrements $|H|$ by merging two components in L. All the possible merge candidates are taken as starting points, and a multiple restart EM determines the best pair to merge. Our initial tests showed that the pair determined by a multiple restart EM is usually among the best pairs determined by a separate EM for each merge candidate (which is computationally much more expensive than a multiple restart EM).

Procedure 3 $Merge(L, \mathbf{D})$

    $\mathcal{L} \leftarrow \emptyset$.
    for all pair of components $\{h_s, h_t\}$ of L **do**
        Obtain model L^* by merging h_s and h_t.
        Add L^* to $\mathcal{L}$.
    end for
    Obtain model L' by running multiple restart EM with $\mathcal{L}$ as starting points and data $\mathbf{D}$.
    Return L'.

---

[5] Both in producing a two-component model and in substituting h_s is L, marginal probabilities of those two components are scaled so that the sum of marginal probabilities of all the components in a model is equal to 1.

Procedure *Adjust* tries to increase the score of L by adjusting its parameters. For an adjustment to be accepted, we require the increase in score to be higher than a positive parameter δ. Otherwise, the algorithm may spend lots of time making insignificant improvements in parameters of L.

First, we try to increase the score of L by decrementing and then incrementing $|H|$. If this does not succeed, then we try to increase the score of L by incrementing and then decrementing $|H|$. Here, when decrementing $|H|$, we consider only those pairs of components where one component has just been obtained by splitting and the other is an "old" component inherited from L.

In the implementation of this procedure, model L' and all models L_s are not computed directly but taken from the last run of *Merge* and the last run of *Split*, because in *LearnLC* the parameter adjustment is performed only when L remains unchanged both in phase 1 and phase 2.

Procedure 4 *Adjust*($L, \mathbf{D}$)

$L' \leftarrow Merge(L, \mathbf{D})$.
$L'' \leftarrow Split(L', \mathbf{D})$.
if $score(L'') > score(L) + \delta$ **then**
    Return L''.
end if
$\mathcal{L} \leftarrow \emptyset$.
for all component h_s of L **do**
    Obtain model L_s as in $Split(L, \mathbf{D})$.
    for all pair of components $\{h_a, h_b\}$ where h_a is a new component and h_b is an old component of L_s **do**
        Obtain model L^* by merging h_a and h_b in L_s.
        Add L^* to $\mathcal{L}$.
    end for
end for
Obtain model L'' by running multiple restart EM with $\mathcal{L}$ as starting points and data $\mathbf{D}$.
if $score(L'') > score(L) + \delta$ **then**
    $L \leftarrow L''$.
end if
Return L.

---

3 Experiments

3.1 Algorithms Tested

In this section, we list the algorithms that are tested in the experiments. Our algorithm, described in Sect. 2.5 and called *SplitMerge*, is compared with the algorithms that use standard starting points for EM. One of them, called *Standard* has almost the same main procedure as *LearnLC* from Sect. 2.5. The difference is that instead of calling *Split*, *Merge*, or *Adjust* it obtains L' by running a multiple restart EM on a model that has correspondingly one component more than L, one component less than L, or the same number of components as L (in the latter case, the score increase by more than δ is also required). Starting points for EM are randomly sampled from the uniform distribution.

Since requiring to change $|H|$ only by one can be too restrictive for a standard approach, we also compare our and standard approaches when the latter has an unfair advantage of knowing $|H|$ from the start. The algorithm called *StandardFixed* uses standard starting points for EM and estimates model parameters when the number of components is fixed. First, it runs a multiple restart EM from a prespecified number of starting points. After that, it repeatedly doubles the number of starting points, runs a multiple restart EM, and takes the resulting model as the current one if the new parameters increase the score. The algorithm is stopped at a prespecified time. Starting points for each EM are randomly sampled from the uniform distribution. $|H|$ is determined by previous runs of *SplitMerge*.

3.2 Setup of Experiments

The setup of the experiments is the following. Function *score* is a well-known BIC score, defined as $BIC(L) = LL(\mathbf{D}|L) - \frac{d}{2}\ln|\mathbf{D}|$, where d is the number of independent parameters in model L, and $|\mathbf{D}|$ is the total number of cases in data $\mathbf{D}$. EM searches for the maximum likelihood parameters. $|\mathcal{L}_0|$ in *Split* and the initial number of starting points for EM in *StandardFixed* are set to 16. The number of starting points for EM in *Standard* is set to 64 in order to make a running time of *SplitMerge* and *Standard* similar. Everywhere in multiple restart EM, $\frac{1}{q}$th of parameterisations giving the highest likelihood are retained after 1, 3, 7, 15, 31, ... iterations of EM, until only one best parameterisation is left. Then EM is run until either the difference between successive values of log-likelihood is less than 0.01 or 200 iterations are reached. $q = 2$ in *Split*, *Standard*, and *StandardFixed*. $q = 10$ in *Merge* and *Adjust*, because the number of merge candidates is high for models with many components. Parameter p from Sect. 2.5 is 0.001. Parameter δ from *Adjust* and *Standard* is 1.

For training data, we selected 20 classification data sets from the UCI Machine Learning Repository [1] and discarded the class information.[6] Continuous variables were converted into binary ones by performing an equal-

---

[6] All the 20 data sets are listed in Table 2. "Credit" stands for "Credit Card Application Approval", "Heart" – for a processed "Cleveland" data set from "Heart Disease", "Image" – for "Image segmentation", "Letter" – for "Letter Recognition", "Pen" – for "Pen-Based Recognition of Handwritten Digits", "Thyroid" – for a data set from "Thyroid Disease" suited for training ANNs, "Wisconsin" – for an original data set from "Wisconsin Breast Cancer". "Satimage", "Shuttle",

frequency binning. When separate training and test data were available, they were joined into a single training data.

The experiments are run on a Java^{TM} 2 platform, 2.8 GHz Intel(R) processor. For each data set, each algorithm (*SplitMerge*, *Standard*, and *StandardFixed*) is run 5 times. For *SplitMerge* and *Standard*, the algorithm is stopped if it does not terminate in 5 hours, and the current model L is returned. For *StandardFixed*, $|H|$ is the same as in the highest scoring model found by *SplitMerge* and the running time is equal to the mean running time of *SplitMerge*.

Table 2. Results.

Data set	Standard			SplitMerge			Standard-Fixed
	score	time	$|H|$	score	time	$|H|$	
Adult	-532344.7 ± 2790.8	300	6	**-520908.5 ± 375.5**	300	11	-520907.8 ± 717.7
Image	-15514.5 ± 35.2	32	14	-15516.5 ± 8.3	28	13	-15503.6 ± 5.5
Letter	-172561.3 ± 331.1	300	19	-172432.8 ± 248.7	300	18	-172314.8 ± 39.5
Mushroom	*-94155.0 ± 1090.5*	300	8	**-82935.1 ± 722.2**	300	17	-83648.2 ± 302.7
Page	-22768.1 ± 39.5	5	14	-22766.6 ± 2.9	5	14	**-22735.1 ± 6.9**
Pen	*-82411.6 ± 827.9*	260	30	**-81454.4 ± 11.5**	300	41	*-81402.1 ± 16.3*
Satimage	*-66602.9 ± 457.0*	300	16	**-64598.4 ± 121.9**	300	25	-64419.4 ± 87.1
Shuttle	*-256561.5 ± 265.9*	5	14	**-256027.6 ± 10.4**	16	21	*-256073.4 ± 28.3*
Spambase	*-83800.0 ± 192.9*	300	11	**-83090.0 ± 74.8**	300	15	-83127.5 ± 81.8
Abalone	-12389.4 ± 0.0	2	5	-12389.5 ± 0.0	1	5	-12389.4 ± 0.0
Audiology	-3403.0 ± 0.1	9	2	-3403.2 ± 0.0	2	2	-3403.1 ± 0.1
Credit	**-7564.7 ± 3.2**	16	5	*-7591.3 ± 22.7*	3	4	-7568.3 ± 3.2
Heart	-2388.1 ± 0.0	1	2	-2388.1 ± 0.0	0	2	-2388.1 ± 0.0
Housing	-3192.6 ± 5.7	2	7	-3198.3 ± 9.4	1	7	-3190.8 ± 5.7
Pima	-3995.6 ± 0.2	1	4	-3995.7 ± 0.0	0	4	-3995.6 ± 0.0
Thyroid	-44562.9 ± 26.1	21	6	-44542.9 ± 0.2	21	7	-44534.6 ± 18.3
Vehicle	-6471.5 ± 7.4	13	10	-6471.1 ± 7.0	9	10	-6467.1 ± 1.3
Voting	-3085.6 ± 0.0	5	5	-3085.6 ± 0.0	2	5	-3085.6 ± 0.0
Wisconsin	-2560.7 ± 0.0	1	3	-2560.9 ± 0.0	0	3	-2560.7 ± 0.0
Yeast	-6213.5 ± 0.0	0	2	-6213.7 ± 0.0	0	2	-6213.5 ± 0.0

3.3 Results

The results are summarized in Table 2. For each data set, the following is shown: the mean score of the final model (with a 95% confidence interval for the mean) for all the three algorithms, and the mean running time (in minutes) with $|H|$ in the highest scoring model for *Standard* and for *SplitMerge* (for *StandardFixed*, "time" and "$|H|$" are the same as for *SplitMerge*). To make the later discussion easier, we partitioned data sets into two groups according to whether $|H|$ for *SplitMerge* is higher than 10 or not. Bold text indicates a significantly better (and italic – a significantly worse) score when comparing

and "Vehicle" are from the Statlog Project. For "Audiology" and "Wisconsin", the identifier variable has been discarded.

pairwise *SplitMerge* with *Standard* and *SplitMerge* with *StandardFixed*. The difference in score is considered to be significant if intervals for the mean do not overlap and the difference in the mean values is higher than 1.

In Figs. 2–5, we display how the mean score changes during time for four data sets. Bars at the end indicate 95% confidence intervals for the mean.

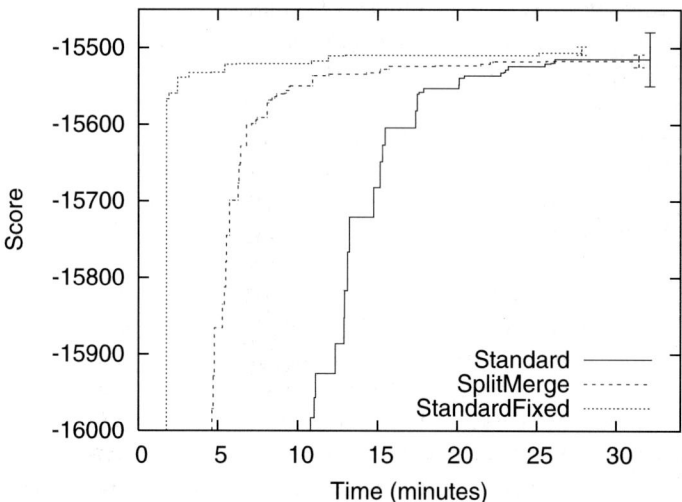

Fig. 2. Score change during time for "Image" data set

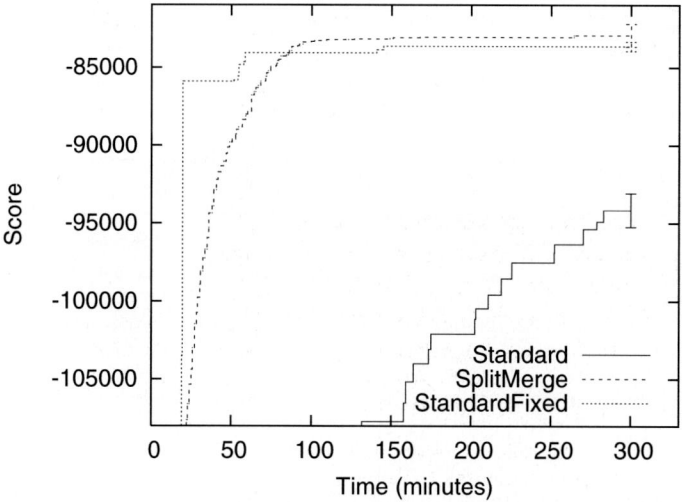

Fig. 3. Score change during time for "Mushroom" data set.

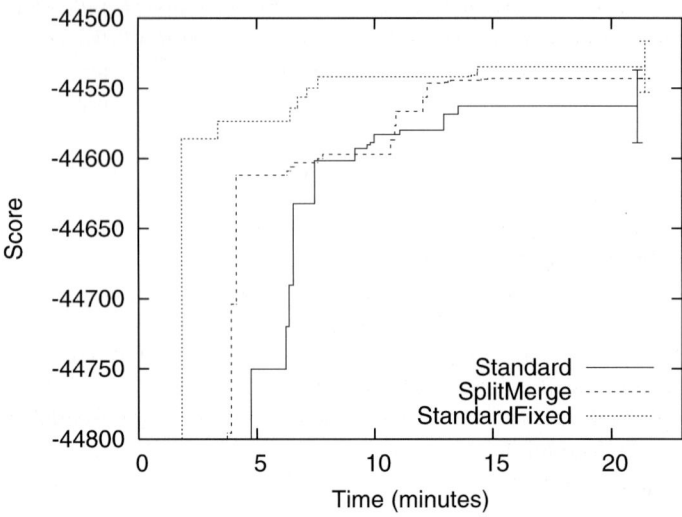

Fig. 4. Score change during time for "Thyroid" data set.

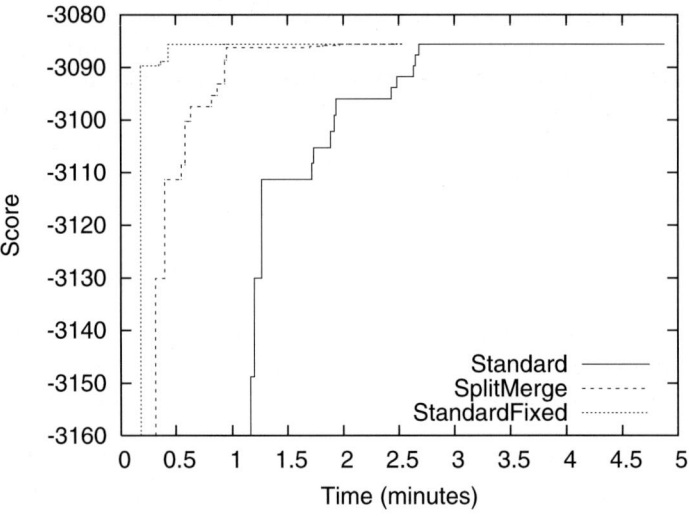

Fig. 5. Score change during time for "Voting" data set.

4 Discussion

As seen from Table 2, *SplitMerge* is significantly better than *Standard* for $|H| > 10$, and except for one data set there is no significant difference for $|H| \leq 10$. Most probably, this is because standard starting points for EM are more likely to lead to local maximum parameters when $|H|$ increases [10]. And when in case of *StandardFixed* the standard approach is from the start given the best $|H|$, it is unable to improve significantly over *SplitMerge* for most of the data sets: *StandardFixed* was significantly better for "Page" and "Pen", and significantly worse for "Shuttle".

For some data sets, the highest scoring models found by different algorithms have different $|H|$. For the data sets where *SplitMerge* is significantly better than *Standard*, $|H|$ determined by *SplitMerge* is much higher than $|H|$ determined by *Standard*. For "Pen" and "Shuttle", *Standard* is simply unable to increase $|H|$. For "Adult", "Mushroom", "Satimage", and "Spambase", *SplitMerge* is faster than *Standard*, and both algorithms are stopped because of a time limit.[7] For "Image", "Letter", "Credit", and "Thyroid", $|H|$ determined by *Standard* and by *SplitMerge* differ by 1. In all of them, the best model found by *Standard* has a higher score than the best model found by *SplitMerge*, even though for "Letter" and "Thyroid" *SplitMerge* performs better on average. This is related to the fact that generally the variance of the score is lower for *SplitMerge* than for *Standard*.

As seen from Figs. 2-5, models found by *SplitMerge* most of the time have higher score than models found by *Standard*. And usually, towards the end *SplitMerge* gets close to *StandardFixed*.

Now we discuss relations between our and other work where the component splitting and merging are used. For continuous data, the component splitting is used by Verbeek et al. [12], where Principal Component Analysis is applied for initialising new components deterministically. In our work, we deal with categorical data and split a component randomly. We did not find a deterministic method that would outperform a random splitting on data that we worked with. Searching for such a method could be a possible future work.

Concerning the component merging, our work has similarities to that of Elidan and Friedman [4]. There, in the context of Bayesian networks, the number of states of a hidden variable is learned by starting with a maximal number of states possible and merging states in a greedy way. The main difference from our method is that for each case in the training data, a "hard" (instead of probabilistic) assignment to a particular state of a hidden variable is maintained. Theoretical properties of such an approach are not known.

Our operation of parameter adjustment is similar to the simultaneous splitting and merging of components used by Ueda et al. [11] to overcome the problem of local maximum parameters in a mixture model with a fixed number of components. The authors require the components involved in splitting to be different from those involved in merging and introduce a criteria for selecting the most promising candidates for splitting and merging. We,

---

[7] We have also performed the experiments where the number of starting points for EM in *Standard* was set to 16 rather than 64. This way, *Standard* increases $|H|$ faster at a cost of worse parameters. In this setup, *Standard* was again significantly worse than *SplitMerge* on 6 data sets.

on the other hand, allow situations where the same component is involved in both splitting and merging. Neither Ueda et al. [11] nor we discuss theoretical properties of such an operation.

In this paper, we have presented a method for learning LC models by splitting and merging components. Even though this method generally does not find global maximum parameters, in a majority of cases it outperforms the standard approach. Since the problem of local maximum parameters generally becomes bigger as the number of unknown parameters to be estimated by EM increases, a promising future work would be to apply the component splitting and merging for learning cardinalities and parameters of hidden variables in Bayesian networks. In recent Ph.D. thesis [6], we have applied component splitting and merging for learning tree-structured Bayesian networks with hidden variables and discussed the extensions to unrestricted Bayesian networks with hidden variables.

Acknowledgments

The author would like to thank Tomáš Kočka, Finn V. Jensen, Nevin L. Zhang, Pedro Larrañaga, and Jose A. Lozano for their help and comments.

References

[1] C. L. Blake and C. J. Merz. UCI repository of machine learning databases, 1998. http://www.ics.uci.edu/~mlearn/MLRepository.html.

[2] D. M. Chickering and D. Heckerman. Efficient approximations for the marginal likelihood of Bayesian networks with hidden variables. *Machine Learning*, 29(2-3):181–212, 1997.

[3] A. Dempster, N. Laird, and D. Rubin. Maximum likelihood from incomplete data via the EM algorithm. *Journal of the Royal Statistical Society*, 39(B):1–38, 1977.

[4] G. Elidan and N. Friedman. Learning the dimensionality of hidden variables. In *Uncertainty in Artificial Intelligence: Proceedings of the Seventeenth Conference*, pages 144–151, 2001.

[5] L. A. Goodman. Exploratory latent structure analysis using both identifiable and unidentifiable models. *Biometrika*, 61:215–231, 1974.

[6] G. Karčiauskas. *Learning with Hidden Variables: A Parameter Reusing Approach for Tree-Structured Bayesian Networks*. PhD thesis, Aalborg University, 2005.

[7] G. Karčiauskas, F. V. Jensen, and T. Kočka. Parameter reusing in learning latent class models. In *Eighth International Symposium on Artificial Intelligence and Mathematics*, 2004.

[8] G. Karčiauskas, T. Kočka, F. V. Jensen, P. Larrañaga, and J. A. Lozano. Learning of latent class models by splitting and merging components. In *Second Workshop on Probabilistic Graphical Models*, 2004.

[9] P. F. Lazarsfeld and N. W. Henry. *Latent Structure Analysis*. Boston: Houghton Mifflin, 1968.

[10] J. Uebersax. A brief study of local maxima solutions in latent class analysis, 2000. http://ourworld.compuserve.com/homepages/jsuebersax/local.htm.

[11] N. Ueda, R. Nakano, Z. Ghahramani, and G. E. Hinton. SMEM algorithm for mixture models. *Neural Computation*, 12(9):2109–2128, 2000.

[12] J. J. Verbeek, N. Vlassis, and B. Krose. Efficient greedy learning of Gaussian mixture models. *Neural Computation*, 15(2):469–485, 2003.

Part IV

Decision Processes

An Efficient Exhaustive Anytime Sampling Algorithm for Influence Diagrams

Daniel Garcia-Sanchez and Marek J. Druzdzel

Decision Systems Laboratory, School of Information Sciences and Intelligent Systems Program, University of Pittsburgh, Pittsburgh, PA 15260
{dgarcia,marek}@sis.pitt.edu

Summary. We describe an efficient sampling algorithm for solving influence diagrams that achieves its efficiency by testing candidate decision strategies against the same set of samples and, effectively, reusing samples for each of the strategies. We show how by following this procedure we not only save a significant amount of computation but also produce better quality anytime behavior. Our algorithm is exhaustive in the sense of computing the expected utility of each of the possible decision strategies.

Keywords: Bayesian networks, algorithms, stochastic sampling

1 Introduction

Influence diagrams (IDs) [18] are acyclic directed graphs modeling decision problems under uncertainty. An ID encodes three basic elements of a decision: (1) available decision options, (2) factors that are relevant to the decision, including how they interact among each other and how the decisions will impact them, and finally, (3) the decision maker's preferences over the possible outcomes of the decision making process. These three elements are encoded in IDs by means of three types of nodes: decision nodes, typically represented as rectangles, random variables, typically represented as ovals, and value nodes, typically represented as diamonds or hexagons. Most popular type of IDs are those in which both the decision options and the random variables are discrete. A decision node in a discrete ID is essentially a list of labels representing decision options. Each random variable is described by a conditional probability table (CPT) containing the probability distribution over its outcomes conditional on its parents. Each value node encodes a utility function that represents a numerical measure of preference over the outcomes of its immediate predecessors. An ID that contains only random variables is called a Bayesian network (BN) [25].

An ID is amenable to an algorithmic analysis that can yield for each possible strategy (an assignment of values to all decision nodes conditional on those chance nodes that are observed before the decision nodes, e.g., [5]) its numerical measure of desirability – expected utility. It is a basic premise of decision theory that a decision maker should maximize his or her gain by choosing the strategy that yields the highest expected utility. Even though this analysis is at its foundations NP-hard [7], there exist several ingenious algorithms that make it feasible for practical ID models. Before we review these algorithms, we will focus on four general characteristic of ID algorithms that will be helpful in making the review more systematic.

While having an exact result is always preferable, in most cases precision can be traded off for computation, dividing ID algorithms into *exact* and *approximate*. The latter class does not give guarantees of optimality, even though it will typically reach or approximate the optimal solution. An important subclass of approximate algorithms are *anytime* algorithms, which are algorithms that regardless of the complexity of the problem have an answer available almost immediately. Subsequent computation, which gradually improves the quality of this answer, can be interrupted at any time. An algorithm is *exhaustive* if it computes the expected utility of each of the possible decision strategies or *focused* if its only goal is identification of the optimal strategy. Focused algorithms are in general more efficient than exhaustive algorithms, but this efficiency comes at a price. The output of a focused algorithm identifies the best next step of a strategy but it does not put it in a context of other possible steps or other strategies. In particular, the output of a focused algorithm does not show whether the suggested decision is a clear winner or a close runner with other, alternative decisions. This is not important in some applications, for example in robotics. However, in applications that include interaction with human decision makers, exhaustive algorithms offer a considerable amount of insight that helps in choosing the superior decision strategy or assists in refining the model. Finally, algorithms can be divided into *direct* and *indirect*, i.e., those that work on IDs directly and those that reduce the IDs solution problem into a series of belief updating problems in its underlying BNs.

There is a sizeable body of literature on the topic of solving IDs. Space limitations prevent us from covering this literature with the exception of those algorithms that are most directly relevant to our paper. The oldest among the direct evaluation methods originates from the work of Olmsted [23], whose algorithm performs a series of node removal and arc reversal operations that successively reduce the ID into a single decision node and a single successor utility node that holds the expected utility of each of the decision options. Shachter [30] has later formalized this algorithm and proven its correctness. Shenoy [32] introduced a new approach that transforms the ID into a valuation network that allows for slightly more efficient calculations. Ndilikilikesha [22] modified Olmsted's algorithm making it computationally as efficient as Shenoy's.

The indirect evaluation methods are inspired by a seminal paper by Cooper [6], who proposed transforming each decision node in an ID into a chance node with uniform distribution and the (single) value node into a chance node whose probability distribution is a linear transformation of the original utility function into the interval [0,1]. Following this transformation, the algorithm instantiates, one by one, each of the strategies, by observing the states of the nodes that represent the original decision nodes, solves the resulting BN using any belief-updating algorithm, and retrieves the expected utility of the current strategy.

Shachter and Peot [28] designed an efficient variation on Cooper's algorithm for IDs that have one decision node and one utility node. After the transformation of the ID into a BN, the algorithm essentially introduces evidence into the value node, solves the BN, and chooses the best decision alternative by comparing the posterior probability of the outcomes of the decision node: the most likely outcome indicates the best decision option. The algorithm is focused, as it identifies only the best decision. A separate, optional step of the algorithm is required to calculate the expected utility of that decision. Zhang [34] describes another focused indirect evaluation algorithm that is an improvement on Shachter and Peot's algorithm. It successively divides the ID into two parts, head and a tail, the tail containing the last decision node and all those other nodes that are relevant to the maximization of that last decision. It then computes the expected utility of the tail and finds the best option for the last decision. The tail, including the last decision node, is then substituted by a single utility node that summarizes all the uncertainty contained in it. The procedure is repeated until no more decision nodes are left. One advantage of this algorithm is that it can be based on any belief-updating algorithm for BNs. Other techniques to solve influence diagrams were offered by Jensen, Jensen and Dittmer [19] and Madsen and Jensen [21].

Another line of research on algorithms for IDs was started by Horvitz [17] who identified a class of inference policies for probabilistic reasoning systems that he called incremental refinement policies. He presented algorithms that improve the accuracy of their solutions as a monotonically increasing function of the allocated resources and the available information. Several algorithms have been developed along these lines. Ramoni [26] proposed an anytime algorithm for solving IDs based on Epistemic Constraint Propagation [27] that incrementally refines the confidence in the expected utility of the decisions. One limitation of this algorithm is that its anytime property holds only for IDs with one decision node. Horsch and Poole [15] developed another anytime algorithm that constructs a tree structure for each decision node in the ID. Each of the constructed tree structures represents a decision function that is improved incrementally. The authors show how improvements to this function lead to the optimal decision function. Similarly to Ramoni's algorithm, the algorithm loses its anytime properties if applied to decision problems with more than one decision node. A refinement of this algorithm, focusing on multi-stage influence diagrams was published later [16]. Charnes and Shenoy [1] propose an approximate sampling algorithm for solving IDs. D'Ambrosio and Burgess

[10] performed experiments comparing a variety of real-time influence diagram algorithms.

This paper shows how sampling algorithms can be utilized very efficiently within an indirect evaluation method. Our approach evaluates all decision strategies on the same set of random samples and allows for saving a significant amount of computation while producing high quality anytime behavior. The algorithm that we propose is suitable for very large decision models for which exact algorithms are not feasible. Recent advances in stochastic sampling algorithms for BNs (e.g., [2; 33]) have made it possible to perform inference in very large models with unlikely evidence. Because our algorithm can be based on any simulation algorithm for BNs, it allows for taking advantage of these developments. The algorithm is general in the sense of admitting full IDs with multiple decision and value nodes and containing observations. As a side-product, the algorithm computes also the posterior probability distribution of every chance node conditional on each of the possible strategies. For example, in a system that helps to decide between medication and surgery, the algorithm will compute not only the expected utility of both choices, but also the probability of the patient dying in each case. We have found that this information is useful in an interactive modeling environment. The algorithm is, to our knowledge, the only algorithm developed for ID that exhibits anytime behavior when the ID has an arbitrary number of decision and utility nodes. Finally, the algorithm is very efficient and this is achieved by reusing random samples. Practically, only one random number per node is generated for each sample of the network. This random number is used to evaluate the probability distribution of the node conditional on each of the strategies.

All random variables used in this paper are multiple-valued, discrete variables. Lower case letters, such as a, b, or c denote random variables. Bold capital letters, such as $\mathbf{A}$, denote sets of variables. Bold capital $\mathbf{E}$ denotes the set of observed nodes and capital $\mathbf{Q}$ denotes the set of other nodes not contained in $\mathbf{E}$. For any node n, Pa(n) denotes the set of parents of n, De(n) denotes the descendants of n, and O(n) denotes the set of outcomes of n.

The remainder of this paper is organized as follows. Section 2 defines some of the terms that we will be using throughout this paper. Section 3 describes the proposed algorithm. Section 4 contains a detailed example that will serve us in explaining the algorithm. Section 5 reports the results of our experiment with the algorithm, and Section 6 contains concluding remarks.

2 Some Definitions

A Bayesian network $\mathcal{B}$ is defined as a pair $\mathcal{B} = (\mathcal{G}, \text{Pr})$, where $\mathcal{G}(\mathbf{C}, \mathbf{A})$ is an acyclic directed graph consisting of a set of nodes $\mathbf{C}$ and a set of directed arcs $\mathbf{A}$. Each node $n \in \mathbf{C}$ has associated a conditional probability distribution, specifying the probability of each outcome $o \in \text{O}(n)$ conditional on Pa(n).

Similarly, we denote the joint probability distribution over a set of nodes $\mathbf{J} \subseteq \mathbf{C}$ as $\Pr(\mathbf{X}_J)$.

Given $\mathcal{B} = (\mathbf{C}, \mathbf{A})$ and sets $\mathbf{J}, \mathbf{K}, \mathbf{L} \subseteq \mathbf{C}$, we say that $\mathbf{X}_L$ is irrelevant [13] to $\mathbf{X}_J$ given $\mathbf{X}_K$ if $\Pr(\mathbf{X}_J|\mathbf{X}_K, \mathbf{X}_L) = \Pr(\mathbf{X}_J|\mathbf{X}_K)$. In other words, $\mathbf{X}_L$ is irrelevant to $\mathbf{X}_J$ given $\mathbf{X}_K$ if, once $\mathbf{X}_K$ has been observed, we cannot learn anything about $\mathbf{X}_J$ by also observing $\mathbf{X}_L$. We extend the notion of irrelevance to decision nodes in the following way. Given an ID model $\mathcal{M} = (\mathbf{C}, \mathbf{A}, \mathbf{D}, \mathbf{U})$ and sets $\mathbf{L} \subseteq \mathbf{D}$ and $\mathbf{J}, \mathbf{K} \subseteq \mathbf{C}$, we say that $\mathbf{L}$ is irrelevant to $\mathbf{X}_J$ given $\mathbf{X}_K$ if $\Pr(\mathbf{X}_J|\mathbf{X}_K, \mathbf{L}) = \Pr(\mathbf{X}_J|\mathbf{X}_K)$. In other words, L is irrelevant to $\mathbf{X}_J$ given $\mathbf{X}_K$ if, once $\mathbf{X}_K$ has been observed, making a decision in L does not impact $\mathbf{X}_J$. Irrelevance for utility nodes $\mathbf{U}$ is defined analogously but with respect to expected utility rather than probability.

An ID $\mathcal{M} = (\mathbf{C}, \mathbf{A}, \mathbf{D}, \mathbf{U})$ is a BN extended with $\mathbf{D}$, a set of decision nodes, and $\mathbf{U}$, a set of utility nodes. The decision nodes are ordered in time, $d_1, \ldots, d_n$, and chance nodes are stratified into sets $\mathbf{E}, \mathbf{I}_1, \ldots, \mathbf{I}_{n+1}$, so that chance nodes $\mathbf{I}_i$ are observed before d_i but after d_{i-1}. The nodes in $\mathbf{E}$ have been observed a-priori while the nodes in $\mathbf{I}_{n+1}$ will never be observed. Let $\mathbf{I} = (\mathbf{I}_1, \ldots, \mathbf{I}_n)$ be nodes that are not evidence but that will be observed before the last decision d_n is made.[1] From here on, we assume, without loss of generality, that $\mathbf{E} = \emptyset$. Dealing with evidence in sampling algorithms is quite well understood and is not the focus of this paper. For convenience, we will sometimes say $n \in \mathcal{M}$ if $n \in \{\mathbf{C} \cup \mathbf{D} \cup \mathbf{U}\}$.

Let us define a partial order Ψ_N over a set of nodes $\mathbf{N} \subseteq \mathcal{M}$ such that, given nodes $n, m \in \mathbf{N}$, if $m \in De(n)$ then n appears in Ψ_N before m. In other words, in Ψ_N a node appears always before any of its descendants.

We say that $d \in (\mathbf{D}, \mathbf{I})$ is an indexing parent of $n \in \mathcal{M}$ if it precedes n in $\Psi_{(\mathbf{D},\mathbf{C},\mathbf{U})}$ and it is relevant to n. We denote by $\mathrm{IP}(n)$ the set of indexing parents of node n. Less formally, the set of indexing parents of a node n is composed of those decision nodes and those chance nodes that will be observed before making a decision, that have an effect on the distribution of the outcomes of n.

A *strategy* is an instantiation of every node $d \in \mathbf{D}$, conditional on $\mathbf{I}$. We will sometimes use the term strategy when making reference to an instantiation of the set of indexing parents of a node.

For each node $n \in \mathcal{M}$ we can define the Cartesian product of the set of outcomes of its indexing parents:

$$\Phi_n = \times_{p \in \mathrm{IP}(n)} O(p) \ .$$

In other words, Φ_n contains all the possible combinations of the outcomes of n's indexing parents. Similarly, Φ is the set of indexing parents of the global utility node

$$\Phi = \times_{p \in \bigcup_{u_i \in \mathbf{U}} \mathrm{IP}(u_i)} O(p) \ .$$

---

[1] We normally indicate this order graphically by having an informational (dashed) arc from every $n \in \mathbf{I}_i$ into d_i.

We will now define the cardinality operator, $\|\Phi_n\|$, in the following way

$$\|\Phi_n\| = \begin{cases} |\Phi_n|, & \Phi_n \neq \emptyset \\ 1, & \Phi_n = \emptyset \end{cases}.$$

This operator, when applied to any set Φ_n, will return the number of elements in Φ_n or 1 if Φ_n is empty. We will now prove a theorem stating that every indexing parent of a node p is also an indexing parent of p's children.

Theorem 1. *Given an ID $\mathcal{M} = (\mathbf{C}, \mathbf{A}, \mathbf{D}, \mathbf{U})$ and nodes $c, p \in \{\mathbf{C}, \mathbf{D}, \mathbf{U}\}$, if $p \in Pa(c)$ then $IP(p) \subseteq IP(c)$.*

Proof. By definition, every node $n \in \text{IP}(p)$ is relevant to p. Since $p \in \text{Pa}(c)$, it follows that n must also be relevant to c and, hence, be an indexing parent of c.

3 The Algorithm

This algorithm solves an ID $\mathcal{M} = (\mathbf{C}, \mathbf{A}, \mathbf{D}, \mathbf{U})$, which means that it calculates the expected utility of each possible strategy and also calculates the distribution of each chance node conditional on each of the strategies. As one of its main features, the algorithm is capable of giving an approximate answer at any time. We describe this algorithm in Fig. 1.

Cooper's transformation in Step [1.1] is a two-step operation consisting of replacing all decision nodes and all utility nodes with chance nodes. Transformation of each utility node $u \in \mathbf{U}$ into a chance node, in particular, is performed by transforming the utility function $V(u) = F(p_1, \ldots, p_n)$, $p_1, \ldots, p_n \in \text{Pa}(u)$ into a probability distribution by linearly mapping the range of F onto the interval [0,1].

In order to achieve [1.2], we can use an algorithm like the one described in Druzdzel & Suermondt [11], Lin & Druzdzel [20] or Shachter [29]. Basically, for a given node $n \in \mathcal{M}$, it traverses the ID, starting from n, marking each d-connected node it finds along the way. Once the traversing is done, those nodes marked are added to the set IP(n).

In Step [3], we iterate only through chance nodes, which means that decision nodes are ignored at this stage. Remember that utility nodes have been transformed into chance nodes in Step [1.1]. The Steps [3.1], [3.2.1], [3.2.2], and [3.2.3] basically follow any forward sampling scheme, such as probabilistic logic sampling [14], likelihood weighting [12; 31], importance sampling [31], Latin hypercube sampling [3], adaptive importance sampling [2], or estimated posterior importance sampling [33]. This produces an unbiased estimator of the distribution of node c. But also note that in [3.2] we are iterating through every $\delta_k \in \Phi_c$ and that in Steps [3.2.1] through [3.2.3] we are assuming 'given δ_k'. The net result of Step [3] is the distribution of each chance node c conditional on each strategy relevant to c:

> **Input:** An ID $\mathcal{M} = (\mathbf{C}, \mathbf{A}, \mathbf{D}, \mathbf{U})$
> **Output:** For every strategy $\delta_k \in \Phi$ expected utility of δ_k.
> (1) For each node $n \in \mathcal{M}$
>   (1.1) If $n \in \mathbf{U}$, apply Cooper's transformation to n and add n to $\mathbf{C}$
>   (1.2) Find IP(n)
>   (1.3) Find $\|\Phi_n\|$
> (2) Determine $\Psi_\mathcal{M}$, a partial order over $\mathcal{M}$
> (3) For each node $c \in \mathbf{C}$ (in $\Psi_\mathcal{M}$ order)
>   (3.1) Generate a random number
>   (3.2) For each strategy $\delta_k \in \Phi_c$ (or once if $\Phi_c = \emptyset$)
>   (3.2.1) Look up the outcome of every node $p \in \text{Pa}(c)$ given δ_k
>   (3.2.2) Instantiate the outcome of c given δ_k
>   (3.2.3) Give a score and annotate the outcome of c given δ_k
> (4) If you wish to terminate the algorithm, proceed to Step 5.
>   Otherwise jump to Step 3.
> (5) Normalize all $\Pr(c), c \in \mathbf{C}$
> (6) For each node $u \in \mathbf{U}$
>   (6.1) For each $\delta_k \in \Phi_u$ (or once if $\Phi_u = \emptyset$)
>   (6.1.1) Compute $EU(u|\delta_k)$, $\delta_k \in \Phi_u$ by reversing Step 1.1
> (7) For each $\delta_k \in \Phi$
>   (7.1) Compute the global expected utility given δ_k.
> (8) Choose the best strategy.

Fig. 1. The basic algorithm.

$$\Pr(c_i|\delta_k), c_i \in \mathbf{C}, \delta_k \in \Phi_c. \qquad (1)$$

In [3.2.1], we say "Look up the outcome of every node $p \in \text{Pa}(c)$ given δ_k." What this "outcome" means depends on the type of the parent. If $p \in \text{IP}(c)$, "outcome" means state of p in the current δ_k. On the other hand, if $p \notin \text{IP}(c)$, "outcome" means the outcome of p that we computed and annotated in Step [3.2.3]. The fact that we are processing nodes in the order specified by $\Psi_\mathcal{M}$ guarantees that p has been dealt with before c, so the outcome of p is available to us. Theorem 1 guarantees that we can always find the state of node p given the current δ_k.

Also, note that we are evaluating in parallel every $\delta_k \in \Phi_c$, reusing the random number generated in [3.1], to compute the distribution of c conditional on every strategy. A complete sample, which is a collection of node outcomes, one for every node, describes a state of the world that we effectively test our decision model on. We use the same state of the world to test each of the possible decision strategies, but we do it in parallel as we continue sampling. This sample reuse achieves two important objectives: (1) it saves a large number of calls to the random number generator, and (2) with every sample generated, it performs the evaluation of all strategies, which is a key to achieving anytime behavior.

The simulation can be interrupted in Step [4], at which point the algorithm yields its best estimate of the expected utilities of each of the strategies.

If we apply an algorithm with a stopping rule, such as the *bounded variance algorithm* [9], the AA algorithm [8], or the AIS-BN-μ / AIS-BN-σ algorithms [4], we may be able to produce an estimate of the error measure for each of the strategies. If the error bounds around the posterior distributions of the transformed utility nodes are small enough, we may even return an exact ordering of the decision strategies.

Step [6] performs reverse of the transformation applied in [1.1] to utility nodes. Utility nodes, indexed by their indexing parents, contain at this point the expected utility of each possible strategy:

$$EU(u_i|\delta_k), \ u_i \in \mathbf{U}, \ \delta_k \in \Phi_u \ . \tag{2}$$

Step [7] is meant to deal with those cases in which **U** has more than one utility node. In this case, a global multi-attribute utility function of the form

$$\text{GEU}(u_1, u_2, \ldots, u_n) = f(U_1(u_1), U_2(u_2), \ldots, U_n(u_n)), \ u_i \in \mathbf{U}$$

should be provided (typically, it is a part of the model). This function is intended to produce a global utility and, for linearly additive utility functions, for example, it is defined as

$$w_1 \, U_1(u_1) + w_2 \, U_2(u_2) + \ldots + w_n \, U_n(u_n) \ , \ u_i \in \mathbf{U} \ .$$

Our algorithm does not depend in any way on the actual form of this multi-attribute utility function. Finally, we identify the strategy $\delta_k \in \Phi$ that maximizes GEU(δ_k).

4 An Example

To facilitate understanding of the algorithm, we are going to introduce a simple ID that we will use to illustrate every step of the algorithm described in the previous section. The ID models the following decision problem.

> A group of terrorists have taken control of an airplane and have threatened to kill all passengers and themselves, if their demands are not met. Information is scarce, the nationality of the assailants is unknown and there exists a possibility of them having explosives, which may make any rescue attempt risky. The authorities are facing two decisions. First, they need to determine whether they should negotiate or not. Second, they need to decide whether to deploy a SWAT team to assault the airplane. Although there is a lot of uncertainty, the authorities have reduced the possible nationalities of the terrorists down to two, *CountryA* and *CountryB*.

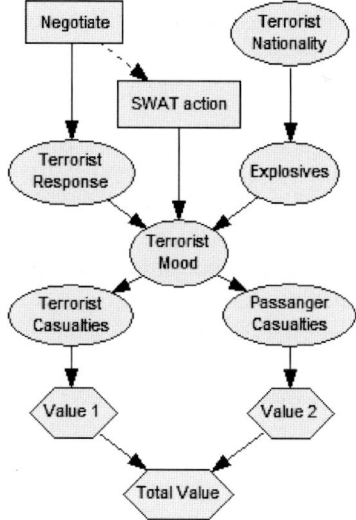

Fig. 2. The hijacked plane ID.

Figure 2 shows the ID for this problem.

Table 1 shows the short labels that we will use in the sequel when referring to the nodes in the model. It also shows the number of outcomes of each of the nodes.

Table 1. The label and the number of outcomes for each node.

Long name	Label	Number of outcomes
Negotiate	*Neg*	2
Terrorists Nationality	*Nat*	2
Swat Action	*Swat*	2
Terrorists Mood	*Mood*	2
Explosives	*Exp*	2
Terrorists Response	*TRes*	3
Terrorist Casualties	*TCas*	2
Passenger Casualties	*PCas*	3
Value 1	*Val1*	–
Value 2	*Val2*	–

We will now walk step-for-step through the algorithm.

Step (1): Initialization

Below we will show, as an example, Cooper's transformation applied to node *Val2*. This node encodes the preferences of the decision-maker with respect to passenger casualties. For simplicity, the decision-maker has considered only three possible outcomes: *All*, *Half*, or *None* of the passengers die and encoded her preferences in the node *Val2* by means of the following utility function:

$$U(Val2|PCas = None) = 100$$
$$U(Val2|PCas = Half) = 25$$
$$U(Val2|PCas = All) = 0$$

We apply a linear transformation to this utility function (please note that utility is invariant to linear transformations) mapping it to the interval $[0, 1]$:

$$\Pr(Val2|PCas = None) = 1.0$$
$$\Pr(Val2|PCas = Half) = 0.25$$
$$\Pr(Val2|PCas = All) = 0$$

After applying similar transformation to *Val1*, we add both *Val1* and *Val2* to the set of chance nodes. Table 2 summarizes the results of computing the indexing parents (Step [1.2]).

Table 2. Some initial values computed.

Node	IP(node)	$\|\Phi_{node}\|$
Neg	∅	–
Nat	∅	1
Swat	∅	–
Mood	{ Neg }	2
Exp	∅	1
TRes	{ Neg, Swat }	4
TCas	{ Neg, Swat }	4
PCas	{ Neg, Swat }	4
Val1	{ Neg, Swat }	4
Val2	{ Neg, Swat }	4

Step (2): Determine Ψ_M, a partial order over $\mathcal{M}$.

There are many valid partial orders for $\mathcal{M}$. We will use the following

$$\Psi_M = (Neg, Nat, Swat, Exp, Mood, TRes, TCas, PCas, Val1, Val2) .$$

Step (3): Take one sample of the network.

Generating a stochastic sample in a BN amounts to randomly instantiating every node to one of its possible states, according to the probability distribution over the node's states, conditional the instantiated states of its parents. In case of forward sampling, this requires every instantiation to be performed in the topological order, i.e., parents are sampled before their children. Supposed we pick the first chance node in Ψ_M, *Nat*, and generate a random number. Since $\Phi_{Nat} = \emptyset$, we do not have any strategies to iterate through. This means that we proceed as in a plain sampling algorithm for BNs: we generate a random number R and then we pick a state indicated by R, depending on the a-priori probability distribution over *Nat*, encoded in its CPT. Let this state be $Nat = CountryA$. We do the same to the next chance node, *Exp*, and generate $Exp = Yes$.

The next chance node in Ψ_M with a non-empty Φ_{Mood} is *Mood*. [3.1] First we generate a random number, say $R = 0.65$. [3.2] Now we start iterating through all $\delta \in \Phi_{Mood}$. We know that $\Phi_{Mood} = \{(Neg=Yes), (Neg=No)\}$, so the first δ is $(Neg=Yes)$. [3.2.1] We need now to find the outcome of each of *Mood*'s parents, only *Neg* in this case. Since $Neg \in \text{IP}(Mood)$, its outcome is taken from the current δ, i.e., $Neg = Yes$.

Assume that the CPT of *Mood* looks as illustrated in Fig. 3–a. [3.2.2] Looking at the probabilities in the first column, which shows $\Pr(Mood|Neg = Yes)$, we choose $Mood = Good$. [3.2.3] Annotate this result in a table similar to the one shown in Fig. 3–b.

In [3.2], we now pick the next element of Φ_{Mood}, which is $\delta = (Neg = No)$, and repeat the above process. In [3.2.1] the current δ tells us that $Neg = No$, since $Neg \in \text{IP}(Mood)$. [3.2.2] Looking at the probabilities in the second column of Fig. 3–a, which shows $\Pr(Mood|Neg = No)$, we find that $Mood = Bad$. [3.2.3] Annotate this result. Figure 3–b shows the state at this point.

Note in Fig. 3–b that we always assign a score of 1 to the sampled outcome. The reason for this, as noted before, is that we assume that there is no prior evidence in the network. In case there is evidence, we need to weigh these scores.

Neg	Yes	No
Good	0.7	0.1
Bad	0.3	0.9

(a)

Neg	Yes	No
Good	1	0
Bad	0	1
OUTCOME	Good	Bad

(b)

Fig. 3. (a) CPT of node *Mood*. (b) Table to score samples and to annotate current outcome. Note that this table is indexed by the indexing parents of *Mood*.

At this point, we have completed sampling from the node *Mood*. There are two remarks that we would like to make at this point. First, note that counting the samples within individual chance nodes allows us to calculate the posterior probability distribution of that node conditional on the strategy applied. We alluded to this earlier, when we talked about evaluating all strategies in parallel. Second, we are reusing the same random number R for all $\|\Phi_{FMood}\|$ strategies, which saves many calls to the random number generator.

We go back to Step 3 and pick the next chance node in Ψ_M, *TRes*. Something slightly more complex happens with this node. [3.1] We generate a new random number, say $R = 0.32$. [3.2] Again, we know that $\Phi_{TRes} = \{(Neg=Yes, Swat=None), (Neg=Yes, Swat=Attack), (Neg=No, Swat=None), (Neg=No, Swat=Attack)\}$. This means that our first strategy is $\delta = (Neg = Yes, Swat = None)$. [3.2.1] The parents of *TRes* are *Mood*, *Swat*, and *Exp*. Since $Swat \in \mathrm{IP}(TRes)$, its outcome is taken, as before, from the current δ, i.e., *Swat=None*. Things get just a little more complex when it comes to finding the outcome of *Mood*, since $Mood \notin \mathrm{IP}(TRes)$. This is because *Mood*, as shown in Fig. 3–b, has two sampled outcomes, not just one. We pick the one whose configuration of indexing parents matches the configuration of the current δ.

Swat	None				Attack			
Exp	No		Yes		No		Yes	
Mood	Good	Bad	Good	Bad	Good	Bad	Good	Bad
Liberate	0.8	0.4	0.8	0.4	0.25	0.1	0.25	0.1
Fight	0.2	0.6	0.15	0.5	0.75	0.9	0.65	0.7
Boom	0	0	0.05	0.1	0	0	0.1	0.2

Fig. 4. CPT of node TResponse.

Indeed, what we want to compute now is $\Pr(TRes|Neg = Yes, Swat = None)$, so it follows that we need to consider the distribution $\Pr(Mood|Neg = Yes, Swat = None)$. Since we have determined in Step [1.1] that *Mood* is independent of *Swat*, $\Pr(Mood|Neg = Yes, Swat = None) = \Pr(Mood|Neg = Yes)$. And since *Neg=Yes* in the current δ, we must pick the outcome of *Mood* from the column indexed by *Neg=Yes* in Fig. 3–b. This yields *Mood=Good*.

We take the same approach to find the outcome of the last parent, *Exp*, but in this case everything seems to be easier since *Exp* is independent from both *Neg* and *Swat* and has only one sampled outcome: *Exp=Yes*. Looking at the third column of Fig. 5, which shows $\Pr(TRes|Swat = None, Exp = Yes, Mood = Good)$, we obtain $TRes = Liberate$.

Neg	Yes		No	
Swat	None	Attack	None	Attack
Liberate	1	0	1	1
Fight	0	1	0	0
Boom	0	0	0	0
OUTCOME	Liberate	Fight	Liberate	Liberate

Fig. 5. Table to score samples and to annotate current outcome of node *TRes*.

[3.2] We now pick the next strategy, $\delta = (Neg=Yes, Swat=Attack)$. [3.2.1] From δ, $Swat=Attack$. Exp is still Yes. Since $Neg=Yes$ again, $Mood=Good$ for the same reasons explained above. [3.2.2] Looking at the seventh column of Fig. 5, which shows $\Pr(TRes|Swat = Attack, Exp = Yes, Mood = Good)$, we obtain $TRes=Fight$. [3.2.3] Annotate and pick next strategy. [3.2] $\delta = (Neg=No, Swat=None)$. [3.2.1] From δ, $Swat=None$. Exp is still Yes. Since now $Neg=No$, Fig. 3–b tells us that $Mood=Bad$. [3.2.2] Looking at the fourth column of Fig. 4, which shows $\Pr(TRes|Swat = None, Exp = Yes, Mood = Bad)$, we obtain $TRes=Liberate$. [3.2.3]. Back to [3.2], the last $\delta = (Neg=No, Swat=Attack)$. Note that we are still reusing the same random number $R = 0.32$, which, looking at column 8 in Fig. 4, yields $TRes=Fight$. We are done with node $TRes$, and Fig. 5 shows the state at this point.

We repeat this process for all nodes in the network. At this point the reader should have an intuitive idea of how the algorithm works. Basically, we choose a state for every chance node until there are no more nodes left in Ψ_M, at which point we would have completed one sample of the whole network.

Step (4): Decide if we should take another sample.

As said before, we need a means to decide whether to terminate the algorithm. The more samples we take, the more accurate the solution will be. If we decide to terminate, we just go on to Step [5] below to do some final calculations.

Step (5): Normalize distributions of chance nodes.

Normalize each of the distributions $\Pr(c_i|\delta_k)$, $c_i \in \mathbf{C}$, $\delta_k \in \Phi_c$ so that they add up to 1. This corresponds to the columns of the tables containing the scores of the outcomes of each node, like those shown in Figs. 3–b and 5.

Step (6): Compute expected utility of each utility node.

Here we just need to reverse the linear transformations of the utility functions that we performed in [1.1]. This step yields $V(Val1|\delta_k)$, $k \in \Phi_{Val1}$ and $V(Val2|\delta_k)$, $\delta_k \delta \in \Phi_{Val2}$.

Step (7): Compute the global expected utility.

Since we have two utility nodes, $Val1$ and $Val2$, we need a function that combines both and returns a global expected utility for the whole model. Let us assume for simplicity that it is a linear combination of the individual utility functions with unit weights, i.e.,

$$\text{GEU}(Val1, Val2) = U_1(Val1) + U_2(Val2).$$

As before, we iterate through all strategies $\delta \in \Phi$ and compute GEU conditional on every strategy. In other words, we compute:

$$\text{GEU}(\delta_k) = U_1(Val1|\delta_k) + U_2(Val2|\delta_k), \delta_k \in \Phi.$$

Step (8): Choose the best strategy.
Choose the strategy $\delta_k \in \Phi$ that maximizes $\text{GEU}(\delta_k)$ calculated in Step [7].

5 Empirical Illustration

It is quite obvious that the proposed algorithm is more efficient in terms of its use of samples and in terms of reducing the variance in results than an algorithm that exhaustively samples all strategies. We illustrate this on HEPAR II [24], a model for diagnosis of liver disorders consisting of 73 nodes and available in Decision Systems Laboratory's model repository at http://genie.sis.pitt.edu/. Since HEPAR II is a BN, we first performed a modification that transformed it into an ID. Basically, HEPAR II consists of nodes representing risk factors, liver disorders, medical tests, and symptoms. The structure of the graph is causal: arcs go from risk factors to disorders and from disorders to tests and symptoms. First, we designated four nodes from among the risk factors to act as decision nodes. These decision nodes represented hypothetical choices a person could have about certain risks factors. For example, alcohol abuse can be considered a risk factor for several liver disorders. Converting this chance node into a decision node represents the choice a person has with respect to drinking. Since each of the transformed decision nodes had two states, we ended up with 16 different strategies to evaluate. The second step was to add utility nodes. We decided to add one utility node to each of the disorder nodes such that the utility of not having a particular disorder was 100 while the utility of having the disorder was 0. For simplicity, we assumed that all disorders as equally severe and created an additive linear multi-attribute utility function with weights $w_i = 1.0$ for each of the individual utility nodes. With a total of eight disorders in the model, the multi-attribute utility function ranged from 0, when a patient suffers from all eight disorders, to 800, when a patient suffers from no disorders at all.

We compared our algorithm against indirect exhaustive algorithm, with both algorithms based on likelihood weighting as their underlying BN algorithm. We ran 1,000,000 samples of the network, taking measurements of intermediate values of expected utility for each of the 16 possible strategies.

Figure 6 shows the expected utility of each of the 16 strategies as a function of the number of samples. In the simulation, the range of expected utilities goes from 500 up to 800 (each horizontal line on the graph represents an increment of 50). This wide range is due to the variations in the estimates that we obtained during the first one hundred samples. Crossing of trajectories means essentially a change in the ranking of the strategies. The expected utilities converge to their exact values (verified by means of an exact algorithm) as the number of samples increases. Figure 7 shows the same plot for the algorithm proposed in this paper. In this simulation, the range of expected utilities goes from 550 up to 800 (each horizontal line also represents an increment of 50). The expected utilities converge to exactly the same values (670 for the worst strategy and 715 for the best one).

Both figures show individual trajectories shifting up and down, but while in the crude exhaustive algorithm each trajectory seems to do so independently of the others, in our algorithm they all shift roughly at the same time

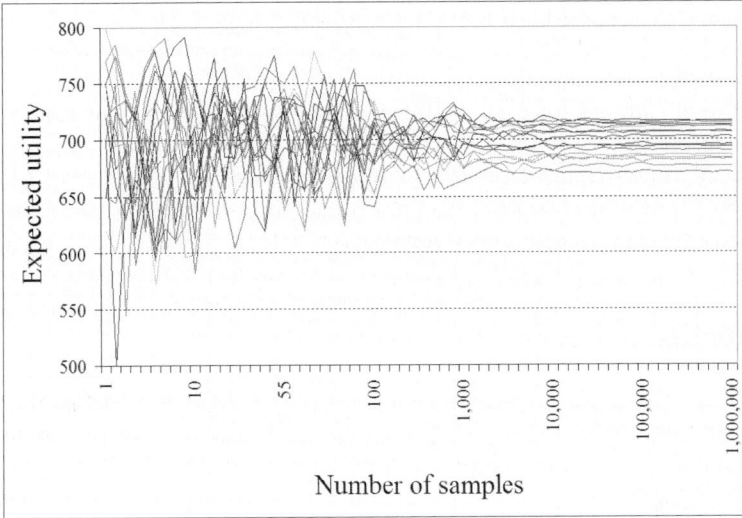

Fig. 6. Expected utility as a function of the number of samples for an indirect exhaustive algorithm based on likelihood weighting.

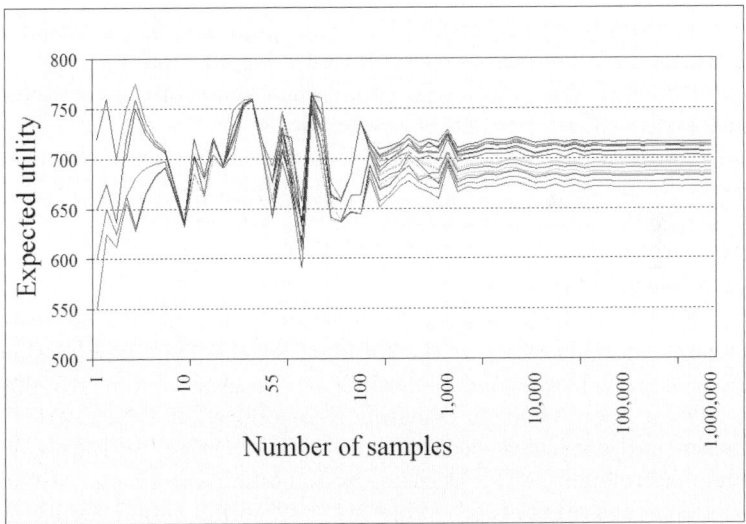

Fig. 7. Expected utility as a function of the number of samples for the exhaustive algorithm based on likelihood weighting proposed in this paper.

and in the same direction. This is a simple consequence of the fact that our simulation tests each strategy against the same sample, i.e., on the same randomly generated state of the world. This greatly reduces variance and our algorithm can rank the strategies after processing a relatively small number of samples. While in Fig. 6 we can find trajectory lines crossing even after as many as 90,000 samples taken, in Fig. 7 the last crossing of trajectory lines occurs at only 900 samples.

Reuse of samples for each of the 16 strategies reduced the number of calls to the random number generator by 16-fold. There was a 20% to 30% computational time overhead related to reuse, resulting in roughly a 12-fold reduction of the computation time of our algorithm compared to the exhaustive algorithm based on likelihood weighting.

6 Conclusion

We introduced an approximate anytime sampling algorithm for IDs that computes the expected utilities of all decision strategies. The algorithm is indirect, in the sense of reducing the problem of solving an ID by first transforming it into a BN. We have shown how by evaluation of each of the strategies on the same set of samples we not only save much computation but also reduce variance and, hence, produce high quality anytime behavior. The proposed algorithm is, furthermore, amenable to parallelization and a possible further increase in speed.

Our algorithm accommodates any forward sampling scheme. A simple empirical test of the algorithm has shown that it rapidly produces the correct order of strategies. We expect that combining this algorithm with stopping rules (e.g., [4]) will lead to efficient algorithms for IDs that give precision guarantees with respect to both the order and numerical expected utilities of strategies.

Acknowledgments

This research was supported by the National Science Foundation under Faculty Early Career Development (CAREER) Program, grant IRI–9624629, and by the Air Force Office of Scientific Research, grant F49620–03–1–0187. All experimental data have been obtained using SMILE, a Bayesian inference engine developed at the Decision Systems Laboratory and available at http://genie.sis.pitt.edu/. While we assume full responsibility for any errors remaining in this paper, we would like to thank our colleagues in the Decision Systems Laboratory for their ideas, comments, and suggestions. Comments from reviewers for the Second European Workshop on Probabilistic Graphical Models (PGM-04), where we presented an earlier version of this paper, helped us to improve the clarity of our presentation.

References

[1] John M. Charnes and Prakash P. Shenoy. Multi-stage Monte Carlo method for solving influence diagrams using local computation. *Management Science*, 50(3):405–418, 2004.

[2] Jian Cheng and Marek J. Druzdzel. BN-AIS: An adaptive importance sampling algorithm for evidential reasoning in large Bayesian networks. *Journal of Artificial Intelligence Research*, 13:155–188, 2000.

[3] Jian Cheng and Marek J. Druzdzel. Latin hypercube sampling in Bayesian networks. In *Proceedings of the 13th International Florida Artificial Intelligence Research Symposium Conference (FLAIRS–2000)*, pages 287–292, Menlo Park, CA, May 2000. AAAI Press/The MIT Press.

[4] Jian Cheng and Marek J. Druzdzel. Confidence inference in Bayesian networks. In *Proceedings of the Seventeenth Annual Conference on Uncertainty in Artificial Intelligence (UAI–2001)*, pages 75–82, San Francisco, CA, 2001. Morgan Kaufmann Publishers, Inc.

[5] Robert T. Clemen. *Making Hard Decisions: An Introduction to Decision Analysis*. Duxbury Press, An Imprint of Wadsworth Publishing Company, Belmont, California, 1996.

[6] Gregory F. Cooper. A method for using belief network algorithms to solve decision-network problems. In *Proceedings of the 1988 Workshop on Uncertainty in Artificial Intelligence, UAI–88*, Minneapolis, Minnesota, 1988.

[7] Gregory F. Cooper. The computational complexity of probabilistic inference using Bayesian belief networks. *Artificial Intelligence*, 42(2–3):393–405, March 1990.

[8] Paul Dagum, Richard Karp, Michael Luby, and Sheldon Ross. An optimal algorithm for Monte Carlo estimation. *SIAM Journal on computing*, 29(5):1481–1496, 2000.

[9] Paul Dagum and Michael Luby. An optimal approximation algorithm for Bayesian inference. *Artificial Intelligence*, 93:1–27, 1997.

[10] Bruce D'Ambrosio and Scott Burgess. Some experiments with real-time decision algorithms. In *Proceedings of the Twelfth Annual Conference on Uncertainty in Artificial Intelligence (UAI–96)*, pages 194–202, San Francisco, CA, 1996. Morgan Kaufmann Publishers.

[11] Marek J. Druzdzel and Henri J. Suermondt. Relevance in probabilistic models: "Backyards" in a "small world". In *Working notes of the AAAI–1994 Fall Symposium Series: Relevance*, pages 60–63, New Orleans, LA (An extended version of this paper is in preparation.), 4–6 November 1994.

[12] Robert Fung and Kuo-Chu Chang. Weighing and integrating evidence for stochastic simulation in Bayesian networks. In M. Henrion, R.D. Shachter, L.N. Kanal, and J.F. Lemmer, editors, *Uncertainty in Artificial Intelligence 5*, pages 209–219. Elsevier Science Publishing Company, Inc., New York, N. Y., 1989.

[13] Dan Geiger, Thomas S. Verma, and Judea Pearl. Identifying independence in Bayesian networks. *Networks*, 20(5):507–534, August 1990.

[14] Max Henrion. Propagating uncertainty in Bayesian networks by probabilistic logic sampling. In L.N. Kanal, T.S. Levitt, and J.F. Lemmer, editors, *Uncertainty in Artificial Intellgience 2*, pages 149–163. Elsevier Science Publishing Company, Inc., New York, N. Y., 1988.

[15] Michael C. Horsch and David Poole. Flexible policy construction by information refinement. In *Proceedings of the Twelfth Annual Conference on Uncertainty in Artificial Intelligence (UAI-96)*, pages 315–324, San Francisco, CA, 1996. Morgan Kaufmann Publishers.

[16] Michael C. Horsch and David Poole. An anytime algorithm for decision making under uncertainty. In *Proceedings of the Fourteenth Annual Conference on Uncertainty in Artificial Intelligence (UAI-98)*, pages 246–255, San Francisco, CA, 1998. Morgan Kaufmann Publishers.

[17] Eric J. Horvitz. Reasoning under varying and uncertain resource constraints. In *Proceedings of the 11th International Joint Conference on Artificial Intelligence, IJCAI-89*, pages 1121–1127, San MAteo, CA, 1989. Morgan Kauffman.

[18] Ronald A. Howard and James E. Matheson. Influence diagrams. In Ronald A. Howard and James E. Matheson, editors, *The Principles and Applications of Decision Analysis*, pages 719–762. Strategic Decisions Group, Menlo Park, CA, 1984.

[19] Frank Jensen, Finn V. Jensen, and Søren L. Dittmer. From influence diagrams to junction trees. In *Proceedings of the Tenth Annual Conference on Uncertainty in Artificial Intelligence (UAI-94)*, pages 367–373, San Francisco, CA, 1994. Morgan Kaufmann Publishers.

[20] Yan Lin and Marek J. Druzdzel. Computational advantages of relevance reasoning in Bayesian belief networks. In *Proceedings of the Thirteenth Annual Conference on Uncertainty in Artificial Intelligence (UAI-97)*, pages 342–350, San Francisco, CA, 1997. Morgan Kaufmann Publishers, Inc.

[21] Anders L. Madsen and Finn V. Jensen. Lazy evaluation of symmetric Bayesian decision problems. In *Proceedings of the Fifteenth Annual Conference on Uncertainty in Artificial Intelligence (UAI-99)*, pages 382–390, San Francisco, CA, 1999. Morgan Kaufmann Publishers.

[22] P. Ndilikilikesha. Potential influence diagrams. *International Journal of Approximate Reasoning*, 11:251–285, 1994.

[23] Scott M. Olmsted. *On Representing and Solving Decision Problems*. PhD thesis, Stanford University, Department of Engineering-Economic Systems, Stanford, CA, December 1983.

[24] Agnieszka Oniśko, Marek J. Druzdzel, and Hanna Wasyluk. Learning Bayesian network parameters from small data sets: Application of Noisy-OR gates. *International Journal of Approximate Reasoning*, 27(2):165–182, 2001.

[25] Judea Pearl. *Probabilistic Reasoning in Intelligent Systems: Networks of Plausible Inference*. Morgan Kaufmann Publishers, Inc., San Mateo, CA, 1988.

[26] Marco Ramoni. Anytime influence diagrams. In *Working notes of the IJCAI–95 workshop on "Anytime Algorithms and Deliberation Scheduling"*, pages 55–62, Montreal, Canada, 1995.

[27] Marco Ramoni and A. Riva. Belief maintenance with probabilistic logic. In *Working notes of the AAAI–93 Fall Symposium on "Automated Deduction in Non-Standard Logics"*, Raleigh, NC, 1993.

[28] R. D. Shachter and M. A. Peot. Decision making using probabilistic inference methods. In *Proceedings of the Eighth Annual Conference on Uncertainty in Artificial Intelligence (UAI–92)*, pages 276–283, San Francisco, CA, 1992. Morgan Kaufmann Publishers.

[29] Ross Shachter. Bayes-Ball: The rational pasttime (for determining irrelevance and requisite information in belief networks and influence diagrams). In *Proceedings of the Fourteenth Annual Conference on Uncertainty in Artificial Intelligence (UAI–98)*, pages 48–487, San Francisco, CA, 1998. Morgan Kaufmann Publishers.

[30] Ross D. Shachter. Evaluating influence diagrams. *Operations Research*, 34(6):871–882, November–December 1986.

[31] Ross D. Shachter and Mark A. Peot. Simulation approaches to general probabilistic inference on belief networks. In M. Henrion, R.D. Shachter, L.N. Kanal, and J.F. Lemmer, editors, *Uncertainty in Artificial Intelligence 5*, pages 221–231. Elsevier Science Publishing Company, Inc., New York, N. Y., 1989.

[32] Prakash P. Shenoy. Valuation-based systems for Bayesian decision analysis. *Operations Research*, 40(3):463–484, 1992.

[33] Changhe Yuan and Marek Druzdzel. An importance sampling algorithm based on evidence pre-propagation. In *Proceedings of the 19th Annual Conference on Uncertainty in Artificial Intelligence (UAI–03)*, pages 624–631, San Francisco, CA, 2003. Morgan Kaufmann Publishers.

[34] N. L. Zhang. Probabilistic inference in influence diagrams. In *Proceedings of the Fourteenth Annual Conference on Uncertainty in Artificial Intelligence (UAI–98)*, pages 514–522, San Francisco, CA, 1998. Morgan Kaufmann Publishers.

Multi-currency Influence Diagrams

Søren Holbech Nielsen, Thomas D. Nielsen, and Finn V. Jensen

Aalborg University, Aalborg, Denmark

Summary. When using the influence diagrams framework for solving a decision problem with several different quantitative utilities, the traditional approach has been to convert the utilities into one common currency. This conversion is carried out using a tacit transformation, under the assumption that the converted problem is equivalent to the original one. In this paper we present an extension of the influence diagram framework. The extension allows for these decision problems to be modelled in their original form. We present an algorithm that, given a linear conversion function between the currencies of the original utilities, discovers a characterisation of all other such functions, which induce the same optimal strategy. As this characterisation can potentially be very complex, we give methods to present it in an approximate way.

1 Introduction

Influence diagrams (IDs) were introduced by [4] as a compact modelling language for decision problems with a single decision maker (DM). When a decision problem is represented using the ID framework, the specification rests on two principal components: A graphical structure for capturing the qualitative part of the domain, and quantitative information in the form of probabilities for representing uncertainty and utilities for representing preferences.

The separation of the qualitative and quantitative part of the ID is one of the appealing properties of IDs when considered as a modelling tool. First, it helps the modeller to focus on structure rather than calculations, and second, the structure emphasises the local relations, which govern the specification of the probabilities. Unfortunately, this locality principle does not completely extend to the specification of the utility function: The utility function is usually specified through a collection of local utility functions, which appear as the additive components of the global utility function. This implies that all local utility functions should appear on the same scale. For instance, in a medical domain money and discomfort would need to be transformed onto a common scale. For decision problems where several parties are affected by the decision making process, the utility functions for the individual stakeholders would also have to be transformed into a single utility function before reasoning can take place. Unfortunately, it is usually difficult to elicit the parameters, which governs such transformations (e.g. what is the monetary cost of one unit of discomfort?).[1] Moreover, the nature of such a transformation has a direct impact on the solution of the ID, but this effect cannot be made transparent when the transformation is tacit. Furthermore, this type of uncertainty, or ignorance, is not easily represented in the model.

In this paper we propose a framework, termed multi-currency IDs (MCIDs), for representing decision problems with local utility functions of different currencies.[2] An MCID can be seen as an ID augmented with currency information for the local utility functions. We propose an algorithm that, based on an MCID representation of a decision problem and a solution corresponding to a given set of currency transformation parameters, provides a characterisation of all combinations of parameters, which would give rise to the same solution – a solution being an optimal strategy composed of a policy for each decision. The result of such an analysis thus provides an indication of how robust the optimal strategy is in terms of the tacit conversion parameters. Regular sensitivity analysis (see e.g. [3] and [8]), on the other hand, provides robustness measures in terms of isolated deviances in one or more of the actual utility values in the ID, with no regard to how the uncertainties in these parameters are related. By encoding each value for which regular sensitivity analysis is to be performed by its own currency, the analysis could be performed by the method we propose here, and it can therefore be seen as a generalization of regular sensitivity analysis.

As the result of the analysis may be quite complex, we provide, in addition to this algorithm, methods for presenting the result to a DM in a comprehensible manner.

Example 1 (A Motivating Example). The ID in Fig. 1 models a decision problem where a doctor is faced with a patient. The health *Health*$_1$ of the patient at the time of the initial consultation is revealed only indirectly by the symptoms *Symptoms* exhibited by the patient. Based on the symptoms, the doctor must decide whether to perform a test *Test*. The test will produce a result *Result*, which the doctor observes before deciding on a treatment (if any) *Treat*. The health of the patient, after a possible treatment has been administered, is represented by the variable *Health*$_2$. As medical supplies are expensive both the *Test* and *Treat* decisions are associated with a monetary cost represented by the utility nodes U_2 and U_3. Furthermore, if a test is performed, it might

---

[1] [12] considers utilities that are defined as a linear combination of cost, insult and risk, for instance.

[2] Even though the word currencies is used here, we are not restricting ourselves to monetary currencies, but consider human lives, spare time etc. as currencies also. Also, in decision problems with several stakeholders, each currency can be seen as the utility for one specific stakeholder.

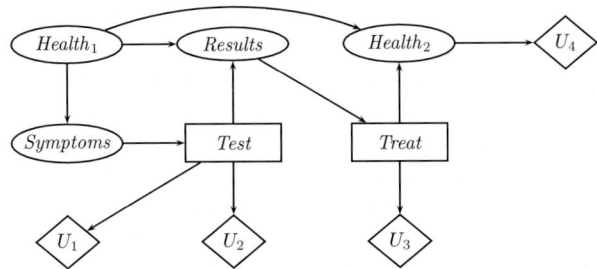

Fig. 1. Example of an ID.

be associated with a degree of pain, risk of mortality, or other side effects on behalf of the patient. This is represented by the utility node U_1. The $Health_2$ variable also has a preferred state (corresponding to the patient being well), which is encoded by the utility node U_4. .

Now, before the doctor calculates an optimal strategy for his decision problem, he needs to transform U_1, U_2, U_3, and U_4 onto a common scale – say, dollars. This involves estimating the monetary equivalents of the patient being ill and of him being subjected to a painful test. However, different transformations might produce differing optimal strategies. Therefore it would be advantageous to know:

1. What choice of conversion parameters has been the basis of the calculated optimal strategy?
2. What other conversion parameters would produce the same optimal strategy? If another stakeholder, such as the patient in this example, disagrees with the parameters, we could guarantee that even though there is disagreement on the exact choice of parameters, the identified set of parameters all render the same strategy optimal.

None of these questions can be answered from the ID alone.

As an example of how decision problems involving several stakeholders can be interpreted as a multi-currency problem, the ID introduced above can be seen as describing a conflict between the interests of the patient and the hospital (and/or the patient's medical insurance company) mediated by the doctor: By letting the patient specify utility values corresponding to utility nodes U_1 and U_4, and the hospital the ones for U_2 and U_3, we can investigate how the decisions of the doctor relate to these two stakeholders; he may be indulgent to please the patient more than the hospital or vice versa.

2 Influence Diagrams

An ID is a directed acyclic graph consisting of *chance* ($\boldsymbol{V_C}$), *decision* ($\boldsymbol{V_D}$), and *utility nodes* ($\boldsymbol{V_U}$), with the two constraints that a node has no children if and only if (iff) it is a utility node, and that there is a directed path encompassing all decision nodes in the diagram. A chance node (drawn as an ellipse) represents a *chance variable*, which is a discrete variable outside the DM's direct control. A decision node (drawn as a rectangle) represents a *decision variable* (or simply a *decision*), which is a discrete variable under the DM's direct control. A utility node (drawn as a diamond) represents a local utility function, and the set of local utility functions constitute the components in an additive factorisation of the global utility function [13]. When the meaning is obvious from the context, we use the terms node and variable interchangeably and do not distinguish between a variable or local utility function and the node representing it.

When speaking of IDs we denote by $pa(V)$ the set of nodes, which are parents of the node V, and by $\mathbf{sp}(V)$ we denote the set of *states* of variable V – states being outcomes for chance variables and decision options for decisions. For a set of variables, $\boldsymbol{S}$, we denote by $\mathbf{sp}(\boldsymbol{S})$ the configurations $\times_{V \in \boldsymbol{S}} \mathbf{sp}(V)$.

An arc in an ID represents either functional dependence, probabilistic dependence, or temporal precedence, depending on the type of node it goes into. In particular, an arc emanating from a node X going into a decision node D is a temporal precedence arc and states that X is observed or decided upon immediately before D is decided upon. *No-forgetting* is assumed so that variables observed or decided upon immediately before previous decisions are remembered at subsequent decisions. A chance variable, which is not a parent of any decision in the ID, is either never observed or observed after the last decision. The temporal precedence arcs impose a partial temporal ordering $\prec$ on $\boldsymbol{V}_C \cup \boldsymbol{V}_D$. Together with the requirement on a directed path through all decisions, this ordering induces a total temporal ordering on $\boldsymbol{V}_D$. For notational convenience, we assume that the decisions are labelled $D_1, \ldots, D_n$, such that $i < j$ implies $D_i \prec D_j$. Furthermore, we use the notation $\boldsymbol{C}_{i-1}$ to mean the set of chance variables observed immediately before deciding on D_i. By $\boldsymbol{C}_n$ we refer to the set of chance variables never observed (or observed after the last decision D_n has been decided upon). In summary we have

$$\boldsymbol{C}_0 \prec D_1 \prec \boldsymbol{C}_1 \prec \cdots \prec D_n \prec \boldsymbol{C}_n ,$$

and no-forgetting amounts to the assumption that for any D_k all observations of variables in $\cup_{i<k} \boldsymbol{C}_i$ and decisions taken for $D_1, \ldots, D_{k-1}$ are remembered when the DM decides on D_k. We define the *past* of decision D to be $\mathbf{past}(D) = \{V \in \boldsymbol{V}_C \cup \boldsymbol{V}_D \mid V \prec D\}$. We encode the quantitative aspects of the modelled decision problem as a set $\boldsymbol{\Phi}$ of conditional probability distributions and a set $\boldsymbol{\Psi}$ of local utility functions:

$$\boldsymbol{\Phi} = \{P(C|pa(C)) \mid C \in \boldsymbol{V}_C\}, \text{ and}$$
$$\boldsymbol{\Psi} = \{U(pa(U)) \mid U \in \boldsymbol{V}_U\} .$$

A pair $(\boldsymbol{\Phi}, \boldsymbol{\Psi})$ is called a *realisation* for the ID. Here, and henceforth, we have used $f(V_1, \ldots, V_k)$ to denote a function of the type $f : \mathbf{sp}(V_1) \times \cdots \times \mathbf{sp}(V_k) \to$

ℝ. Such a function is called a *potential*; we distinguish between *probability potentials*, denoted by ϕ's, and *utility potentials*, denoted by ψ's. Furthermore, the set of variables $\{V_1, \ldots, V_k\}$ is referred to as the *domain* of f, denoted $\mathrm{dom}(f)$.

Given an ID and a decision D a function, $\delta_D : \mathbf{sp}(\mathbf{past}(D)) \to \mathbf{sp}(D)$, is called a *policy* for D. A collection of policies for each decision in an ID,

$$\mathbf{\Delta} = \{\delta_D : \mathbf{sp}(\mathbf{past}(D)) \to \mathbf{sp}(D) \mid D \in \mathbf{V}_D\},$$

is called a *strategy* for the ID. Given a policy δ_D for a decision D we define the *chance variable policy*[2] for D $P_{\delta_D}(D|\mathbf{past}(D))$ as $P_{\delta_D}(d|\mathbf{c}) = 1$ if $\delta_D(\mathbf{c}) = d$ and 0 otherwise. An *optimal strategy* $\mathbf{\Delta}^*$ for an ID is a strategy that fulfills

$$\mathbf{\Delta}^* = \arg\max_{\mathbf{\Delta}} \sum_{\mathbf{c} \in \mathbf{sp}(\mathbf{V}_C \cup \mathbf{V}_D)} \left(\prod_{D \in \mathbf{V}_D} P_{\delta_D}(\mathbf{c}) \prod_{\phi \in \mathbf{\Phi}} \phi(\mathbf{c}) \sum_{\psi \in \mathbf{\Psi}} \psi(\mathbf{c}) \right). \qquad (1)$$

The individual policies in an optimal strategy are referred to as *optimal policies*. To denote that a policy for a decision D is a part of an optimal strategy, we write it as δ_D^*. The quantity that is maximised in (1) is the *expected utility* of the decision problem given the strategy $\mathbf{\Delta}$, and it is denoted $\mathrm{eu}(\mathbf{\Delta})$. Not all variables in the past of a decision D are necessarily relevant for D. Therefore, we call a variable V *required* for D, if there exists a realisation and a configuration $\mathbf{c}$ over the variables in $\mathbf{past}(D) \setminus \{V\}$, such that $\delta_D^*(\mathbf{c}, v_i) \neq \delta_D^*(\mathbf{c}, v_j)$ for two states v_i and v_j in $\mathbf{sp}(V)$. The set of required variables for a decision D we denote by $\mathbf{req}(D)$.[3] We may then redefine a policy to be a function $\delta_D : \mathbf{sp}(\mathbf{req}(D)) \to \mathbf{sp}(D)$.

When optimal policies are to be identified, it is usually easier to work with a recursive expression for the maximum expected utility instead of (1). An example is the *variable elimination* algorithm [5]: Define

$$\mathbf{\Phi}_{(n)} = \{\phi \in \mathbf{\Phi} \mid \mathrm{dom}(\phi) \cap (\mathbf{C}_i \cup \{D_i\}) \neq \varnothing\},$$

and similarly for $\mathbf{\Psi}_{(n)}$. Then

$$\delta_{D_n}^*(\mathbf{c}) = \arg\max_{d \in \mathbf{sp}(D_n)} \sum_{\mathbf{e} \in \mathbf{sp}(\mathbf{C}_n)} \prod_{\phi \in \mathbf{\Phi}_{(n)}} \phi(\mathbf{c}, d, \mathbf{e}) \sum_{\psi \in \mathbf{\Psi}_{(n)}} \psi(\mathbf{c}, d, \mathbf{e}), \qquad (2)$$

and we set

$$\phi_n(\mathbf{c}) = \sum_{\mathbf{e} \in \mathbf{sp}(\mathbf{C}_n)} \prod_{\phi \in \mathbf{\Phi}_{(n)}} \phi(\mathbf{c}, \delta_{D_n}^*(\mathbf{c}), \mathbf{e}), \qquad (3)$$

and

$$\psi_n(\mathbf{c}) = \left(\sum_{\mathbf{e} \in \mathbf{sp}(\mathbf{C}_n)} \prod_{\phi \in \mathbf{\Phi}_{(n)}} \phi(\mathbf{c}, \delta_{D_n}^*(\mathbf{c}), \mathbf{e}) \sum_{\psi \in \mathbf{\Psi}_{(n)}} \psi(\mathbf{c}, \delta_{D_n}^*(\mathbf{c}), \mathbf{e}) \right) / \phi_n(\mathbf{c}), \qquad (4)$$

---

[3] [9] provides an operational method for determining $\mathbf{req}(D)$ for any decision D in an ID.

for all configurations **c** over $\mathbf{req}(D_n)$. For all $i < n$ we recursively define

$$\boldsymbol{\Phi}_{(i)} = \{\phi \in \boldsymbol{\Phi} \cup \{\phi_{i+1}, \ldots, \phi_n\} \mid dom(\phi) \cap (\boldsymbol{C}_i \cup \{D_i\}) \neq \emptyset\} \setminus \cup_{j>i} \boldsymbol{\Phi}_{(j)} \quad (5)$$

and similarly for $\boldsymbol{\Psi}_{(i)}$ with $\psi_{i+1}, \ldots, \psi_n$. We then get

$$\delta_{D_i}^*(\mathbf{c}) = \arg\max_{d \in \mathbf{sp}(D_i)} \sum_{\mathbf{e} \in \mathbf{sp}(\boldsymbol{C}_i)} \prod_{\phi \in \boldsymbol{\Phi}_{(i)}} \phi(\mathbf{c}, d, \mathbf{e}) \sum_{\psi \in \boldsymbol{\Psi}_{(i)}} \psi(\mathbf{c}, d, \mathbf{e}) , \quad (6)$$

and set

$$\phi_i(\mathbf{c}) = \sum_{\mathbf{e} \in \mathbf{sp}(\boldsymbol{C}_i)} \prod_{\phi \in \boldsymbol{\Phi}_{(i)}} \phi(\mathbf{c}, \delta_{D_i}^*(\mathbf{c}), \mathbf{e}) , \quad (7)$$

and

$$\psi_i(\mathbf{c}) = \left(\sum_{\mathbf{e} \in \mathbf{sp}(\boldsymbol{C}_i)} \prod_{\phi \in \boldsymbol{\Phi}_{(i)}} \phi(\mathbf{c}, d, \mathbf{e}) \left(\sum_{\psi \in \boldsymbol{\Psi}_{(i)}} \psi(\mathbf{c}, d, \mathbf{e}) \right) \right) / \phi_i(\mathbf{c}) , \quad (8)$$

for all configurations **c** over $\mathbf{req}(D_i)$. We may then write the maximum expected utility of the ID as

$$\mathrm{eu}(\boldsymbol{\Delta}^*) = \sum_{\mathbf{e} \in \mathbf{sp}(\boldsymbol{C}_0)} \prod_{\phi \in \boldsymbol{\Phi}_{(0)}} \phi(\mathbf{e}) \sum_{\psi \in \boldsymbol{\Psi}_{(0)}} \psi(\mathbf{e}) .$$

Given an ID and a realisation, an optimal strategy may be found through the use of any one of a number of algorithms including [5; 7; 10], and [11], which all utilise the distributive and associative law on the expressions in (2) to (8).

3 Multi-currency Influence Diagrams

From the summations of utility potentials in (2) to (8), it is clear that these must be of the same type, i.e. defined over the same currency; in the ID framework this is tackled by transforming the different currencies into one currency during construction of the ID. We now introduce a framework capable of handling decision problems involving utilities of several currencies. We call models in this framework Multi-currency Influence Diagrams (MCIDs).

Basically, an MCID is just an ID where each of the utility nodes is annotated with the currency of the corresponding local utility function. Formally, the syntax and semantics of IDs described in Sect. 2 carry over to MCIDs, except for the requirement that the local utility functions must be an additive decomposition of the global utility function. By assuming some arbitrary, but fixed, order of the currencies in the MCID $s_1, \ldots, s_m$ we may refer to the currency of a local utility function by a natural number $i \in \{1, \ldots, m\}$.

In what follows we refer to the number of different currencies of an MCID as the *dimension* of the MCID, and throughout we assume this to be m. A realisation of an MCID is therefore a tuple $(\boldsymbol{\Phi}, \boldsymbol{\Psi}_1, \ldots, \boldsymbol{\Psi}_m)$, where $\boldsymbol{\Psi}_i$ is the set of local utility functions of currency i. We require that, for each $\boldsymbol{\Psi}_i$, its elements form an additive decomposition of a utility function, which encodes the same preference ordering as if the DM had disregarded all consequences measured in currencies different from i, and that there is some linear combination of these utility functions describing the decision makers preferences in full. Note that it follows that an MCID of dimension 1 is an ID, and hence from this point on we assume m to be larger than 1.

A strategy for an MCID is the same as for an ID: A prescription of choices given the required variables in the past. However, if we want to compute an optimal strategy for an MCID, we need a method for comparing amounts of one currency with amounts of another. This can be seen from the following example.

Example 2 (A Simple Example of an MCID). Consider a simple two-dimensional MCID, over the currencies A and B, with only a single binary decision D and two local utilities U_1 and U_2 defined as $U_1(D = d_1) = 1A$, $U_1(D = d_2) = 5A$, $U_2(D = d_1) = 2B$, and $U_2(D = d_2) = 1B$.

Both choices of D can be optimal choices depending on how much the DM values amounts of currency A relative to amounts of currency B. If $D = d_1$ should be an optimal strategy then $\mathrm{eu}(D = d_1) \geq \mathrm{eu}(D = d_2)$, which is equivalent to

$$1A + 2B \geq 5A + 1B \iff -4A + B \geq 0 ,$$

If we regard the currency name A as a real variable, representing the DM's degree of appreciation of amounts of A, and similarly for currency name B, then $-4 \cdot A + B$ corresponds to an amount of appreciation equivalent to -4 A's and one B. The set of all values for A and B, where $-4 \cdot A + B \geq 0$, then corresponds to the possible attitudes of the DM rendering d_1 the optimal choice of the decision problem modelled by the MCID. The state space of $A \times B$, viz. $\mathbb{R}^2$, is thus partitioned into two regions, each corresponding to an optimal strategy.

To sum up, we see that the payoff of following a specific strategy is an element of $\mathbb{R}^m$ (e.g. $(5, 1)$ in Ex. 2) rather than a scalar value as is the case with strategies for IDs. The means we use for comparing amounts of different currencies are called *currency mappings*:

Definition 1. *Let $\mathcal{I}$ be an MCID of dimension m and $\boldsymbol{\alpha} = (\alpha_1, \ldots, \alpha_m)$ a point in $\mathbb{R}^m$, then $\boldsymbol{\alpha}$ is a currency mapping (CM) for $\mathcal{I}$.*

The semantics of a CM $\boldsymbol{\alpha}$, *reflecting a DM's preferences*, is that, for any two amounts x_i and x_j of currencies i and j, respectively, we have that $\alpha_i x_i$ equals $\alpha_j x_j$ iff the DM values x_i of currency i as much as x_j of currency j. We also say that the DM *adheres* to $\boldsymbol{\alpha}$. This way α_i becomes a measure of appreciation for the DM of one unit of currency i. In a multi-stakeholder setting, α_i becomes a weight of importance attributed to satisfying the i'th stakeholder compared to the other stakeholders. In any case, the objective of the DM is to maximize the *expected global utility* of a strategy $\boldsymbol{\Delta}$ given by

$$\boldsymbol{\alpha} \cdot \mathrm{eu}(\boldsymbol{\Delta}) = \sum_{i=1}^{m} \alpha_i \sum_{\mathbf{c} \in \mathrm{sp}(V_C \cup V_D)} \left(\prod_{D \in V_D} P_{\delta_D}(\mathbf{c}) \prod_{\phi \in \boldsymbol{\Phi}} \phi(\mathbf{c}) \sum_{\psi_i \in \boldsymbol{\Psi}_i} \psi_i(\mathbf{c}) \right). \quad (9)$$

It follows that we assume the preferential relationship between currencies is a linear one. Thus, in Ex. 2, a DM adhering to the CM $(3,2)$, would have an expected global utility of $3 \cdot 1 + 2 \cdot 2 = 7$ for choosing d_1 and $3 \cdot 5 + 2 \cdot 1 = 17$ for choosing d_2. In order to maximize the expected global utility he should therefore choose d_2.

We use bold face Latin letters ($\mathbf{f}$, $\mathbf{q}$, etc.) to denote arbitrary points in $\mathbb{R}^m$, and Greek letters ($\boldsymbol{\alpha}$ and $\boldsymbol{\beta}$) to denote points when we want to emphasise that they are CMs. In general, we use q_i to refer to the i'th coordinate of a point $\mathbf{q}$. We shall not distinguish between a point $\mathbf{q}$ and the corresponding vector going from the origin to $\mathbf{q}$. In what follows we furthermore use $\mathbf{f} \cdot \mathbf{q}$ to denote the scalar product $\sum_i f_i q_i$.

In this paper we assume, without loss of generality, that each element of a currency mapping is positive, i.e. a CM is an element of $\mathbb{R}_+^m$ rather than $\mathbb{R}^m$, where $\mathbb{R}_+$ denotes the set of strictly positive reals. This assumption implies that everybody should be able to agree on whether amounts of each currency, i, is beneficial to be had or not, and that no-one would contest the relevance of amounts of any currency. If a currency i is disadvantageous to be had (meaning that α_i should be negative) we expect the modelled decision problem to have negative utilities specified for positive amounts of i, as is usually done when costs are specified in IDs.

If for a strategy, $\boldsymbol{\Delta}$, we have that $\boldsymbol{\alpha} \cdot \mathrm{eu}(\boldsymbol{\Delta})$ is greater than or equal to $\boldsymbol{\alpha} \cdot \mathrm{eu}(\boldsymbol{\Delta}')$ for all other strategies $\boldsymbol{\Delta}'$, it means that a DM adhering to $\boldsymbol{\alpha}$ appreciates the expected utility of following $\boldsymbol{\Delta}$ at least as much as that of following any other strategy, and we consequently say that $\boldsymbol{\Delta}$ is optimal given the CM $\boldsymbol{\alpha}$. As can be seen from (9), and the distributive law of the scalar product, the optimal strategy for $\boldsymbol{\alpha}$ is equivalent to the one obtained by solving an ID resulting from multiplying the individual utilities in the MCID with $\boldsymbol{\alpha}$ beforehand. We denote an optimal strategy for an MCID given a CM $\boldsymbol{\alpha}$ as $\boldsymbol{\Delta}_{\boldsymbol{\alpha}}^*$. As several strategies might give rise to the same expected utility, the set of all optimal strategies corresponding to $\boldsymbol{\alpha}$ is denoted as $\overline{\boldsymbol{\Delta}_{\boldsymbol{\alpha}}^*}$.

4 Support Analysis of MCIDs

Clearly, if we are given a CM $\boldsymbol{\alpha}$ in addition to an MCID we can solve it by means of simply converting the MCID into an ID, through multiplying each local utility function of currency i by α_i, and then solving the resulting ID. This simple solution method allows for optimal strategies to be computed for any $\boldsymbol{\alpha}$, and the distinction between MCIDs and CMs emphasises the assumptions leading to the results. However, we would not be closer to answering the second question in the motivating example, viz. what other currency mappings would lead to the same optimal strategy? In order to do this, we must render the effect of $\boldsymbol{\alpha}$ on the optimal strategy transparent. We would then be able to reason about the universality of the applicability of that optimal strategy. We obtain this transparency by postponing the conversion of utilities until it is needed, so that we can analyse the requirements these conversions bestow upon those CMs giving rise to the same optimal strategy.

4.1 Preliminaries

Given an MCID and a strategy $\boldsymbol{\Delta}$, we call the set $\mathbf{su}(\boldsymbol{\Delta}) = \{\boldsymbol{\beta} \in \mathbb{R}_+^m \mid \boldsymbol{\Delta} \in \overline{\boldsymbol{\Delta}_{\boldsymbol{\beta}}^*}\}$ the *support* of $\boldsymbol{\Delta}$. Intuitively, $\mathbf{su}(\boldsymbol{\Delta})$ is the set of CMs for which $\boldsymbol{\Delta}$ is an optimal strategy. We refer to the process of calculating the support of an optimal strategy as performing support analysis of the MCID. In Ex. 2 we actually found the support of both strategies $D = d_1$ and $D = d_2$ (the two partitions of $\mathbb{R}_+^2$ defined by $-4A + B = 0$). Later it will become apparent that any such support can be described as an intersection of partitions of $\mathbb{R}_+^m$ – each partition described by a linear inequality.

As mentioned in the beginning of this section, we postpone the conversion of utility potentials until needed. Hence, we introduce a new type of potential, which can represent utility functions of several currencies:[4] A *multi-currency utility potential* (MCUP) of dimension m over the variables in a set $\boldsymbol{S}$ is a function $\theta : \mathbf{sp}(\boldsymbol{S}) \to \mathbb{R}^m$ attributing to each configuration $\mathbf{c}$ over the variables in $\boldsymbol{S}$ a measure of utility $(\theta(\mathbf{c}))_i$ of each currency i. In Ex. 2 we could have exchanged the two utility functions U_1 and U_2 with the two MCUPs θ_1 and θ_2, where $\theta_1(d_1) = (1,0)$, $\theta_1(d_2) = (5,0)$, $\theta_2(d_1) = (0,2)$, $\theta_2(d_2) = (0,1)$. For a DM whose preferences are reflected by the CM $\boldsymbol{\beta}$, $\boldsymbol{\beta} \cdot \theta(\boldsymbol{S})$ is a utility potential that for any configuration, $\mathbf{c}$, over variables in $\boldsymbol{S}$ yields the value $\beta_1(\theta(\mathbf{c}))_1 + \cdots + \beta_m(\theta(\mathbf{c}))_m$, which to the DM is equivalent to the amounts $(\theta(\mathbf{c}))_1, \ldots, (\theta(\mathbf{c}))_m$ of currencies $1, \ldots, m$, respectively. When adding MCUPs or multiplying probability potentials onto them, we simply treat each dimension of the MCUP as a regular utility potential, and perform the operation on each dimension separately. For instance, in Ex. 2 we have that $\theta_1 + \theta_2$ is the MCUP θ_+ where $\theta_+(d_1) = (1,2)$ and $\theta_+(d_2) = (5,1)$, which to a DM adhering to $\boldsymbol{\beta} = (3,2)$ would be equivalent to a utility potential ψ, where $\psi(d_1) = 7$ and $\psi(d_2) = 17$.

4.2 Support Analysis

We are now ready to give a procedure for performing support analysis of an MCID. We describe the method first, and give an example of its application

---

[4] Such potentials are not part of the MCID framework as such, but rather data structures used by the proposed method for doing support analysis.

afterwards. We assume the existence of an optimal strategy Δ_α^* determined by some initial currency mapping α as well as a realisation $(\Phi, \Psi_1, \ldots, \Psi_m)$, and we look for the support of Δ_α^* for this realisation. The method is inspired by that of Lazy evaluation presented in [7] and basically traces the steps of this method while recording the requirements on α for Δ_α^* to be an optimal strategy. The method consists of an initialisation phase and an identification phase. The initialisation phase consists of two steps: First, two empty sets Ξ and Θ are generated, where Ξ will hold inequalities defining $\mathbf{su}(\Delta_\alpha^*)$, and Θ is a container for MCUPs used in the identification phase. Second, for each currency i each potential ψ in Ψ_i is converted to a MCUP θ and put into Θ, such that $\theta_k = \psi$ if $k = i$ and 0 otherwise, where 0 denotes the function yielding the zero value for all input.

The identification phase follows the Lazy evaluation method, except for the steps normally carried out when a variable is eliminated (see Algorithm 1). The major difference between these steps and the corresponding steps in Lazy evaluation is that we do not perform a maximisation to uncover an optimal strategy. Instead we look for a set of linear inequalities (Step 4) that need to be fulfilled if Δ_α^* is to be optimal. We refer to the inequalities in Ξ as *constraints*, since they constrain the support set.

Algorithm 1: *The elimination steps of the identification phase. The strategy Δ_α^* is assumed to be given apriori.*

---

1. Let V be the variable to be eliminated, and let Φ_V denote the set of probability potentials in Φ with V in their domain, and Θ_V denote the set of MCUPs in Θ with V in their domain.
2. Let
$$\phi_V = \prod_{\phi \in \Phi_V} \phi, \quad \text{and} \quad \theta_V = \sum_{\theta \in \Theta_V} \theta \,.$$
3. If V is a chance variable, then set
$$\Phi \leftarrow (\Phi \setminus \Phi_V) \cup \left\{ \sum_V \phi_V \right\},$$
and
$$\Theta \leftarrow (\Theta \setminus \Theta_V) \cup \left\{ \frac{\sum_V (\phi_V \theta_V)}{\sum_V \phi_V} \right\}.$$
4. If V is a decision variable, then set
$$\Phi \leftarrow (\Phi \setminus \Phi_V) \cup \{\phi(V = v) \mid \phi \in \Phi_V\},$$
where v is some arbitrary state of V^5, and
$$\Theta \leftarrow (\Theta \setminus \Theta_V) \cup \{\theta_V(\delta_V^*)\} \,,$$

---

[5] Note that any potential in Φ_V must be constant over V

where δ_V^* is the appropriate element of $\boldsymbol{\Delta}_{\boldsymbol{\alpha}}^*$. For each configuration $\mathbf{c}$ over the variables in $\mathbf{req}(V)$ and state $v \neq \delta_V^*(\mathbf{c})$ of V, set

$$\Xi \leftarrow \Xi \cup \{\mathbf{f}_{\mathbf{c},v} \cdot \boldsymbol{\gamma} \geq 0\}, \tag{10}$$

where $\mathbf{f}_{\mathbf{c},v}$ denotes $\theta(\mathbf{c}, \delta_V^*(\mathbf{c})) - \theta(\mathbf{c}, v)$.

Before proceeding, we illustrate the workings of the algorithm by an example:

Example 3. We specify the ID from Ex. 1 as an MCID. We assume that the state space of the variables are as follows:

$$\mathbf{sp}(Health_1(H_1)) = \{bad(b_1), healthy(h_1)\},$$
$$\mathbf{sp}(Symptoms(S)) = \{symptoms(sy), none(no)\},$$
$$\mathbf{sp}(Test(Te)) = \{test(te), no\ test(nte)\},$$
$$\mathbf{sp}(Results(R)) = \{positive(po), negative(ne), no\ result(nr)\},$$
$$\mathbf{sp}(Treat(Tr)) = \{treat(tr), no\ treatment(ntr)\}, \text{ and}$$
$$\mathbf{sp}(Health_2(H_2)) = \{bad(b_2), healthy(h_2)\}.$$

The three currencies we work with are, and are ordered as *comfort(c)*, *dollars($)*, and *health(h)*. The realization that we work with is defined by the following parameters:

$P(b_1) = 0.5$, $P(sy|b_1) = 0.95$, $P(sy|h_1) = 0.1$, $P(nr|nte, \cdot) = 1$,
$P(nr|te, \cdot) = 0$, $P(po|te, b_1) = 0.99$, $P(po|te, h_1) = 0.01$,
$P(b_2|b_1, tr) = 0.001$, $P(b_2|b_1, ntr) = 0.999$, $P(b_2|h_1, tr) = 0$,
$P(b_2|h_1, ntr) = 0.0001$, $U_1(te, nte) = (-1c, 0)$, $U_2(te, nte) = (-\$1000, 0)$,
$U_3(tr, ntr) = (-\$10000, 0)$, and $U_4(b_2, h_2) = (-1h, 0)$.

We assume that the doctor is adhering to a CM $\boldsymbol{\alpha} = (10, 1, 100000)$ meaning that he regards discomfort on behalf of the patient as bad as the loss of \$10 and the death of the patient as bad as a loss of \$100000. The CM corresponds to a strategy prescribing *treatment* either if a *test* was conducted, and a *positive* test result was gotten, or *no test* was conducted and there was *symptoms*, and *no treatment* otherwise.

In the initialization part of the algorithm, we construct an empty set Ξ and convert the four utility potentials into a set of four MCUPs

$$\Theta = \{\theta_1(te, nte) = ([-1, 0, 0], [0, 0, 0]), \theta_2(te, nte) = ([0, -1000, 0], [0, 0, 0]),$$
$$\theta_3(tr, ntr) = ([0, -10000, 0], [0, 0, 0]), \theta_4(b_2, h_2) = ([0, 0, -1], [0, 0, 0])\}.$$

Next we perform variable elimination in an order that respects the $\prec$-ordering of the MCID: First $Health_2$, then $Health_1$, $Treat$, $Results$, $Test$, and lastly

Symptoms. Marginalizing out *Health*$_2$, according to Steps 2 and 3 results in dropping $P(H_2|H_1, Tr)$ from the set of probability potentials and replacing θ_4 with a new MCUP:

$$\theta_{H_1,Tr} = \frac{\sum_{H_2} P(H_2|H_1,Tr) \cdot \theta_4}{\sum_{H_2} P(H_2|H_1,Tr)},$$

in which $\theta_{H_1,Tr}(b_1, tr) = [0, 0, -10^{-3}]$, $\theta_{H_1,Tr}(b_1, ntr) = [0, 0, -0.999]$, $\theta_{H_1,Tr}(h_1, tr) = [0, 0, 0]$, and $\theta_{H_1,Tr}(h_1, ntr) = [0, 0, -10^{-4}]$. Further marginalizing out *Health*$_1$ (again according to Steps 2 and 3) replaces all probability potentials with

$$\phi_{R,S,Te} = \sum_{H_1} P(H_1) \cdot P(S|H_1) \cdot P(R|H_1, Te)$$

and $\theta_{H_1,Tr}$ with

$$\theta_{R,S,Te,Tr} = \frac{\sum_{H_1} P(H_1) \cdot P(S|H_1) \cdot P(R|H_1, Te) \cdot \theta_{H_1,Tr}}{\phi_{S,R,Te}}.$$

This last potential is shown in Table 1.

Table 1. $\theta_{R,S,Te,Tr}$

		te		nte	
		tr	ntr	tr	ntr
po	sy	$[0, 0, -9.99 \cdot 10^{-4}]$	$[0, 0, -9.98 \cdot 10^{-1}]$	$[0, 0, 0]$	$[0, 0, 0]$
	no	$[0, 0, -8.46 \cdot 10^{-4}]$	$[0, 0, -8.45 \cdot 10^{-1}]$	$[0, 0, 0]$	$[0, 0, 0]$
ne	sy	$[0, 0, -8.76 \cdot 10^{-4}]$	$[0, 0, -8.76 \cdot 10^{-2}]$	$[0, 0, 0]$	$[0, 0, 0]$
	no	$[0, 0, -5.61 \cdot 10^{-7}]$	$[0, 0, -6.60 \cdot 10^{-4}]$	$[0, 0, 0]$	$[0, 0, 0]$
nr	sy	$[0, 0, 0]$	$[0, 0, 0]$	$[0, 0, -9.05 \cdot 10^{-4}]$	$[0, 0, -9.04 \cdot 10^{-1}]$
	no	$[0, 0, 0]$	$[0, 0, 0]$	$[0, 0, -5.26 \cdot 10^{-5}]$	$[0, 0, -5.27 \cdot 10^{-2}]$

When marginalizing *Treat*, we construct the MCUP $\theta^+_{R,S,Te,Tr} = \theta_3 + \theta_{R,S,Te,Tr}$ (shown in Table 2), according to Step 2, and by marginalizing *Treat* out of this, according to Step 4 and Δ^*_α, we end up with a new MCUP $\theta_{R,S,Te}$ (not shown), but also add 12 constraints to Ξ – one for each configuration over *Results*, *Symptoms*, and *Test*. For instance, for the configuration (po, sy, te), we construct the constraint ($\boldsymbol{\gamma} = (\gamma_c, \gamma_\$, \gamma_h)$)

$$((0, -10000, -9.99 \cdot 10^{-4}) - (0, 0, -9.98 \cdot 10^{-1}))\boldsymbol{\gamma} \geq 0$$
$$-10000\gamma_\$ + 9.97 \cdot 10^{-1}\gamma_h \geq 0,$$

which is satisfied by $\boldsymbol{\alpha}$, as can easily be verified. After the other 11 constraints have been calculated the algorithm continues with elimination of the remaining variables.

Table 2. $\theta^+_{R,S,Te,Tr}$

		te		nte	
		tr	ntr	tr	ntr
po	sy	$[0, -10^4, -9.99 \cdot 10^{-4}]$	$[0, 0, -9.98 \cdot 10^{-1}]$	$[0, -10^4, 0]$	$[0, 0, 0]$
	no	$[0, -10^4, -8.46 \cdot 10^{-4}]$	$[0, 0, -8.45 \cdot 10^{-1}]$	$[0, -10^4, 0]$	$[0, 0, 0]$
ne	sy	$[0, -10^4, -8.76 \cdot 10^{-4}]$	$[0, 0, -8.76 \cdot 10^{-2}]$	$[0, -10^4, 0]$	$[0, 0, 0]$
	no	$[0, -10^4, -5.61 \cdot 10^{-7}]$	$[0, 0, -6.60 \cdot 10^{-4}]$	$[0, -10^4, 0]$	$[0, 0, 0]$
nr	sy	$[0, -10^4, 0]$	$[0, 0, 0]$	$[0, -10^4, -9.05 \cdot 10^{-4}]$	$[0, 0, -9.04 \cdot 10^{-1}]$
	no	$[0, -10^4, 0]$	$[0, 0, 0]$	$[0, -10^4, -5.26 \cdot 10^{-5}]$	$[0, 0, -5.27 \cdot 10^{-2}]$

As presented here, the algorithm presupposes that $\boldsymbol{\Delta}^*_\alpha$ has been computed beforehand. Alternatively, the Lazy evaluation algorithm itself can easily be interleaved by inserting the following step prior to Step 4 in Algorithm 1:

* For each configuration **c** over the variables in **req**(V) set

$$\delta^*_V(\mathbf{c}) = \arg\max_{v \in \mathbf{sp}(V)} \boldsymbol{\alpha} \cdot \theta_V(\mathbf{c}, v) .$$

Although the method, as presented here, follows the structure of the Lazy evaluation method, it can easily be adapted to follow the structure of any other solution method that is based on the expressions in (2) to (8). We conjecture that any such adaptation would identify the support set as long as the constraints are identified and stored. With or without this modification, though, we have the following important result:

Theorem 1. *Let $\boldsymbol{\beta}$ be in $\mathbb{R}^m_+$, $\boldsymbol{\Delta}$ a strategy, and $\boldsymbol{\Xi}$ the result of running the method described above on $\boldsymbol{\Delta}$. Then $\boldsymbol{\beta}$ is an element of $\mathbf{su}(\boldsymbol{\Delta})$ iff $\boldsymbol{\beta}$ satisfies all inequalities in $\boldsymbol{\Xi}$.*

Proof. The "if" part of the theorem is obvious. We therefore only show the "only if" part, viz. that if $\boldsymbol{\beta}$ fails to satisfy at least one constraint in $\boldsymbol{\Xi}$, then $\boldsymbol{\Delta}$ cannot be an optimal strategy for a DM adhering to $\boldsymbol{\beta}$, and hence that $\boldsymbol{\beta}$ is not an element of $\mathbf{su}(\boldsymbol{\Delta})$.

Without loss of generality, assume that $f \in \boldsymbol{\Xi}$ is the first constraint not satisfied by $\boldsymbol{\beta}$ that is identified by the algorithm. This happens during elimination of some decision D_i in Step 4 of the algorithm, with some MCUP θ_{D_i} having been calculated during Step 2. For some configuration **c** over $\text{dom}(\theta_{D_i}) \setminus \{D_i\}$ and state $v \neq \delta_{D_i}$ we must have that f is

$$(\theta_{D_i}(\mathbf{c}, \delta_{D_i}(\mathbf{c})) - \theta_{D_i}(\mathbf{c}, v)) \cdot \boldsymbol{\gamma} \geq 0 .$$

Since $\boldsymbol{\beta}$ fails to satisfy this constraint, it follows that

$$(\theta_{D_i}(\mathbf{c}, \delta_{D_i}(\mathbf{c})) - \theta_{D_i}(\mathbf{c}, v)) \cdot \boldsymbol{\beta} \not\geq 0 ,$$

which is equivalent to

$$\theta_{D_i}(\mathbf{c}, \delta_{D_i}(\mathbf{c})) \cdot \boldsymbol{\beta} < \theta_{D_i}(\mathbf{c}, v) \cdot \boldsymbol{\beta} \,.$$

We construct the strategy $\boldsymbol{\Delta}'$ obtained from $\boldsymbol{\Delta}$ by letting $\delta'_{D_i}(\mathbf{c}) = v$ and leaving all other policies intact. We then have that $\mathrm{eu}(\boldsymbol{\Delta}') \cdot \boldsymbol{\beta}$ is strictly greater than $\mathrm{eu}(\boldsymbol{\Delta}) \cdot \boldsymbol{\beta}$, and hence that $\boldsymbol{\Delta}$ cannot be an optimal strategy for a DM adhering to $\boldsymbol{\beta}$.

Glancing over the constraints identified in Ex. 3 it is clear that each constraint $f : a\gamma_1 + b\gamma_2 + c\gamma_3 \geq 0$ defines a hyperplane in $\mathbb{R}^3$ given by the equation $a\gamma_1 + b\gamma_2 + c\gamma_3 = 0$. As the zero vector $\mathbf{0} = (0,0,0)$ is a satisfying solution to each equation, it follows that all these hyperplanes must pass through the origin of $\mathbb{R}^3$, and hence that the points satisfying all constraints must lie in a bottomless pyramid extending from the origin, as illustrated in Fig. 2(a). A corresponding visualization for a two-dimensional MCID is shown in Fig. 2(b). That the support of a strategy extends indefinitely from the origin as a pyramid, also makes sense from a purely semantical point of view: If a CM is scaled by multiplying each entry by some positive constant, the relative difference in appreciation between amounts of the individual currencies, for a DM adhering to this CM, stays the same. Hence, if a CM renders some strategy optimal, that strategy should also be optimal for any positively scaled version of this CM. From a more formal point of view, we also have that, for any CM $\boldsymbol{\alpha}$ and two strategies $\boldsymbol{\Delta}_i$ and $\boldsymbol{\Delta}_j$, if $\boldsymbol{\alpha} \cdot \mathrm{eu}(\boldsymbol{\Delta}_i) \geq \boldsymbol{\alpha} \cdot \mathrm{eu}(\boldsymbol{\Delta}_j)$ then it must necessarily be the case that $c\boldsymbol{\alpha} \cdot \mathrm{eu}(\boldsymbol{\Delta}_i) \geq c\boldsymbol{\alpha} \cdot \mathrm{eu}(\boldsymbol{\Delta}_j)$ for any $c > 0$, and hence that the pyramidal forms of support areas are what we should expect.

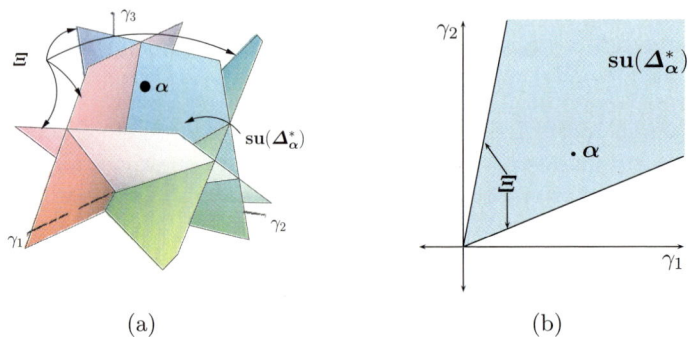

(a)   (b)

Fig. 2. Supports for a three-dimensional (**a**) and a two-dimensional MCID (**b**).

5 Finding a Minimal Support Set

The procedure described above finds the support of a strategy for a given CM and stores it as a set of constraints $\varXi$. The cardinality of $\varXi$ is given by

$$|\varXi| = \sum_{D \in V_D} |\mathbf{sp}(\mathbf{req}(D))|(|\mathbf{sp}(D)| - 1) \ ,$$

and storing it requires an amount of memory proportional to $m \cdot |\varXi|$. This size can be problematic for larger decision problems – both in terms of memory requirements and in terms of representing the resulting $\varXi$ to a human DM in a comprehensible manner.

We have devised a method for keeping the size of $\varXi$ minimal during the analysis. By minimal we mean that $\varXi$ does not include a constraint f, such that the set $\varXi \setminus \{f\}$ defines the same volume as $\varXi$. Conversely, if such a constraint f is in $\varXi$, we call it *superfluous*. Traditionally, such a task would be carried out by repeated applications of linear programming, but this can be done more efficiently when the number of dimensions m is significantly smaller than the number of decisions in the MCID. This seems to be the case in most examples of multi-currency decision problems in the literature, see e.g. [1] and [12].

The approach is purely geometric and rests on the already mentioned fact that all constraints in $\varXi$ define a pyramid extending from the origin of $\mathbb{R}^m$ (see Figs. 2(a) and 3(a)), and a black box view of the support analysis as a simple constraint generating process. For sake of clarity, we first describe the method informally in the intuitively understood three dimensional setting, and then give a formal presentation of the more general m-dimensional setting.

As the support of a strategy is a pyramid extending from the origin, we can represent the volume defined by $\varXi$ as a set of lines extending from the origin, each corresponding to the intersection of two hyperplanes defined by constraints in $\varXi$ (see Fig. 3(a), which illustrates the support defined by the initial set of constraints $\gamma_1 \geq 0$, $\gamma_2 \geq 0$, and $\gamma_3 \geq 0$, as well as an additional constraint f). We refer to such lines as *edges* of the pyramid, and we denote the set of them as $\boldsymbol{E_\varXi}$. When a new constraint g is to be added to $\varXi$ (see Fig. 3(b)) it may be superfluous. With the alternative representation this question can be restated as whether the points on any of the edges in $\boldsymbol{E_\varXi}$ fail to satisfy g. If this is the case, g is not superfluous. In Fig. 3(b) we have that there was originally a pyramid defined by the edges e_1, e_2, e_3, and e_4, but as all points (but the origin) on e_1 and e_2 fail to satisfy the new constraint g, it is not superfluous. Therefore it is added to $\varXi$, which now defines a new pyramid whose edges are e_3, e_4, e_5, and e_6. Thus $\boldsymbol{E_\varXi}$ is updated by dropping e_1 and e_2 and adding e_5 and e_6. This means that the original constraint $\gamma_1 \geq 0$ no longer participates in defining any of the edges in $\boldsymbol{E_\varXi}$, and *it* is therefore superfluous now. Consequently, $\gamma_1 \geq 0$ must be dropped from $\varXi$ to keep it minimal (see Fig. 3(c)). That is, in this alternative representation, we can also detect when old constraint are rendered superfluous by new constraints.

There is one main hurdle to be overcome by an implementation: Whenever a new constraint h is identified, the edges in $\boldsymbol{E_\varXi}$ need to be checked for points not satisfying h. By storing the pyramid as a list of edges sorted according to their angular distance to $\boldsymbol{\alpha}$ (see Fig. 4(a)), and calculating the minimal angular

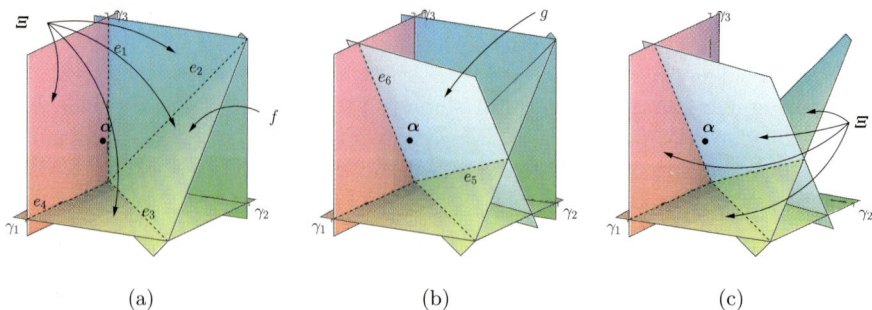

Fig. 3. Keeping Ξ minimal when a new constraint g is added. Edges are shown as dashed lines.

distance from $\boldsymbol{\alpha}$ to the hyperplane defined by h, we can quickly determine those edges that can contain points not satisfying h, i.e. those with an angular distance greater than the distance to the hyperplane defined by h. If no such edges exist (as in Fig. 4(b)), h is clearly superfluous. If such edges do exist, they have to be checked against h one at the time, and only if all of them satisfy h, it must be superfluous. However, as the set of constraints grow, the girth of the pyramid becomes smaller, and we would therefore expect that more new constraints fail the initial check of being closer to $\boldsymbol{\alpha}$ than any of the edges in Ξ, and that this further check by enumeration can be avoided.

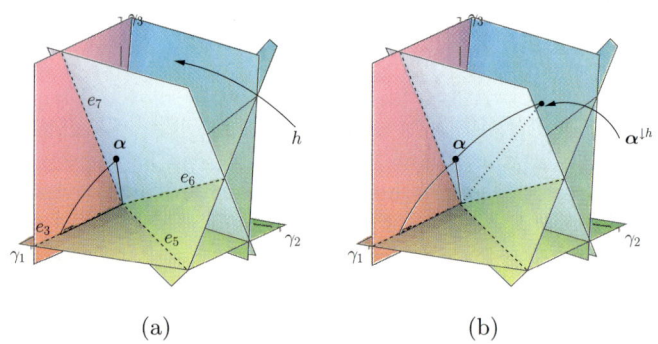

Fig. 4. Rejecting a constraint, h, because of its angular distance to $\boldsymbol{\alpha}$.

To describe the approach precisely, we formalize and generalize the discussion above: First, if $f(\boldsymbol{\gamma}) = \mathbf{f} \cdot \boldsymbol{\gamma} \geq 0$ is a constraint in Ξ and $\boldsymbol{\beta}$ is some point in $\mathbb{R}^m$, then we use $f(\boldsymbol{\beta})$ to state that $\mathbf{f} \cdot \boldsymbol{\beta} \geq 0$ and $\neg f(\boldsymbol{\beta})$ to state that $\mathbf{f} \cdot \boldsymbol{\beta} < 0$. Furthermore, we denote the hyperplane $\{\boldsymbol{\beta} \in \mathbb{R}^m \mid \mathbf{f} \cdot \boldsymbol{\beta} = 0\}$ as $\boldsymbol{H}(f)$, and talk of Ξ as consisting of hyperplanes when it introduces no ambiguity. A constraint f in Ξ is then defined to be *superfluous* in Ξ if there exists no point $\boldsymbol{\beta}$ in $\mathbb{R}^m$, such that $\neg f(\boldsymbol{\beta})$ and $g(\boldsymbol{\beta})$ for all $g \neq f$ in $\boldsymbol{\Xi}$. If no constraint in a set of constraints $\boldsymbol{\Xi}$ is superfluous, we say that $\boldsymbol{\Xi}$ is *minimal*.

The *angular distance* from $\boldsymbol{\beta}$ to $\boldsymbol{\alpha}$ is given by

$$\angle\boldsymbol{\beta}\boldsymbol{\alpha} = \cos^{-1} \frac{\boldsymbol{\beta} \cdot \boldsymbol{\alpha}}{\|\boldsymbol{\beta}\| \cdot \|\boldsymbol{\alpha}\|} .$$

Here and henceforth $\|\mathbf{q}\|$ denotes the Euclidean length of the vector $\|\mathbf{q}\|$. Moreover, we say that the set of constraints $\boldsymbol{\Xi}$ constitutes a *pyramid*, if for all points $\boldsymbol{\beta}$ in $\mathbb{R}^m$, where the angular distance to $\boldsymbol{\alpha}$ is greater than $90°$, there exists at least one constraint f in $\boldsymbol{\Xi}$, such that $\neg f(\boldsymbol{\beta})$. The previously introduced assumption of only considering CMs in $\mathbb{R}^m_+$, we restate as an inclusion of the m constraints $\gamma_i \geq 0$ in $\boldsymbol{\Xi}$ from the outset of the support analysis. This ensures that $\boldsymbol{\Xi}$ is a pyramid at the beginning of the support analysis.

The triangle inequality ensures that the points in a pyramid having the largest angular distance to $\boldsymbol{\alpha}$ must lie at the intersection of a number of hyperplanes in $\boldsymbol{\Xi}$. We need to consider intersections of exactly $m-1$ hyperplanes. The reason is that an intersection of k non-parallel hyperplanes in $\mathbb{R}^m$ describe an $m-k$ dimensional subspace of $\mathbb{R}^m$, and the points in this subspace will have varying angular distances to $\boldsymbol{\alpha}$ unless the subspace has dimension 1, and this is only the case when k is 1. If part of such an intersection lies within $\boldsymbol{\Xi}$, we refer to that part as an *edge*. An edge determined by constraints $f_1, \ldots, f_{m-1}$ we represent by a pair $e = (\boldsymbol{I}, \mathbf{p})$, where $\boldsymbol{I} = \{f_1, \ldots, f_{m-1}\}$ and $\mathbf{p} \in \mathbb{R}^m_+ \cap \boldsymbol{H}(f_1) \cap \cdots \cap \boldsymbol{H}(f_{m-1})$. The set of edges of a pyramid $\boldsymbol{\Xi}$ is denoted $\boldsymbol{E}_{\boldsymbol{\Xi}}$, and we assume it to be kept sorted such that $(\boldsymbol{I}_i, \mathbf{p}_i)$ is stored before $(\boldsymbol{I}_j, \mathbf{p}_j)$ only if $\angle\mathbf{p}_i\boldsymbol{\alpha} \leq \angle\mathbf{p}_j\boldsymbol{\alpha}$. Finally, for any constraint f, we denote by $\boldsymbol{\alpha}^{\downarrow f}$ the projection of $\boldsymbol{\alpha}$ onto the subspace $\boldsymbol{H}(f)$. With these terms specified, we can now present the proposed method in Algorithm 2.

Algorithm 2: *Takes as input a minimal pyramid $\boldsymbol{\Xi}$ with a sorted set of edges $\boldsymbol{E}_{\boldsymbol{\Xi}}$ and a constraint f. Outputs a new minimal pyramid $\boldsymbol{\Xi}'$, describing the same volume as $\boldsymbol{\Xi} \cup \{f\}$, along with a sorted set of edges $\boldsymbol{E}_{\boldsymbol{\Xi}'}$.*

---

1. Partition $\boldsymbol{E}_{\boldsymbol{\Xi}}$ into two sets $\boldsymbol{E}_+$ and $\boldsymbol{E}_-$, such that each edge in $\boldsymbol{E}_+$ has angular distance to $\boldsymbol{\alpha}$ less than or equal to $\angle\boldsymbol{\alpha}\boldsymbol{\alpha}^{\downarrow f}$.
2. Move each edge $(\boldsymbol{I}, \mathbf{p})$ in $\boldsymbol{E}_-$, where $f(\mathbf{p})$, from $\boldsymbol{E}_-$ to $\boldsymbol{E}_+$.
3. If $\boldsymbol{E}_- = \emptyset$ then stop and return $\boldsymbol{\Xi}' = \boldsymbol{\Xi}$ and $\boldsymbol{E}_{\boldsymbol{\Xi}'} = \boldsymbol{E}_{\boldsymbol{\Xi}}$.
4. Let the new set of constraints be

$$\boldsymbol{\Xi}' = \{f\} \cup \bigcup_{(\boldsymbol{I}, \mathbf{p}) \in \boldsymbol{E}_+} \boldsymbol{I},$$

and the set of constraints possibly defining new edges be

$$\boldsymbol{\Xi}_N = \left(\boldsymbol{\Xi}' \cap \bigcup_{(\boldsymbol{I}, \mathbf{p}) \in \boldsymbol{E}_-} \boldsymbol{I} \right) \cup \{f\} .$$

Then let

$$\boldsymbol{E}_{\boldsymbol{\Xi}'} = \boldsymbol{E}_+ \cup \{(\boldsymbol{I}, \mathbf{p}_{\boldsymbol{I}}) \mid \boldsymbol{I} \subseteq \boldsymbol{\Xi}_N, \ f \in \boldsymbol{I}, \ |\boldsymbol{I}| = m-1, \ \text{and} \ g(\mathbf{p}_{\boldsymbol{I}}) \ \forall g \in \boldsymbol{\Xi}'\},$$

where $\mathbf{p}_{\boldsymbol{I}}$ is the unique point in $\mathbb{R}^m_+ \cap (\cap_{h \in \boldsymbol{I}} \boldsymbol{H}(h))$ for which $\|\mathbf{p}_{\boldsymbol{I}}\| = 1$.
5. Return $\boldsymbol{\Xi}'$ and $\boldsymbol{E}_{\boldsymbol{\Xi}'}$.

Proposition 1. *Let $\boldsymbol{\Xi}$ constitute a minimal pyramid and f be a constraint not in $\boldsymbol{\Xi}$. Then the constraints in $\boldsymbol{\Xi} \cup \{f\}$ define the same region as those in $\boldsymbol{\Xi}'$ obtained from using Algorithm 2 to add f to $\boldsymbol{\Xi}$. Furthermore, $\boldsymbol{\Xi}'$ is a minimal pyramid.*

The complexity of inserting an element in the sorted set $\boldsymbol{E}_{\boldsymbol{\Xi}}$ is $O(\log|\boldsymbol{E}_{\boldsymbol{\Xi}}|)$, and we have a maximum of $\binom{|\boldsymbol{\Xi}|}{m-2}$ new edges to add to $\boldsymbol{E}_{\boldsymbol{\Xi}}$ in Step 4, in effect yielding a worst-case complexity of $O(\binom{|\boldsymbol{\Xi}|}{m-2} \log|\boldsymbol{E}_{\boldsymbol{\Xi}}|)$ for insertion of a constraint. This is an improvement with respect to the complexity of $(|\boldsymbol{\Xi}|+1)m^{|\boldsymbol{\Xi}|+1}$, offered by using simplex repeatedly [6], especially when the dimension m is low.

6 Presenting the Support to a Human DM

Given that $\boldsymbol{\Xi}$ has been identified as in Fig. 5(a), we provide two compact abstractions, which are useful for presenting the support to a DM.[6]

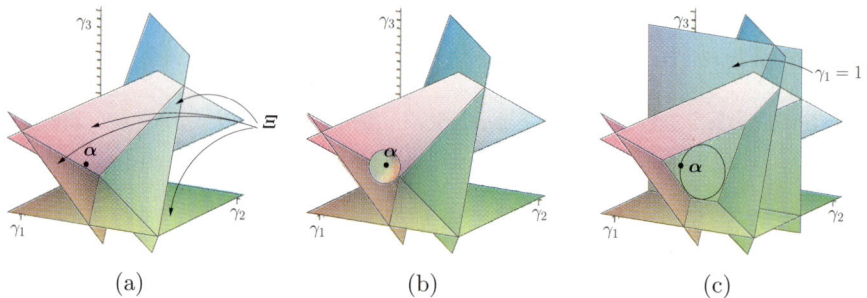

Fig. 5. Three ways of representing the result of support analysis.

An immediate approach to representing $\boldsymbol{\Xi}$ in a compact manner is to define an m-dimensional ball centered at $\boldsymbol{\alpha}$ with radius equal to the minimum Euclidean distance from $\boldsymbol{\alpha}$ to any one of the hyperplanes defined by constraints in $\boldsymbol{\Xi}$ (see Fig. 5(b)).

The approach allows for a highly compact representation of $\boldsymbol{\Xi}$ during computation too: Only a single scalar value (the radius of the ball) needs to

---

[6] These techniques do not presuppose that $\boldsymbol{\Xi}$ is minimal.

be stored. Whenever a constraint closer to $\boldsymbol{\alpha}$ is identified, we replace the old radius with the distance from $\boldsymbol{\alpha}$ to this new constraint. Unfortunately this representation can be a rather crude abstraction, as seen in Fig. 5(b).

Another, more accurate, representation technique is to present $\boldsymbol{\Xi}$ by the largest ball that will fit into the support defined by it. That is, abandon $\boldsymbol{\alpha}$ as a fixed center of the representation. As $\boldsymbol{\Xi}$ describes an infinite pyramid, no such largest ball exists, though, so we propose to make a cut through the computed pyramid $\boldsymbol{\Xi}$, by forcing one of the γ_is to be 1, and then representing the resulting intersection by the largest ball, which can fit into it (see Fig. 5(c)). The first step of this approach corresponds to identifying a base currency i that all other currencies are compared to.

For the second step to be successfully completed it is necessary that the intersection of the pyramid and the cut is a bounded volume. This is the case if the hyperplane defined by $\gamma_i = 1$ intersects all hyperplanes in $\boldsymbol{\Xi}$. This is equivalent to that there exists no f in $\boldsymbol{\Xi}$, such that the hyperplane defined by f is parallel to the hyperplane defined by $\gamma_i = 1$. As all hyperplanes pass through the origin, it is sufficient to choose an i where the constraint $\gamma_i \geq 0$ is not in $\boldsymbol{\Xi}$. If no such i exists, additional linear constraints on parameters will have to be put into $\boldsymbol{\Xi}$, e.g. $\gamma_i < k$ for some currency i and positive constant k.

Once a base currency i has been chosen, each constraint $f : f_1\gamma_1 + \cdots + f_m\gamma_m \geq 0$ in $\boldsymbol{\Xi}$ is replaced with the constraint $f_1\gamma_1 + \cdots + f_{i-1}\gamma_{i-1} + f_i + f_{i+1}\gamma_{i+1} + \cdots + f_m\gamma_m \geq 0$, corresponding to f's effect on points on the hyperplane defined by $\gamma_i = 1$. The resulting set of constraints we denote $\boldsymbol{\Xi}^{\gamma_i=1}$. To find the largest ball enclosed in the volume of $\mathbb{R}^{m-1}$ defined by the constraints in $\boldsymbol{\Xi}^{\gamma_i=1} = \{f^1, \ldots, f^k\}$ we solve the linear program

$$f^1(\mathbf{y}), \cdots, f^k(\mathbf{y}), z \leq \text{dist}(\mathbf{y}, \boldsymbol{H}(f^1)), \cdots, z \leq \text{dist}(\mathbf{y}, \boldsymbol{H}(f^k)),$$

where $\mathbf{y}$ is in $\mathbb{R}^{m-1}$ and $\text{dist}(\mathbf{y}, \boldsymbol{H}(f^i))$ denotes the Euclidean distance from $\mathbf{y}$ to the hyperplane $\boldsymbol{H}(f^i)$. The function that is to be optimised is z, and upon resolution the ball centered at $\mathbf{y}$ having radius z is the largest ball inscribed in the part of the support corresponding to γ_i being 1.

Acknowledgments

We would like to thank Martin Raussen of the Department of Mathematical Sciences at Aalborg University for useful discussions and two anonymous reviewers for helpful comments and suggestions. Furthermore, we are grateful for access to the Hugin software, which has been used in the implementation of the algorithms described in the paper. The implementation can be downloaded from `http://www.cs.aau.dk/~holbech/mcids.tgz`

References

[1] Magnus Boman, Paul Davidsson, and Håkan L. Younes. Artificial decision making under uncertainty in intelligent buildings. In *Proceedings of the Fifteenth Conference on Uncertainty in Artificial Intelligence*, pages 65–70, 1999.

[2] Gregory F. Cooper. A method for using belief netwoks as influence diagrams. In *Proceedings of the Fourth Conference on Uncertainty in Artificial Intelligence*, pages 55–63, 1988.

[3] James C. Felli and Gordon B. Hazen. Do sensitivity analysis really capture problem sensitivity? an empirical analysis based on information value. *Risk, Decision and Policy*, 4(2):79–98, 1999.

[4] Ronald A. Howard and James E. Matheson. Influence diagrams. *Readings on the Principles and Applications of Decision Analysis*, pages 720–763, 1984.

[5] Frank Jensen, Finn V. Jensen, and Søren L. Dittmer. From influence diagrams to junction trees. In R. Lopez de Mantaras and D. Poole, editors, *Proceedings of the Tenth Conference on Uncertainty in Artificial Intelligence*, pages 367–373. Morgan Kaufmann, 1994.

[6] V. Klee and G. J. Minty. How good is the simplex algorithm? In *Inequalities III*, pages 159–175. Academic Press Inc., 1972.

[7] Anders L. Madsen and Finn V. Jensen. Lazy evaluation of symmetric Bayesian decision problems. In *Proceedings of the Fifteenth Conference on Uncertainty in Artificial Intelligence*, pages 382–390, 1999.

[8] Thomas Dyhre Nielsen and Finn Verner Nielsen. Sensitivity analysis in influence diagrams. *IEEE Transactions on Systems, Man, and Cybernetics*, 33(2):223–234, 2003.

[9] Ross D. Shachter. Bayes-ball: The rational pastime (for determining irrelevance and requisite information in belief networks and influence diagrams. In G. F. Cooper and S. Moral, editors, *Uncertainty in Artificial Intelligence: Proceedings of the Fourteenth Conference*, pages 480–487. Morgan Kaufmann.

[10] Ross D. Shachter. Evaluating influence diagrams. *Operations Research*, 34:871–882, 1986.

[11] Prakash P. Shenoy. Valuation-based systems for Bayesian decision analysis. *Operations Research*, 40:463–484, 1991.

[12] Claus Skaaning. A knowledge acquisition tool for Bayesian-network troubleshooters. In *Proceedings of the Sixteenth Conference on Uncertainty in Artificial Intelligence*, pages 549–557, 2000.

[13] Joseph A. Tatman and Ross D. Shachter. Dynamic programming and influence diagrams. *IEEE Transactions on Systems Management and Cybernetics*, 20:265–279, 1990.

Parallel Markov Decision Processes

L. Enrique Sucar

ITESM Cuernavaca, Reforma 182-A, Temixco, Morelos, Mexico

Summary. We propose a framework for solving complex decision problems based on a partition in simpler problems that can be solved independently, and then combined to obtain an optimal, global solution. Each aspect of the problem is represented as an MDP and solved independently. At any state of the problem, each MDP sends its value for each possible action (its Q value) and an "arbiter" selects the action with the greatest combined value. In contrast to previous approaches for hierarchical MDPs, in our approach all the MDPs work in *parallel*, so we obtain a reactive system based on a decision theoretical framework. We present an algorithm for solving parallel MDPs and prove it obtains the global optimum, assuming an additive value. We present experimental results in two cases: (i) a simulated robot navigation problem, (ii) a real robot in a message delivery task.

1 Introduction

Our work is motivated by planning under uncertainty in robotics. Consider a mobile robot that has to perform a complex task in an uncertain environment. To accomplish its goal, the robot has to do several subtasks simultaneously, such as finding the shortest route to certain location, and at the same time, avoid obstacles and maintain its location in the map. It might also need to recognize objects in the environment and interact with people. A popular approach to solve this problem in robotics is based on Brooks *subsumption architecture* [4], in which several processes can sense and act in *parallel*. The conflicts that could arise between the different *behaviors* are usually solved by a fixed priority structure. However, this way of task coordination has several drawbacks: (i) as the number of subtasks increases, defining the priority structure becomes very difficult, (ii) the priority is fixed, and can not change depending on the current situation. We consider an alternative approach based on decision–theoretic planning, in which the priority of the subtasks can be decided dynamically such that the *best* action can be taken at each decision point.

Markov Decision Processes (MDPs) [2; 16] have developed as a standard method for representing uncertainty in decision-theoretic planning. They are simple for domain experts to specify, or can be learned from data. They are the subject of much current research, and have many well studied properties including exact and approximate solution and learning techniques. However, if we represent the robot task coordination problem as a single MDP, we have to consider all possible combinations of all the possible simultaneous actions. This implies an explosion in the action–state space and thus an important increase in complexity for solving the MDP. It also becomes much more difficult to specify or learn the model. Given that each subtask is usually implemented as a separate software module, it is more natural to try to view each subtask as a different MDP and then in some way combine their policies to obtain the optimal global policy.

In this paper we propose and approach for solving several subtasks represented as MDPs and combine the results to obtain the optimal global policy, that we call *Parallel MDPs*. Each aspect of the problem is represented as an MDP an solved independently. At any state of the problem, each MDP sends its value for each possible action (its Q value) and an "arbiter" selects the action with the greatest combined value. So we have the advantages of both, reactive and decision–theoretic planning: a reactive system based on a decision theoretical framework. We present an algorithm for solving parallel MDPs and prove it obtains the global optimum, assuming an additive value. We illustrate our approach with a simulated robot navigation problem, and with a real robot solving a complex task.

The paper is organized as follows. In the next section we present a formal definition of MDPs, the standard techniques for their solution and factored and abstract MDP representations. We review related work on hierarchical and loosely coupled MDPs. We then give a formal definition of parallel MDPs, and an algorithm for their solution. We present preliminary results in two robotics applications. We conclude with a summary and directions for future work.

2 Markov Decision Processes

A Markov Decision Processes (MDP) [16] models a sequential decision problem, in which a system evolves in time and is controlled by an agent. The system dynamics is governed by a probabilistic transition function that maps states an actions to states. At each time, the agents receives a reward that depends on the current state and the applied actions. Thus, the main problem is to find a control strategy or *policy* that maximizes the expected reward over time. A graphical model representation of an MDP is depicted in Figure 1.

Formally, an MDP is a tuple $M =< S, A, \Phi, R >$, where S is a finite set of states $\{1, ..., n\}$. A is a finite set of actions. $\Phi : A \times S \to \Pi(S)$ is the state transition function specified as a probability distribution. The probability of

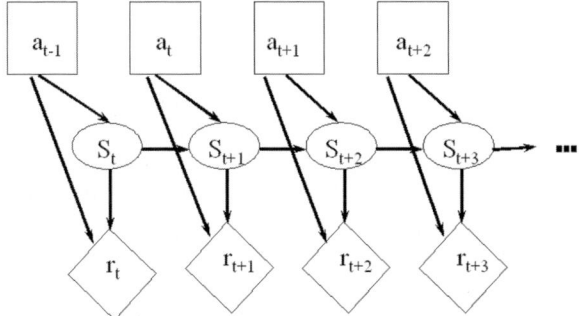

Fig. 1. A probabilistic graphical model representation of an MDP. The figure depicts the actions (a_i), states (s_i) and rewards (r_i) for several time periods, and the dependencies between them under the Markov assumption.

reaching state s' by performing action a in state s is written $\Phi(a, s, s')$. $R : S \times A \to \Re$ is the reward function. $R(s, a)$ is the reward that the system receives if it takes action a in state s.

A policy for an MDP is a mapping $\pi : S \to A$ that selects and action for each state. Given a policy, we can define its finite-horizon value function $V_n^\pi : S \to \Re$, where $V_n^\pi(s)$ is the expected value of applying the policy π for n steps starting in state s. The value function is defined inductively with $V_0^\pi(s) = R(s, \pi(s))$ and $V_m^\pi(s) = R(s, \pi(s)) + \Sigma_{s' \in S} \Phi(\pi(s), s, s') V_{m-1}^\pi(s')$. Over an infinite horizon, a discounted model is used to have a bounded expected value, where the parameter $0 \le \gamma < 1$ is the *discount factor*, used to discount future rewards at a geometric rate. Thus, if $V^\pi(s)$ is the discounted expected value in state s following policy π forever, we must have $V^\pi(s) = R(s, \pi(s)) + \gamma \Sigma_{s' \in S} \Phi(\pi(s), s, s') V_{m-1}^\pi(s')$, which yields a set of linear equations in the values of $V^\pi()$.

A solution to an MDP is a policy that maximizes its expected value. For the discounted infinite–horizon case with any given discount factor $\gamma \in [0, 1)$, there is a policy V^* that is optimal regardless of the starting state that satisfies the *Bellman* equation [2]:

$$V^*(s) = max_a \{R(s, a) + \gamma \Sigma_{s' \in S} \Phi(a, s, s') V^*(s')\} \qquad (1)$$

Two popular methods for solving this equation and finding an optimal policy for an MDP are: (a) value iteration and (2) policy iteration [16].

In policy iteration, the current policy is repeatedly improved by finding some action in each state that has a higher value than the action chosen by the current policy for the state. The policy is initially chosen at random, and the process terminates when no improvement can be found. This process converges to an optimal policy [16].

In value iteration, optimal policies are produced for successively longer finite horizons until they converge. It is relatively simple to find an optimal policy over n steps $\pi_n^*(.)$, with value function $V_n^*(.)$ using the recurrence relation:

$$\pi_n^*(s) = arg\ max_a \{R(s,a) + \gamma \Sigma_{s' \in S} \Phi(a,s,s')V_{n-1}^*(s')\}, \qquad (2)$$

with starting condition $V_0^*(.) = 0\ \forall\ s \in S$, where V_m^* is derived from the policy π_m^* as described earlier.

An alternative formulation of an MDP can be given in terms of the action–value function or *Q function*. This function, $Q^\pi(s,a)$ gives the expected cumulative reward of performing action a in state s and following policy π thereafter. The optimal action value function $Q^*(s,a)$, gives the expected cumulative reward of performing action a in state s and following an optimal policy thereafter. Note that given $Q^*(s,a)$, we can obtain $V^*(s)$ by maximizing over the actions.

The main drawback of the MDP approach is that the solution complexity is polynomial on size of the state–action space, and this can be *very large* for most applications. There are two main approaches to deal with complexity: state space abstraction and problem decomposition.

2.1 Factored and Abstract MDPs

Traditional MDP solution techniques have the drawback that they require an explicit state space, limiting their applicability to real-world problems. Factored representations address this drawback via compactly specifying the state-space in factored form. In a factored MDP, the set of states is described via a set of random variables $X = \{X_1, .., X_n\}$, where each X_i takes on values in some finite domain $Dom(X_i)$. The framework of dynamic Bayesian networks (DBN) [6] gives us the tools to describe the transition model function concisely. For each action, a two-stage DBN specifies the transition model. An even more compact representation can be obtained by representing the transition tables and value functions as decision trees [3] or algebraic decision diagrams [12].

A further reduction in complexity can be obtained by state abstraction and aggregation techniques [3]. Dean and Givan [5] describe an algorithm that partitions the state space into a set of blocks such that the each block is *stable*; that is, it preserves the same transition probabilities as the original model. Although this algorithm produces an exact partition, this could still be too complex. In many applications an approximate model could be sufficient to construct near–optimal policies. Other approaches consider problem decomposition, in which an MDP is partitioned in several problems that are solved independently and then pieced together [3], called hierarchical MDPs.

2.2 Hierarchical MDPs

Hierarchical MDPs accelerate the solution of complex problems by defining different subtasks that correspond to intermediate goals, solving for each subgoal, and then combining these subprocesses to solve the overall problem. Hierarchical MDP approaches include MAXQ [7] and HAM [15], among others. Most of these approaches assume that the domain hierarchy is given; a notable exception that learns the decomposition is HEXQ [11]. Although our approach also considers a decomposition of the problem, it is a different one. Hierarchical MDPs provide a *sequential* decomposition, in which different subgoals are solved in sequence to reach the final goal. That is, at the execution phase, only one task is *active* at a given time. In Parallel MDPs, the subtasks are concurrent, so these are executed in parallel to solve the global task.

The previous work most closely related to our approach are loosely coupled MDPs [13]. They consider several independent subprocesses whose reward and transition functions are independent of each other, but these are coupled due to common resource constraints. Their solution is divided in two phases. In the first, off–line phase, value functions are calculated for the individual subtasks. In the second, on–line phase, the value functions are used to calculate the next action for each process. To choose the actions, they use an iterative procedure based on a heuristic allocation of resources to each task. There are important differences in the types of problems we are solving and also in the solutions. We are interested in problems in which the subtasks have a common goal, but each consider a different aspect of the problem. Each MDP has the same action set, and only one of these actions can be executed at any time; while in loosely coupled MDPs each subtask executes an action at each state. So our solution selects a single action for each state, and this is computed directly.

3 Parallel MDPs

Parallel MDPs (PMDPs) are a set of discrete time Markov decision processes that are executed in *parallel*. At each time period an action is selected from each MDP, and an *arbiter* selects an action among the ones proposed by each process. These processes share a common state and action space (that is, each MDP has access to the same set of actions), but have different rewards. A graphical reprasentation of a parallel MDP is shown in Figure 2, considering two MDPs. Following we define a parallel MDP and give an algorithm for its solution.

Definition 1: A parallel MDP is a set of K Markov decision processes, $P_1, P_2, ..., P_k$, such that each process, P_i is an MDP. All the MDPs share the same state space, action set and transition function, that is $S_1 = S_2 = ... = S_k$; $A_1 = A_2 = ... = A_k$; $\Phi_1(a, s, s') = \Phi_2(a, s, s') = ... = \Phi_k(a, s, s')$. Each MDP has a different reward function, $R_1, R_2, ..., R_k$. The total reward, RT, is the sum of the individual rewards: $RT = R_1 + R_2 + ... + R_k$

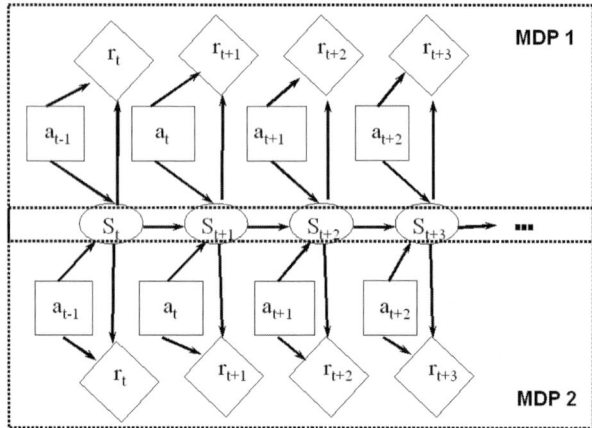

Fig. 2. Graphical representation of a parallel MDP. The two MDPs share the same state space but have different rewards. The action set is shown separated for both MDPs, but it is assumed that each one (MDP 1 and MDP 2) have the same set of actions.

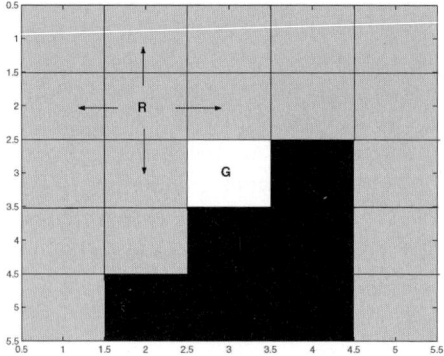

Fig. 3. Grid world. This simulated environment consists of cells that can be free (gray), obstacle (black) and a goal (white). The robot (R) can move from each cell to its 4 neighbors as shown.

For example, assume we have a simulated robot in a "grid world", as depicted in Figure 3. The robot has to go from the its actual position to the goal, and at the same time avoid obstacles. So we can define two MDPs: (i) a *Navigation MDP*, for going to the goal, (ii) an *obstacle avoidance MDP*, for avoiding obstacles. They both have the same state space (each cell in the grid) and actions (move up, down, right or left); but a different reward. The Navigation MDP gets a positive reward when it arrives to the goal, while the Obstacle Avoidance MDP gets a negative reward when it collides with an obstacle. So this case corresponds to a Parallel MDP. In the results section we present how this example is solved with our approach.

Next we will give an algorithm to obtain the optimal policy for a Parallel MDP, but first we present a theorem that gives the theoretical bases for the algorithm.

Theorem 1: Given a Parallel MDP, and assuming additive reward and value functions, the optimal value V_I^* and optimal policy π_I^* are:

$$V_I^*(s) = max_a \sum_{i \in K} Q_i^*(s,a) \quad (3)$$

$$\pi_I^*(s) = argmax_a \sum_{i \in K} Q_i^*(s,a) \quad (4)$$

Where $Q_i^*(s,a)$ is the optimal Q value obtained by solving each individual MDP.

Proof: The solution to the parallel MDP is given by the *Bellman* equation [2]:

$$V^*(s) = max_a \{RT(s,a) + \gamma \Sigma_{s' \in S} \Phi(a,s,s') V^*(s')\} \quad (5)$$

Where the total reward RT is given by:

$$RT = R_1(s,a) + R_2(s,a) + ... + R_k(s,a) \quad (6)$$
$$= \Sigma_{i=1}^{K} R_i(s,a)$$

Given the additive value assumption, the total value can also be written in terms of the individual MDP values:

$$V^*(s') = V_1^*(s') + V_2^*(s') + ... + V_k^*(s') \quad (7)$$
$$= \Sigma_{i=1}^{K} V_i^*(s')$$

Substituting 6 and 7 in 5 we obtain:

$$V^*(s) = max_a \{\Sigma_{i=1}^{K} R_i(s,a) \quad (8)$$
$$+ \gamma \Sigma_{s' \in S} \Phi(a,s,s') \Sigma_{i=1}^{K} V_i^*(s')\}$$

which can be written as:

$$V^*(s) = max_a \{\Sigma_{i=1}^{K} [R_i(s,a) + \gamma \Sigma_{s' \in S} \quad (9)$$
$$\Phi(a,s,s') V_i^*(s')]\}$$

By definition:

$$Q_i^*(s,a) = R_i(s,a) + \gamma \Sigma_{s' \in S} \Phi(a,s,s') V_i^*(s') \quad (10)$$

so 9 can be written as:

$$V^*(s) = max_a \{\Sigma_{i=1}^{K} [Q_i^*(s,a)]\} \quad (11)$$

And by the definition of the Q function, $Q^*(s,a) = \Sigma_{i=1}^{K} [Q_i^*(s,a)]$, so:

$$\pi^*(s) = argmax_a\{\Sigma_{i=1}^{K}[Q_i^*(s,a)]\} \qquad (12)$$

Q.E.D.

By using Theorem 1, we know give an algorithm for solving a Parallel MDP:

1. Solve each individual MDP by using value iteration [16]) and obtain the optimal Q_i values per action–state:

$$Q_i^*(s,a) = \{r_i(s,a) + \gamma\Sigma_{s'\in S}\Phi(a,s,s')V_i^*(s')\}$$

2. Compute the global optimal Q:

$$Q^*(s,a) = \Sigma_{i=1}^{K}Q_i^*(s,a)$$

3. Obtain the optimal policy:

$$\pi^*(s) = argmax_a\{\Sigma_{i=1}^{K}Q_i^*(s,a)\}$$

We can think that, at any state of the problem, each MDP sends its value for each possible action (its Q value) and the "arbiter" selects the action with the greatest combined value.

Using a flat state representation, there is not reduction in complexity by using a parallel MDP, with respect to solving the compound problem (including all the subtasks) as a single MDP. However, there are several important advantages:

- In a similar way as hierarchical MDPs [7], parallel MDPs facilitate state abstraction. For instance, in the robot grid example, Navigation only needs to consider its position with respect to the goal; while Obstacle Avoidance might just include a local map with the distance to obstacles. By using these abstractions, an important reduction in the number of states can be achieved.
- The subtask decomposition is also helpful for learning. Each subtask Q function can be learn individually using reinforcement learning [17], and then combined using our algorithm. In the grid world example, the robot can independently learn to go to the goal and to avoid obstacles, which is *easier* than learning both tasks simultaneously.
- From a practical perspective, in many areas such as robotics, there are different software programs for solving different aspects of the problem, such as path planning, obstacle avoidance, localization, etc. So it is better to consider each module as a task and combine their policies, instead of designing a single MDP for the complete task.

Next we illustrate our approach with an example of a simulated robot in the grid world.

4 Experimental Results

4.1 Navigation in a Simulated Environment

To test our approach, we consider a simulated robot in the grid world (see Figure 3). The state is given by the coordinates (X, Y) of robot's location in the grid. At each state the robot can move to its 4 neighbor locations, so there are 4 possible actions: *up, right, down, left*. The uncertainty in the actions is given by assuming that the robot has an 80% chance of going to the desired location, and 10% to each of the two adjacent locations. So this defines the transition function. In this experiments we consider two tasks: navigation and obstacle avoidance. The reward for navigation is a fixed amount for reaching the goal state and zero otherwise. The obstacle avoidance subtask receives a negative reward for going into a cell with an obstacle, and zero otherwise.

We tested with different grid sizes, goal positions and obstacle distributions, and compared the resulting policy of: (a) a parallel MDP with two subtasks, navigation and obstacle avoidance, and (b) a single MDP that considers both aspects. To illustrate the results we choose a small grid so the optimal value function and policy can be shown graphically. Figure 4 shows this test case. We show the results with the parallel MDP for two different position of the goal: $(5, 5)$ and $(1, 1)$. Figure 5 depicts the value and policy functions for the goal at $(5, 5)$. The policy is represented in terms of the gray level of each cell, lighter represents a higher value. The policy is shown as arrows that correspond to the optimal action per state (cell). We can see that both, the value and the policy are optimal for this case.

Figure 6 shows the value function for the goal position at $(1, 1)$. Again, this corresponds to the optimal value function. In these two experiments, we solve the navigator task two times, one for each goal, but the obstacle avoidance only once.

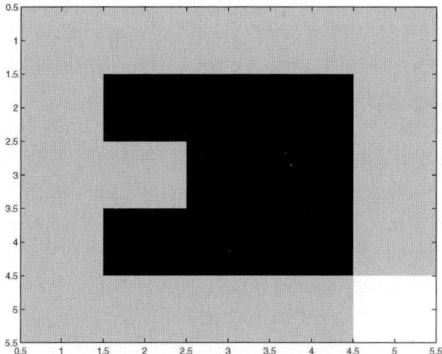

Fig. 4. Test case used in the experiments. The black cells are obstacles and the white cell is the goal (in this case at $5, 5$).

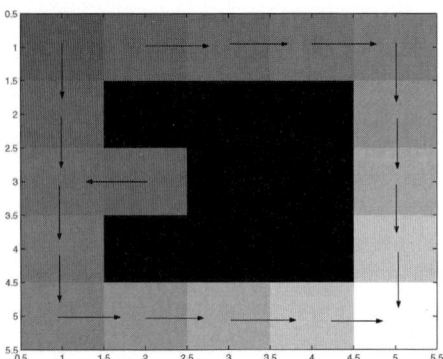

Fig. 5. Value function and policy for the goal at 5, 5. For each free cell, we show the value (as a gray level) and the optimal action (arrow).

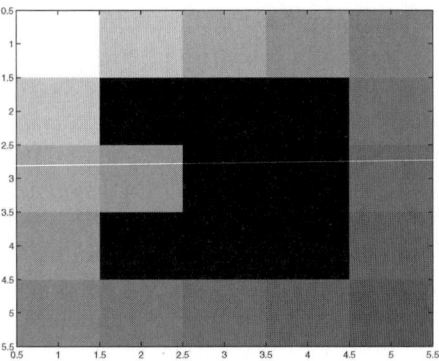

Fig. 6. Value function for the goal at 1, 1. For each free cell, we show the value (as a gray level).

In the previous examples, the results are the same for the parallel MDPs and for the single MDP. However, although in many cases the PMDP approach gives an optimal policy, we have found that there are some cases in which this is not true. A counter example is given in Figure 7. The combined value function obtained with the parallel MDP approach is shown in Figure 8. In this case the states with non-optimal policies are at the left of the obstacle, where the action selected by the PMDP is shown (arrows). The underlying problem is that the MDPs are not independent, they interact through a common state and action space.

We notice that in general the combined value function (and corresponding policy) is correct for *most* of the states (cells), and it fails in just a few of them. This suggests a two phase approach to improve the policy obtained initially by the parallel MDP:

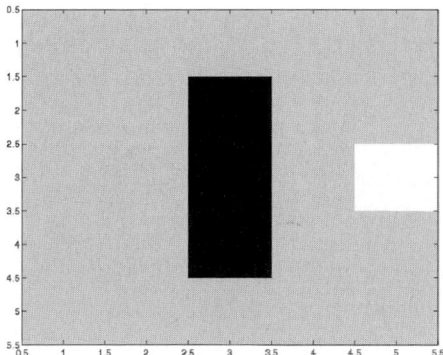

Fig. 7. Counter example. The black cells are obstacles and the white cell is the goal.

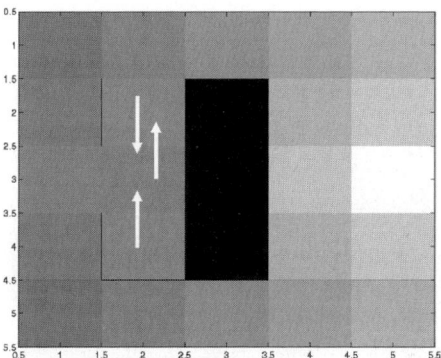

Fig. 8. Parallel MDP solution to the example in Figure 7. The value function is shown for each free cell (as a gray level). Notice that the actions for the cells at the left of the obstacle are incorrect, as shown by the arrows.

1. An approximate optimal solution is obtained considering separate MDPs with the parallel MDP approach.
2. The initial solution is refined considering the complete problem, taking as starting point the solution (value function and policy) obtained in the first phase.

We are currently working in the development of the refinement phase.

4.2 Message Delivery with Homer

We have tested a variant of parallel MDPs with a real robot in a message delivery task. In this case we consider simultaneous actions and no conflicts, so each MDP selects its action independently and all are executed concurrently, coordinated implicitly with common state variables. By assuming no conflicts, there is, in principle, no need for an arbiter. In practice, we give priority to one of the tasks. A detailed description of the application of multiple MDPs for task coordination for a service robot is given in [10]. Next we just present a brief summary of the main results.

HOMER [8], the Human Oriented MEssenger Robot, is a mobile robot that communicates messages between humans in a workspace. The message delivery task is a challenging domain for an interactive robot. It presents all the difficulties associated with uncertain navigation in a changing environment, as well as those associated with exchanging information and taking commands from humans using a natural interface. For this task we use 3 MDPs: the navigator, the dialogue manager and the gesture generator. Together they coordinate 10 behaviors [9] for accomplishing the message delivery task.

Our robot, HOMER, shown in Figure 9(a), is a Real World Interface B-14 robot, and has a single sensor: a Point Grey Research Bumblebee^{TM} stereo vision camera [14].

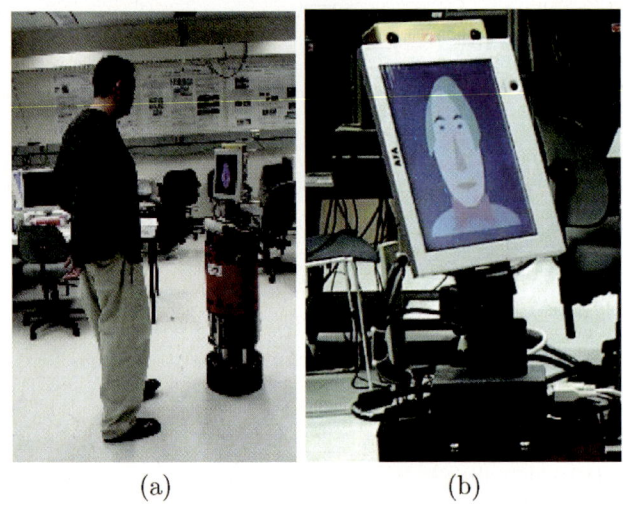

Fig. 9. (a) HOMER the messenger robot interacting with a person and (b) closeup of HOMER's head.

HOMER's message delivery task consists of accepting messages, finding recipients and delivering messages. In his quiescent state, HOMER explores the environment looking for a message sender. A potential sender can initiate an interaction with HOMER by calling his name, or by presenting herself to the robot. HOMER asks the person for her name (sender), the recipient's name, and the message. During the interactions, HOMER uses speech recognition, speech generation and gesture generation to communicate with people. Once HOMER has a message to deliver, he must find the recipient. This requires some model of the typical behavioral patterns of people within HOMER's workspace. We use a static map of person locations, which is updated when new information is obtained about the presence or absence of persons. This map allows HOMER to assess the most likely location to find a person at any time. Navigation to that location is then attempted. If the location is not reachable, HOMER finds another location and re-plans. If the location is reached, then HOMER attempts to find a potential receiver using face and voice detection. Upon verifying the receivers name, HOMER delivers the message. During the entire process, HOMER will localize in the map if necessary, or it will go *home* to recharge if its battery is low.

The message delivery task can be divided in 3 subtasks, each one controlled by an MDP. The **N**avigator controls the navigation and localization of the robot, the **D**ialogue Manager controls the interaction with people using speech, and the **G**esture Generator controls the interaction with people using gestures performed by an animated face. Each MDP includes the relevant variables as its state space, and controls several behaviors through its actions. The complete state is represented by 13 variables. The goal of the message delivery task is encoded in the reward function: a *small* reward for receipt of a message, a *big* reward for message delivery, and a negative reward for a low battery. The Dialogue and Gesture planners only include rewards for message receipt and delivery, while the navigator includes all three.

We solved the 3 MDPs using SPUDD [12] and generated the optimal policies for each one. During concurrent execution of the policies, potential conflicts are avoided by simply giving priority to the Navigator. Thus, if HOMER is navigating to a location, such as *home*, it does not stop for an interaction.

We ran several experiments with HOMER. Each experiment involves the robot receiving and delivering a message by visiting locations as necessary. Initially, we performed a guided exploration task in order to build all the necessary maps for navigation and localization. We also manually specified a list of possible users and the most likely areas they inhabit. HOMER then ran autonomously for the message delivery task. Although we did not try to prove optimality for this approach, empirically, our simple solution method works well for these examples.

5 Conclusions and Future Work

We have presented a framework for solving complex planning problems by dividing them into several subtasks represented as MDPs. We obtain the optimal policy for each subtask, and then combine the results to obtain the optimal global policy. At any state of the problem, each MDP sends its value for each possible action (its Q value) and an "arbiter" selects the action with the greatest combined value. So we have the advantages of both, reactive and decision–theoretic planning: a reactive system based on a decision theoretical framework. We present an algorithm for solving parallel MDPs and prove it obtains the global optimum, assuming an additive value. Initial experiments with a simulated robot in the grid world and with a real message delivery robot show good results.

We are currently working in improving the initial non-optimal policy by considering a two phase approach, in which the solution of the parallel MDP is considered as an initial solution to a second phase which considers the global problem. We are also interested in extending the framework to consider concurrect actions with conflicts, and apply it to complex decision problems for service robots.

References

[1] Arkin, C.E. (1998) *Behavior-based Robotics.* MIT Press, Cambridge, Mass.
[2] Bellman, R.E. (1957) *Dynamic Programming.* Princeton U. Press, Princeton, N.J.
[3] Boutilier, C., Dean, T., Hanks, S. (1999) Decision-theoretic planning: structural assumptions and computational leverage. *Journal of Artificial Intelligence Research* 11:1–94.
[4] Brooks, R. (1986) A robust layered control system for a mobile robot. *IEEE Journal of Robotics and Automation* 2(1):14–23.
[5] Dean, T., Givan, R. (1997) Model minimization in markov decision processes. In: *Proceedings of the 14th National Conference on Artificial Intelligence*, 106–111.
[6] Dean, T., Kanasawa, K. (1989) A model for reasoning about persistence and causation. *Computational Intelligence* 5:142–150.
[7] Dietterich, T. G. (2000) Hierarchical reinforcement learning with the maxq value function decomposition. *Journal of Artificial Intelligence Research* 13:227–303.
[8] Elinas, P., Hoey, J., Little, J.J. (2003) Human oriented messenger robot. In: *Proc. of AAAI Spring Symposium on Human Interaction with Autonomous Systems*, AAAI Stanford, CA.
[9] Elinas, P., Little, J.J. (2002) A robot control architecture for guiding a vision-based mobile robot. In: *Proc. of AAAI Spring Symposium in Intelligent Distributed and Embedded Systems*, AAAI Stanford, CA.
[10] Elinas, P., Sucar, L.E., Reyes, A., Hoey, J. (2004) A decision theoretic approach for task coordination in social robots. In: *Proc. IEEE International Workshop on Robot and Human Interactive Communication (RO-MAN)* IEEE, pp. 679–684.
[11] Hengst, B. (2002) Discovering hierarchy in reinforcement learning with hexq. In: *Proceedings of the National Conf. on Artificial Intelligence.* AAAI.

[12] Hoey, J., St-Aubin, R., Hu, A., Boutilier, C. (1999) SPUDD: Stochastic planning using decision diagrams. In: *Proceedings of International Conference on Uncertainty in Artificial Intelligence (UAI '99)*, Stockholm.

[13] Meuleau, N., Hauskrecht, M., Kim, K.E., Pashkin, L., Kaelbling, L., Dean, T., Boutilier, C. (1998) Solving very large weakly coupled Markov decision processes. In: *Proc. AAAI*.

[14] Murray, D., Little, J.J. (2000) Using real-time stereo vision for mobile robot navigation. *Autonomous Robots* 8:161–171.

[15] Parr, R., Russell, S. (1997) Reinforcement learning with hierarchies of machines. In: *Advances in Neural Information Processing Systems (NIPS)*, volume 9. MIT Press.

[16] Puterman, M.L. (1994) *Markov Decision Processes: Discrete Stochastic Dynamic Programming.*. Wiley, New York, NY.

[17] Sutton, R., Barto, A. G. (1998) *Introduction to reinforcement learning.* MIT Press, Cambridge, MA.

Part V

Applications

Applications of HUGIN to Diagnosis and Control of Autonomous Vehicles

Anders L. Madsen[1] and Uffe B. Kjærulff[2]

[1] HUGIN Expert A/S, Gasværksvej 5, DK-9000, Aalborg, Denmark
[2] Aalborg University, Department of Computer Science, Fredrik Bajers Vej 7E, DK-9220, Aalborg, Denmark

Summary. We present an application of HUGIN to solve problems related to diagnosis and control of autonomous vehicles. The application is based on a distributed architecture supporting diagnosis and control of autonomous units. The purpose of the architecture is to assist the operator or piloting system in managing fault detection, risk assessment, and recovery plans under uncertainty. To handle uncertainty, we focus on the use of probabilistic graphical models (PGMs) as implemented in the HUGIN tool.

We describe the application of PGMs to three problems of diagnosis and control of autonomous vehicles. Based on the HUGIN tool, limited memory influence diagrams (LIMIDs) are used to represent and solve complex problems of diagnosis and control of autonomous ground and underwater vehicles. In particular, we describe how battery monitoring and control problems related to an underwater and a ground vehicle are solved and how to solve the problem of assessing the quality of a sonar image related to an underwater vehicle.

1 Introduction

The HUGIN tool [1; 5; 13] supports construction and deployment of complex statistical models known as probabilistic graphical models (PGMs) for reasoning and decision making under uncertainty. We describe how a particular kind of PGMs, known as limited memory influence diagrams (LIMIDs), have been applied to solve complex reasoning and decision making problems related to autonomous ground vehicles and autonomous underwater vehicles.

The ADVOCATE project (acronym for *Advanced On-board Diagnosis and Control of Autonomous Systems*), which was formed in 1997 under the IST program of the European Commission, had as one of its objectives to increase the performance of unmanned underwater vehicles in terms of availability, efficiency, and reliability of the systems and in terms of safety for the systems themselves as well as for their environments. The aim of the follow-up project, ADVOCATE II, was partly to design and develop a general-purpose software architecture for Autonomous Underwater Vehicles (AUVs) and Autonomous

Ground Vehicles (AGVs) and partly to develop software-based systems to increase the degree of automation, efficiency, and reliability of the vehicles. The interest of such a concept from the market point of view was demonstrated by a market study.

The latter objective was reached by adding artificial intelligence (AI) into existing and new control software to diagnose and recover from any dysfunction or failure situation of the system. To improve the management of uncertainty in AUVs and AGVs it was decided that the ADVOCATE II architecture should allow for easy incorporation and merging of different AI techniques in a highly modular fashion.

Three end-user partners were involved in the ADVOCATE II project: University of Alcalá designs piloting modules for AGVs for surveillance applications, Ifremer designs AUVs for scientific applications, and ATLAS Elektronik designs AUVs and semi-AUVs for industrial applications. Each end-user partner presented diagnosis and control problems related to a single vehicle. Several such problems were presented for each vehicle involving different kinds of dysfunctions and failures:

- Thruster or motor failure diagnosis and recovery in case of abnormal behaviour of the vehicle due to thrusters or motors.
- Sensor malfunction diagnosis and recovery on sensor state in order to account for failure situations in case of corrupt sensor signals due, for instance, to noise.
- Power consumption diagnosis and recovery monitoring the level of remaining energy in a vehicle battery in order to avoid mission abortion.
- Motion diagnosis and recovery monitoring and assessing the motion characteristics of a vehicle.

Three different AI techniques have been applied for solving these diagnosis and control problems, namely probabilistic graphical models (PGMs), neuro-symbolic systems (NSSs), and fuzzy logic (FL). In this chapter, we focus on the development and application of PGMs for a selected set of the problems. We discuss how LIMIDs have been used to represent and solve complex problems of diagnosis and control of AGVs and AUVs. In particular, we describe how battery monitoring and control problems related to an AUV and an AGV are solved and how a sonar image quality assessment problem related to an AUV is solved.

In Sect. 2, we briefly present the HUGIN tool used for constructing and executing the LIMIDs. Section 3 introduces the problem domain of semi-autonomous ground and underwater vehicles. The ADVOCATE II communication architecture is described in Sect. 4. Section 5 presents some preliminaries and notation on the LIMID representation used to model and solve the diagnosis and control problems. The knowledge extraction process and method developed as part of the ADVOCATE II project is described in Sect. 6. Section 7 describes the models developed to solve the diagnosis and control problems, while in Sect. 8 we discuss how the developed LIMIDs are solved.

Section 9 describes model integration and validation, and presents the results of real world trials. Finally, Sect. 10 ends the chapter with a discussion of our work.

2 The HUGIN Tool

The HUGIN tool [1; 5; 13] is a general purpose tool for constructing and deploying probabilistic graphical models (PGMs) such as Bayesian networks and influence diagrams.

In the HUGIN tool, inference in PGMs is performed through message passing in a secondary computational structure known as a junction tree [7]. The junction tree is constructed from the PGM through processes known as moralization and triangulation [12]. The nodes of a junction tree are sometimes referred to as *cliques*. To each clique (containing a subset of the variables of the PGM) is associated tables representing joint probability and utility functions over variables of the clique. The messages passed between cliques represent joint probability and utility functions over variables common to both the sending and the receiving clique.

The HUGIN tool consists of a graphical user interface (HUGIN Graphical User Interface) and an inference engine (HUGIN Decision Engine). The HUGIN Decision Engine has Application Programming Interfaces (APIs) for four different programming languages: C, C++, Java, and Visual Basic for Applications.

The core functionality of the HUGIN Decision Engine is implemented in the C programming language according to the ANSI C standard. This makes the HUGIN Decision Engine highly efficient and portable. Interfaces for C++, Java and Visual Basic for Applications are constructed on top of the core implementation. The HUGIN Graphical User Interface is implemented in Java, which makes it highly portable.

The HUGIN Decision Engine has been deployed on a large number of different platforms ranging from PDAs to multiprocessor mainframes.

3 Problem Domain

The transition of autonomous vehicles from experimental research tools to real applications increases the need for reliable and safe performance of the vehicles. This includes detection, avoidance, and recovery from any dysfunction.

Both the AGVs and the AUVs considered in the ADVOCATE II project are supplied with energy from batteries. This poses the problem of monitoring the remaining energy level of the battery and providing diagnosis and recovery actions in order to manage the mission parameters related to the energy consumption and to avoid unnecessary mission aborts. One AUV is equipped with an advanced object detection and avoidance system. This system works well in situations where obstacles can be detected by sonar. Hence, it is important to assess the quality of the sonar image and to suggest recovery actions to improve the sonar image quality or to suggest reductions in speed in case of poor image quality.

The ADVOCATE II consortium consisted of eight partners from five different European countries. Each partner served different roles in the consortium. The roles were robot manufactures and end-users, technology providers, marketing and communication specialists, and project coordinator. HUGIN Expert A/S served the role as technology provider and developer of intelligent modules.

3.1 The DeepC AUV

ATLAS Elektronik is developing a new type of underwater vehicle operating with autonomous mission durations of up to 60 hours. This vehicle is referred to as DeepC (see Fig. 1).

Fig. 1. The DeepC underwater vehicle.

The long mission durations impose the need for advanced AI techniques to detect, avoid, and recover from any dysfunction. All end-users and ATLAS Elektronik in particular were faced with problems, which could not easily be solved by existing systems. The current approach to handle mission faults is to abort the mission, which is, however, very expensive.

As a fully autonomous system, the DeepC vehicle has to rely on its sensors to survive operationally. The DeepC is equipped with an advanced object detection and avoidance system. The object detection system consists of a mechanically scanning, forward looking sonar and its control electronics. This system works well when the sonar image is of sufficient quality. The problem considered is to construct a model for assessing the sonar image quality and for suggesting actions to avoid object collisions.

Applications of HUGIN   317

3.2 The VORTEX AUV

Ifremer has developed the remotely operated underwater vehicle VORTEX (see Fig. 2), which for our purpose is functionally considered as an AUV, as it allows programming of autonomous complex missions.

Fig. 2. The VORTEX underwater vehicle.

The motivation for equipping the vehicle with AI technology is much the same as for the DeepC, including optimization of the mission plan, diagnosis of abnormalities, recovery planning in case of abnormalities, avoidance of mission abortion (which is very expensive), and avoidance of vehicle loss.

3.3 The BART AGV

To put the ADVOCATE II concept into practice also for autonomous ground vehicles, University of Alcalá was deploying a telesurveillance application using the BART AGV (see Fig. 3). Two independent actuators powered by an on-board battery drive the vehicle.

Fig. 3. The BART autonomous ground vehicle.

In order to increase the probability of mission success in case of energy problems or in case the vehicle gets stalled, the overall mission of the vehicle as well as other navigational issues can be managed using ADVOCATE II technology. The ADVOCATE II system provides the AGV with intelligent diagnosis capabilities and ability to recommend optimal recovery actions resulting in more reliable and safer operations. The diagnosis and recovery capabilities are concerned with aspects of navigation, energy system, sensors, actuators, etc.

4 ADVOCATE II Architecture

The ADVOCATE II architecture is a distributed architecture based on a generic communication protocol. The architecture is modular and easy to evolve and adapt to future piloting systems.

The purpose of the architecture is to assist the operator or piloting system in managing fault detection, risk assessment, and recovery plans under uncertainty. The generic communication protocol is based on SOAP/XML technology implementing HTTP for communication between different types of modules (see Fig. 4).

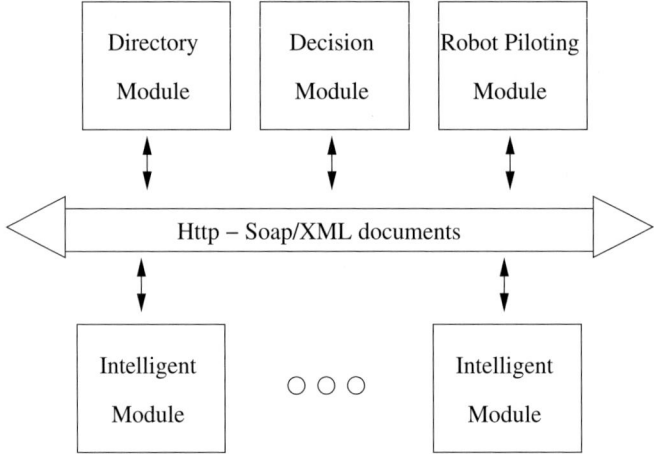

Fig. 4. The communication architecture.

The architecture is generic, open, and modular consisting of a set of interacting modules including a decision module and a set of intelligent modules. The decision module communicates with the intelligent modules to request and obtain diagnosis and recovery action proposals based on data obtained from the robot piloting module.

The architecture is designed to allow easy integration of different AI techniques into preexisting systems. The decision to support the simultaneous use of multiple AI techniques was made to allow these techniques to collaborate on the task of reasoning and making decisions under uncertainty. This raises the question of how to most efficiently integrate different AI techniques into new and existing systems. We have found that this is most efficiently done through an open and generic architecture with a sophisticated communication interface.

The architecture consists of four different types of modules.

4.1 Robot Piloting Module

The robot piloting module manages mission plans and communicates directly with the sensors and actuators of the vehicle. This module also implements recovery plans received from the decision module into actions on the vehicle.

Each end-user partner of the ADVOCATE II project was responsible for the piloting module corresponding to the end-user vehicle.

4.2 Decision Module

The decision module manages the diagnosis and recovery action process. This includes integration of information provided by different intelligent modules, user validation of diagnosis and recovery actions when required by the system, and translation of recovery actions into recovery plans.

The decision module communicates with the intelligent modules receiving diagnoses and recovery actions, the robot piloting module, and the user.

4.3 Intelligent Module

The role of an intelligent module is to provide possible diagnoses, suggestions for recovery actions, or both. An intelligent module encapsulates a knowledge base to a specific problem domain and an inference engine.

A diagnosis on an operational vehicle corresponds to identification of system state while a recovery action corresponds to performing a sequence of actions on the vehicle (e.g., to avoid collision or to recover from any dysfunction).

The intelligent module communicates with the robot vehicle piloting module and the decision module. The robot vehicle piloting module supplies the intelligent module with data. These data are used in conjunction with the knowledge base to generate diagnoses and recovery actions. The diagnosis or recovery action is communicated to the decision module.

There may be multiple intelligent modules connected to the ADVOCATE II architecture. Multiple intelligent modules may consider the same or different problems related to the vehicle. Each intelligent module implements the communication protocol defined for the ADVOCATE II architecture.

In each application considered, the HUGIN Decision Engine is used for both reasoning and decision making under uncertainty, i.e., both to make a diagnosis and to generate recovery actions.

4.4 Directory Module

The architecture is organized around the directory module. The directory module is the central point of communication in the sense that it maintains a list of registered and on-line modules.

5 Limited Memory Influence Diagram

Limited memory influence diagrams are used to represent and solve the selected set of diagnosis and recovery action problems. An influence diagram [3] is a compact and intuitive probabilistic graphical model for reasoning about decision making under uncertainty. It is a graphical representation of a decision problem involving a sequence of interleaved decisions and observations. In essence, an influence diagram is a Bayesian network [2; 6; 14; 15] augmented with decision variables and preference (or utility) functions.

An influence diagram $\mathcal{N} = (\mathcal{X}, \mathcal{G}, \mathcal{P}, \mathcal{U})$ over random variables $\mathcal{X}_C$ and decision variables $\mathcal{X}_D$ such that $\mathcal{X} = \mathcal{X}_C \cup \mathcal{X}_D$ consists of an acyclic, directed graph $\mathcal{G} = (V, E)$ over nodes V, connected by directed links $E \subseteq V \times V$, a set of conditional probability distributions $\mathcal{P}$, and a set of utility functions $\mathcal{U}$. The nodes V of $\mathcal{G}$ represent the random variables, decision variables, and utility functions of $\mathcal{N}$.

Each decision variable, D, represents a specific point in time under the model of the problem domain where the decision maker has to make a decision. The decision options or alternatives are the finite set of states $(d_1, \ldots, d_n)$ of D where n is the size of the state space of D. The usefulness of a decision option d_i is measured by local utility functions associated with D or one of its descendants in $\mathcal{G}$.

Mathematically, an influence diagram is a compact representation of a joint expected utility (EU) function:

$$\mathrm{EU}(V) = \prod_{X \in \mathcal{X}_C} P(X \mid \pi(X)) \sum_{u \in \mathcal{U}} u,$$

where $\pi(X)$ denotes the immediate predecessors (or parents) of variable X in $\mathcal{N}$.

To solve an influence diagram $\mathcal{N}$ is to determine an optimal strategy for the decision maker to follow. The strategy consists of one decision policy δ_D for each decision variable $D \in \mathcal{X}_D$. A policy δ_D is a mapping from the requisite past of D (i.e., past observations that may impact the choice of decision option for D) to the state space of $D \in \mathcal{X}_D$.

In the graphical representation of an influence diagram, random variables are represented as ovals, decision variables as rectangles, and utility functions as diamond-shaped nodes. A link into a node representing a random variable represents a probabilistic dependence relation, a link into a utility node identifies a domain variable of the corresponding utility function, and a link into a node representing a decision variable specifies that the parent is observed prior to the decision is made. Links into decision nodes are referred to as informational links.

An influence diagram supports the representation and solution of sequential decision problems under the no-forgetting assumption (i.e., assuming perfect recall of all observations and decisions made in the past that are influential in a given decision situation). A LIMID [11] is an influence diagram relaxing the no-forgetting assumption to a limited memory assumption. This implies that all information available to the decision maker must be specified using informational links for each decision. This is contrary to an influence diagram where some informational links may be implicitly assumed present.

Figure 5 shows a LIMID representation of a simple decision problem involving two decisions D_1 and D_2.

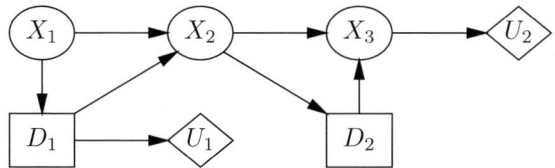

Fig. 5. A decision problem with two decisions.

If the graph in Fig. 5 is interpreted as an influence diagram, then the domain of the decision policy for D_2 will consist of X_1, D_1, and X_2 (due to the no-forgetting assumption). If, on the other hand, the graph in Fig. 5 is interpreted as a LIMID, then the domain of the decision policy for D_2 will consist only of X_2. In a LIMID, all informational links are shown explicitly.

In the HUGIN tool, an influence diagram is solved by message passing in a so-called strong junction tree [4]. A LIMID, on the other hand, is solved by message passing in an ordinary junction tree. The difference in computational complexity between a strong junction tree and an ordinary junction tree can be tremendous [11].

An OO LIMID is an extension of a LIMID with support for object-oriented constructions [10], considering a LIMID as a class of which instances can exist in several other classes (LIMIDs). Thus, in addition to the elements of a LIMID, an OO LIMID contains instance nodes. An instance node represents an instantiation (or realization) of one LIMID class within another LIMID class following the object-oriented paradigm. In graphical representations of LIMIDs, instance nodes are represented using box-shaped nodes with rounded corners. The interface of a class is its input and output variables (indicated as nodes with a gray outer part, where input nodes have a dashed black border); see Fig. 9 for an example.

6 Knowledge Extraction Methodology

Unfortunately, the construction of a PGM can be a labour intensive task with respect to both knowledge acquisition and formulation. LIMIDs are not exceptional in this respect. The knowledge acquisition and formulation process associated with building the three LIMID models in the ADVOCATE II project involved knowledge engineers and domain experts located in four different countries. The knowledge engineers and domain experts had limited possibilities for face-to-face meetings and the domain experts had limited knowledge of LIMIDs. Therefore, a knowledge acquisition scheme had to be developed that did not rely on familiarity with terminology of probabilistic graphical models and direct contact with the knowledge engineers.

The scheme is based on building a problem hierarchy for an overall problem. The problems (or causes) of the hierarchy relate to the states of the different parts of a vehicle and its environment.

Figure 6 shows such a cause hierarchy related to the energy problem of the BART AGV. The causes of the hierarchy are grouped into causes that qualify as satisfactory explanations of the overall problem and causes that do not. The first group of causes are referred to as *permissible diagnoses*. The subset of these that can actually be identified based on available information are referred to as *possible diagnoses*. Possible diagnoses are marked with a "+" in Fig. 6, and permissible diagnoses that are not possible are marked with a "−".

The cause hierarchy acts as a road-map for describing the relevant diagnostic information and the possible recovery actions. A cause of a sub-tree of the cause hierarchy that does not contain any possible diagnoses is unlikely to

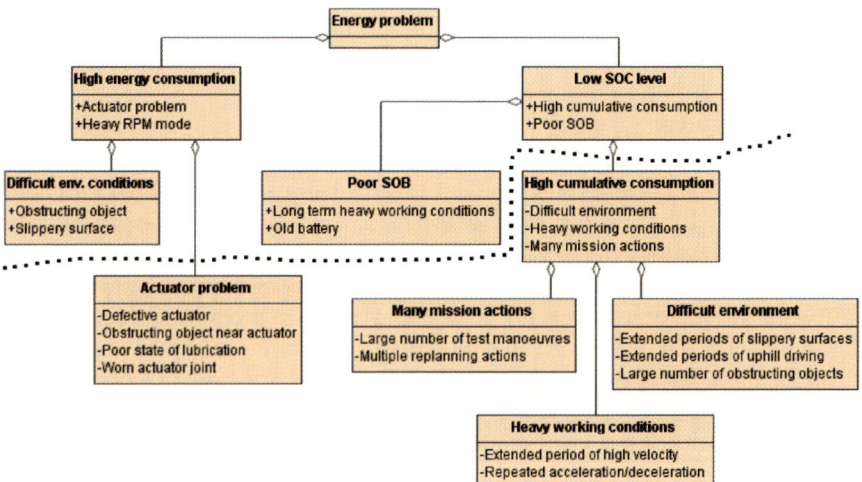

Fig. 6. Cause hierarchy for the BART AGV energy problem.

provide relevant diagnostic information or error recovery information. Thus, if there are no observable manifestations of the cause strong enough to identify a possible diagnosis for the cause, we need not worry about it when eliciting the diagnostic and error recovery information. In particular, none of the causes below the dotted line in Fig. 6 contain any possible diagnoses. The domain expert provides the relevant diagnostic information and the recovery actions in matrix form with one row for each cause "above the dotted line" and one column for each kind of diagnostic information (i.e., background information and symptoms) and one column for possible recovery actions.

The qualitative knowledge elicited following such a scheme provides a sufficient basis for a knowledge engineer to construct the structure of a PGM, on the basis of which the quantitative knowledge can then be elicited.

For a detailed description of the knowledge acquisition scheme, we refer the reader to [9].

7 Models

Using the knowledge extraction method described in Sect. 6, one LIMID model for each of the vehicles has been developed in collaboration with the end-user. For reasons of space limitations, we include only a subset of the cause hierarchies and models developed.

7.1 The VORTEX AUV

The purpose of the PGM intelligent module of the VORTEX is to assess the status of the energy consumption of the actuators and the payload systems of the AUV. The payload systems consist of various sensors for scientific investigations. More concretely, the task of the module is to compute the probabilities of the various possible root causes of unexpected high energy consumption and the expected utilities of the various recovery actions given the information available.

There are two different aspects (or sub-causes) of "Energy consumption problem", namely "High energy consumption" indicating that the current level of energy consumption is significantly higher than recommended, and "Low state of charge (SOC)" indicating either an abnormally high level of cumulative energy consumption or a poor state of the battery (SOB). These two aspects relate to, respectively, the *present energy consumption* and the *cumulative energy consumption*. The present energy consumption is defined as the average consumption over the last 10 seconds.

To identify the cause of low SOC as a high cumulative energy consumption, the model should either be dynamic, capable of representing phenomena evolving over time, or rely on a measurement of the cumulative total consumption and an indication of the recommended cumulative total consumption at any given point of the mission. We decided to go with the latter approach, as a dynamic model would result in serious computational complexity problems. Also, given that periodic requests are issued from the decision module to the VORTEX intelligent module, determining that the cumulative energy consumption is high is a straightforward task that might as well be performed by the decision module itself.

Figure 7 shows the resulting LIMID. There are four groups of random variables in Fig. 7: Ten diagnosis variables, eleven background information variables, nine symptom variables, and eighteen auxiliary variables. The ten diagnosis variables represent the following distinct root causes of an energy consumption problem of the VORTEX: *Old battery*, *Long-term heavy working conditions*, *Poor SOB*, *Cold battery*, *High cumulative energy consumption*, *Obstructing object*, *Strong currents*, *Fast acceleration*, *Actuator problem*, and *Unhealthy payload*. The posterior probability distributions for these diagnoses are computed on the basis of information provided through the twenty evidence variables (symptom measurements and background information).

The domain experts identified a group of nine different actions that can be performed in response to energy problems: *Mission action* (e.g., "Continue", "Reduce velocity", "Abort mission", etc.), *Test SOB*, *Replace battery*, *Back/forth manoeuvre* (i.e., to escape from an obstructing object), *Check payload sensors*, etc.

Except that *Replace battery* must be preceded by a *Test SOB* action there are no natural orderings among the actions. Also, observations will be provided for all twenty evidence variables (symptom measurements and background information) before any decisions are going to be made. These two facts imply

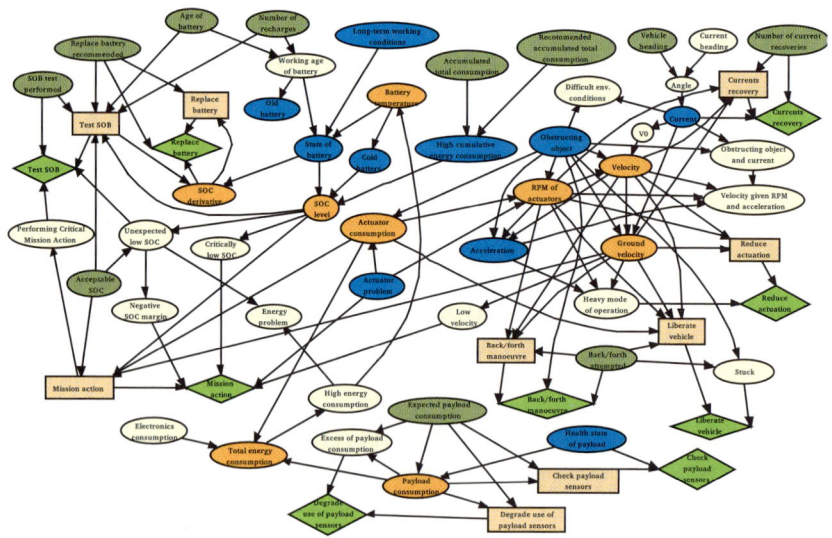

Fig. 7. The VORTEX LIMID.

that the model is not naturally represented as an influence diagram. Also, it would make exact inference absolutely intractable.

The LIMID framework therefore offers an ideal representation of this combined diagnosis and decision problem. In fact, the size of the junction tree for the network in Fig. 7 is only about 30K (measured as the sum of the sizes of the clique tables). We should note, however, that the "limited memory" aspect of the model contributes significantly to this fact, as there are observed variables that belong to the "relevant past" [16] of some decision variables that do not appear as parents of these variables. For example, according to the model in Fig. 7 the observed variables *RPM of actuators* and *Velocity* appear to be relevant for the *Mission action* decision, but there are no information links from these variables to the decision variable, as the *Actuator consumption* and the *Ground velocity* variables are assumed to cater for their influences.

Despite the "limited memory" aspect of the model in Fig. 7, preliminary evaluations of the model provided satisfactory results.

7.2 The BART AGV

The purpose of the PGM intelligent module of the BART AGV is very similar to that of the VORTEX AUV. The fact that the BART carries no payload systems and the obvious difference that the BART is an AGV and the VORTEX an AUV, give rise to some differences in the two models, but for the most part, the BART model shown in Fig. 8 constitutes a subset of the VORTEX model. After some adjustments of the model, a preliminary evaluation of the

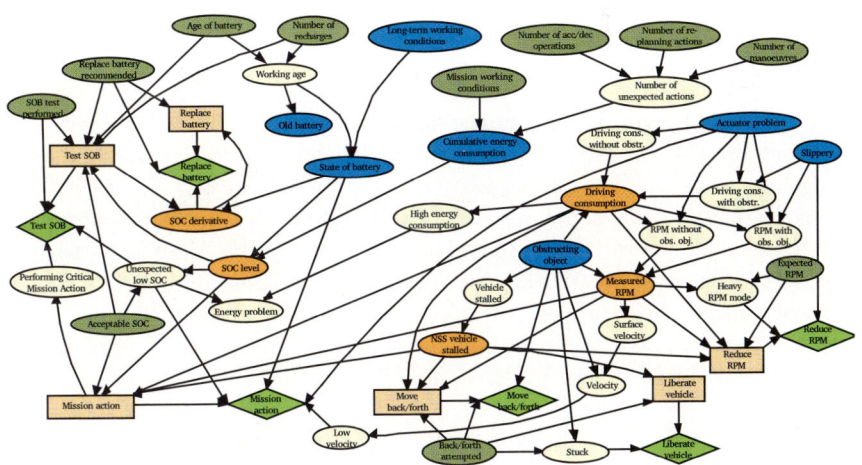

Fig. 8. The BART LIMID.

model showed an almost complete agreement between expert diagnoses and recommendations and those provided by the model.

7.3 The DeepC AUV

The purpose of the PGM intelligent module of the DeepC is to assess the quality of the sonar image and in the case of bad sonar image quality to suggest appropriate actions to avoid damage to the vehicle or even loss of vehicle.

The assessment of the sonar image quality is based on the computation of three sonar image quality indicators. The quality indicators are determined by the robot piloting module and fed into the model as evidence. The sonar image quality indicators are pixel entropy, pixel mean value, and pixel substance, see [8] for details. From the above description of the problem domain, it is clear that the amount of disturbance in the sonar image and the presence of objects is time dependent.

The main modeling challenges were to capture the dynamics of the process (how the quality indicators relate to the position and behaviour of the vehicle, the noise sources, and the quality of the image), to address the inherent infinite horizon problem, and to maintain a computationally efficient model (small cliques and policies).

We model the problem as a discrete time, finite horizon partially observed Markov decision process. The model is dynamic in the sense that it models the behaviour of the system over (discrete) time. The state of the system at any given point in time is partially observed as sensor readings are available, but not all entities of the problem domain are observed. The process is represented as an OO LIMID.

The top-level LIMID class $\mathcal{N}$ contains three instantiations, $\mathcal{M}_i$, $\mathcal{M}_{i+1}$, and $\mathcal{M}_{i+2}$, of the class $\mathcal{M}$ shown in Fig. 9; see Fig. 10 where $\mathcal{M}_i$ has label Ti. The input variables are located at the top of Fig. 9, while the output variables are located at the bottom. In $\mathcal{N}$, the output variables of $\mathcal{M}_i$ are connected to the input variables of the subsequent time-slice $\mathcal{M}_{i+1}$, and similarly for $\mathcal{M}_{i+1}$ and $\mathcal{M}_{i+2}$.

Each instantiation of $\mathcal{M}$ represents the system at a given point in time. The model $\mathcal{N}$ represents the system at three consecutive time steps with an 8 seconds interval, which is the time the image analysis component needs to analyze a single sonar image.

To avoid combinatorial explosion and thereby maintain computational efficiency, the model specifies that *Altitude*, *Depth*, *Pitch*, and *Speed* are observed prior to the decision *Recovery Action*, but not the image quality indicators. The values computed for the image quality indicators are inserted as evidence and subsequently policies are recomputed. Hence, the policy for the decision in the next time-slice will only depend on the most recent observations on *Altitude*, *Depth*, *Pitch*, and *Speed*. Since the image quality indicators are observed

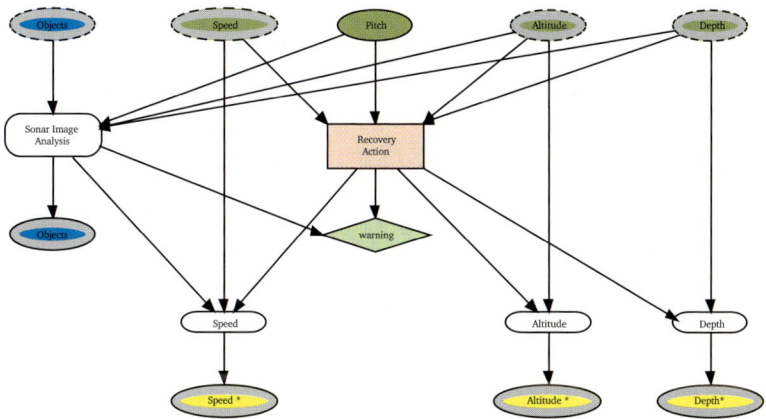

Fig. 9. The generic time-slice model for sonar image assessment

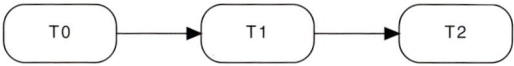

Fig. 10. The top-level LIMID class for sonar image assessment is a time-sliced model of three instances of the model (class) shown in Fig. 9.

each time a sonar image is analyzed, we need to resolve the LIMID with the observations on the image quality indicators entered as evidence.

The decision has a potential impact on speed, altitude, and depth of the vehicle. Deciding on a recovery action changing any of these properties will impact the quality of the next sonar image. The probability of a collision is modeled in the class instance *Speed*.

The instance *Sonar Image Analysis*, which is an instance of the network class shown in Fig. 11, models the sonar image assessment process. The three image quality indicators are represented in this class by the variables *Entropy*, *Mean*, and *Substance*. The quality indicators are influenced by the presence of disturbance or objects in the sonar image. Disturbance may be caused by reverberation or noise.

The hierarchical construction of the LIMID enforced by the object-oriented paradigm has simplified the knowledge acquisition phase considerably as it is easy to focus on well-defined subparts of the LIMID in isolation. Using class instances, it is a simple task to create and maintain multiple instances of the same LIMID class. Furthermore, it is a simple task to change the class of an instance to another class. This is particularly useful in the knowledge acquisition phase where each LIMID class has been revised and updated multiple times.

For further details on the DeepC model, see [8].

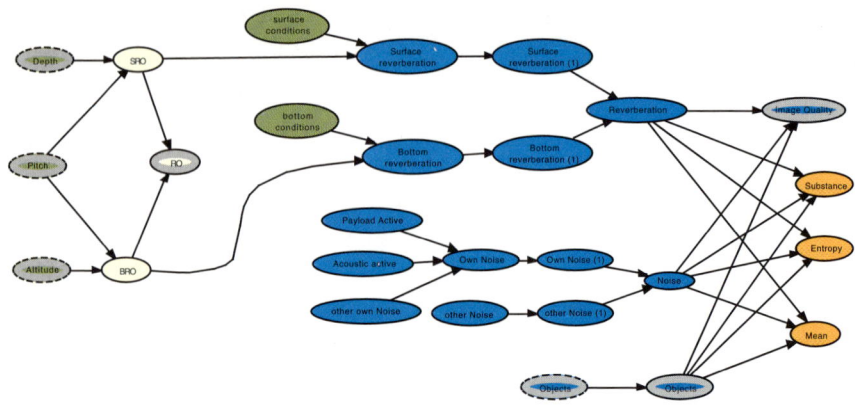

Fig. 11. The model (class) representing the sonar image assessment process.

8 Solving Models

A LIMID is solved using a message passing procedure in a junction tree as briefly described in Sect. 5. This procedure proceeds by iteratively passing messages in a junction tree representation of the LIMID. The procedure for solving LIMIDs differs from algorithms solving influence diagrams [4; 11]. The procedure for solving a LIMID is an iterative procedure working on a junction tree representation.

The complexity of solving a LIMID depends heavily on the structure of the junction tree, i.e., the sizes of cliques in the junction tree. The structure of the junction tree is determined by the connectivity of the graph of the LIMID, which in turn is determined by the number of nodes and links of the graph. The state space sizes of variables and the number of links determine the number of parameters of the model. Table 1 shows the number of chance and decision variables as well as the number of parameters to be specified in the three LIMID models constructed. For the DeepC model the number of variables in the instantiated network is shown.

Table 1. Number of variables and parameters of the three models.

Model	Chance variables	Decision variables	Total	Parameters
DeepC	121	3	124	3,556
VORTEX	57	9	66	3,475
BART	49	6	51	901

From Table 1 it can be seen that the DeepC model contains the largest number of variables while the VORTEX and BART models have approximately the same number of variables. The VORTEX model has a higher number of parameters though.

Table 2 shows the total clique state space of the optimal junction trees used for solving each LIMID where the optimality criterion is clique state space size. The total clique state space size is defined as $\sum_{C \in \mathcal{C}} \prod_{X \in C} \|X\|$, where $\mathcal{C}$ denotes the set of cliques of the junction tree and $\|X\|$ denotes the number of states of variable X.

Table 2. Total clique state space size of the three models.

Model	Influence Diagram	LIMID
DeepC	$> 2^{32} - 1$	108,841
VORTEX	$> 2^{32} - 1$	26,099
BART	14,017,742	39,322

Table 2 also shows the total clique state space size when each model is solved as an influence diagram. The table shows that the DeepC and VORTEX models when considered as influence diagrams cannot be solved on a standard 32 bits PC platform (each vehicle is equipped with a standard 32 bits PC platform running either Windows or Linux operating systems). Thus, applying probabilistic graphical models to diagnosis and control of autonomous vehicles would be infeasible without the use of LIMIDs.

9 Integration, Validation, and Real-World Trials

Each LIMID model is encapsulated in an intelligent module of the ADVOCATE II architecture (see Fig. 4). The intelligent module takes care of the communication with the decision module and robot piloting modules. Module integration was performed using a special-purpose integration tool, which has greatly simplified the integration process as it allows developers to integrate their modules into an architecture consisting of a mix of mock-up modules, man-machine interfaces, and real modules. This was very helpful as module developers were located in different countries with different working hours and with limited possibilities for face-to-face meetings.

The validation of each module is equivalent to validation of the knowledge bases (models). The validation of a model was performed by a careful investigation of the performance and behaviour of the system based on a selected set of test scenarios.

Extensive prototyping has helped to ensure appropriate performance and behaviour of each module. The test cases used to validate and measure the performance of a model covers all important, critical situations that possibly can occur in realistic operations. Finally, domain expert(s) and the knowledge engineers have evaluated the model against the test cases iteratively.

The usefulness of each LIMID was evaluated by a sequence of trials where each LIMID was deployed as an intelligent module as part of the ADVOCATE II architecture on the vehicles described above. In each trial the LIMID was deployed in an instantiation of the architecture consisting of a directory module, a decision module, a robot piloting module, and an intelligent module.

The DeepC model had an 88.4% accuracy on image quality assessment on the test set consisting of 1048 sonar images. The sonar images were recorded at the ATLAS test pond in Bremen where the real sea trials were performed.

The BART model had a 93.4% accuracy on 36 selected test cases. The test cases where hand generated to reflect operation of BART at the University of Alcalá campus.

No real trial data was available for the VORTEX model. Preliminary results on simulated data were promising, but the data is of insufficient quality for a proper validation.

10 Discussion

We have presented an application of HUGIN to solve reasoning and decision making problems under uncertainty related to diagnosis and control of autonomous vehicles. The application is based on the ADVOCATE II architecture. The ADVOCATE II architecture is a distributed architecture supporting diagnosis and control of autonomous vehicles.

The main objective of the ADVOCATE II project was to develop an architecture to allow the implementation of intelligent modules for AGVs and AUVs in order to increase their reliability and efficiency.

The performance of the ADVOCATE II architecture is constrained by (soft) real-time requirements. This implies that the performance of the communication protocol and the intelligent modules needs to be very high. This has implied a high focus on computational performance in the model construction.

Not only does the communication architecture enable efficient integration of different AI technologies into new and existing systems, but it also allows various AI techniques to interact (through the decision module and robot piloting module, though). This option of interactions has been used to dedicate PGMs to certain types of problems and to have variables in a PGM represent the output of an NSS intelligent module (e.g., variable "NSS vehicle stalled" of the BART LIMID; see Fig. 8). In addition, different AI techniques may be used to solve the same problem. This will often be the case for mission critical error handling. In our case, this raises the issue of a common scale of measurement of the usefulness of actions. We have chosen to use normalized expected utility as the measurement of usefulness of recovery actions.

Even though the framework of PGMs have been available for more than 15 years, it is our experience that the efficient use of these models in certain domains still requires some research and development. It is often a problem for the PGM technology that only a few very convincing success stories are available to the public. Often a PGM captures all or almost all the knowledge a company has in a particular business area. This implies that the company is not interested in sharing this model or even sharing the knowledge that such a model exists. The results of the ADVOCATE II project, however, add to the increasing number of successful applications of PGMs available to the public.

One of our key experiences from research and development projects is that even though graphical models are intuitive, they are difficult to build for inexperienced domain experts. In the ADVOCATE II project, it was necessary to develop a new methodology to solve the knowledge acquisition task. We have developed a knowledge elicitation and formulation method, which is applicable in general to problems of reasoning and decision making on complex machinery such as AGVs and AUVs.

The LIMID representation [11] and the object-oriented knowledge representation paradigm [10] implemented in the HUGIN tool [1; 5] have been two major cornerstones of the success of Bayesian modeling in the ADVOCATE II project. Still a lot of work remains to develop the object-oriented framework further.

Acknowledgment

We acknowledge the partners of ADVOCATE II: University of Alcalá (Spain), Getronics (France), Technical University of Madrid (Spain), ATLAS Elektronik GmbH (Germany), Ifremer (France), HUGIN Expert A/S (Denmark), Innova S.p.A (Italy), and e-motive (France). The EU Commission supported the ADVOCATE II project under grant IST-2001-34508. Visit the project web-site for more information on ADVOCATE II:

> http://www.advocate-2.com

Please visit the company web-site for more information on applications using HUGIN software:

> http://www.hugin.com

References

[1] S. K. Andersen, K. G. Olesen, F. V. Jensen, and F. Jensen. HUGIN — a Shell for Building Bayesian Belief Universes for Expert Systems. In *Proceedings of IJCAI'89*, pages 1080–1085, 1989.

[2] R. G. Cowell, A. P. Dawid, S. L. Lauritzen, and D. J. Spiegelhalter. *Probabilistic Networks and Expert Systems*. Springer-Verlag, 1999.
[3] R. A. Howard and J. E. Matheson. Influence Diagrams. In *The Principles and Applications of Decision Analysis*, volume 2, chapter 37, pages 721–762. 1981.
[4] F. Jensen, F. V. Jensen, and S. L. Dittmer. From Influence Diagrams to Junction Trees. In *Proceedings of 10th Conference on UAI*, pages 367–373, 1994.
[5] F. Jensen, U. B. Kjærulff, M. Lang, and A. L. Madsen. HUGIN - The Tool for Bayesian Networks and Influence Diagrams. In *Proceedings of PGM'02*, pages 212–221, 2002.
[6] F. V. Jensen. *Bayesian Networks and Decision Graphs*. Springer-Verlag, 2001.
[7] F. V. Jensen, S. L. Lauritzen, and K. G. Olesen. Bayesian updating in causal probabilistic networks by local computations. *Computational Statistics Quarterly*, 4:269–282, 1990.
[8] J. Kalwa and A. L. Madsen. Sonar image quality assessment for an autonomous underwater vehicle. In *Proceedings of ISORA'04*, 2004.
[9] U. B. Kjærulff and A. L. Madsen. A methodology for acquiring qualitative knowledge for probabilistic graphical models. In *Proceedings of IPMU'04*, pages 143–150, 2004.
[10] D. Koller and A. Pfeffer. Object-oriented Bayesian networks. In *Proceedings of UAI'97*, pages 302–313, 1997.
[11] S. L. Lauritzen and D. Nilsson. Representing and solving decision problems with limited information. *Management Science*, 47:1238–1251, 2001.
[12] S. L. Lauritzen and D. J. Spiegelhalter. Local computations with probabilities on graphical structures and their application to expert systems. *Journal of the Royal Statistical Society, Series B*, 50(2):157–224, 1988.
[13] A.L. Madsen, M. Lang, U. Kjærulff, and F. Jensen. The Hugin Tool for Learning Bayesian Networks. In *Proceedings of 7th European Conference on Symbolic and Quantitative Approaches to Reasoning with Uncertainty*, pages 594–605, 2003.
[14] R. E. Neapolitan. *Learning Bayesian Networks*. Prentice Hall, 2003.
[15] J. Pearl. *Probabilistic Reasoning in Intelligence Systems*. Series in Representation and Reasoning. Morgan Kaufmann Publishers, 1988.
[16] R. Shachter. Efficient Value of Information Computation. In *Proceedings of UAI'99*, pages 594–601, 1999.

Biomedical Applications of Bayesian Networks

Peter J.F. Lucas

Institute for Computing and Information Sciences, Radboud University Nijmegen, Toernooiveld 1, 6525 ED Nijmegen, The Netherlands,
peterl@cs.ru.nl

Summary. The central role played by uncertainty in medical decision making explains why medicine was amongst the first areas where applications based on Bayesian networks were developed. Biomedical research is firmly grounded on statistical methods; new methods are adopted only slowly by the field. During the past decade, however, Bayesian networks have become important tools for building decision-support systems in medicine and are now steadily becoming main stream. More recently, Bayesian networks have also been adopted as analytic tools in human biology, mainly in research that aims to elucidate the biological mechanisms underlying disease. In this paper, we review some of the applications in both medicine and human biology and we make an attempt to unravel some of the characteristics of Bayesian networks in biomedicine.

1 Introduction

Researchers that develop computer-based systems for use within medicine and human biology are normally confronted by the complexity and uncertain nature of processes in these areas. In many cases, the situation is even worse as many of the processes in medicine and human biology are only partly known. Whether partly or fully known, experience has shown that Bayesian networks with their associated methods are geared to reasoning with uncertainty in a way closely resembling medical doctors [45]. As uncertainty is one of the essential features of medical management of patients, consisting of the establishment of a diagnosis, treatment of identified disease of the patient, prediction of treatment outcome, and monitoring of the efficacy of the treatment, it may not come as a surprise that all of these areas have been explored using probabilistic methods, in particular by means of Bayesian networks. Examples of the deployment of Bayesian networks outside the direct clinical environment include the use of Bayesian networks for the construction of disease models in epidemiology and within bioinformatics for the interpretation of microarray gene expression data.

As the analysis of data in medicine and human biology, whether the data is obtained from clinical trials or from experimental research, is firmly based on statistics, one of the problems originally faced by researchers working in the area of probabilistic graphical models was that probabilistic graphical models, if at all familiar to the biostatistician, were considered to be unproven technology in comparison to traditional techniques, such as logistic regression. Researchers interested in applying probabilistic graphical models to biomedical problems were, therefore, working outside main-stream biomedical research. This situation has slowly changed in the last decade.

However, whereas researchers interested in probabilistic approaches to reasoning with uncertainty have moved beyond Bayesian networks, studying probabilistic graphical models based on chain graphs and maximal ancestral graphs, [54], biomedical applications are still mainly restricted to acyclic directed graph representations, i.e., Bayesian networks. This may be due to the fact that in medicine and biology it is considered to be important that a probabilistic graphical model can be given an interpretation in terms of cause-effect relationships. It is unclear whether or not this will remain so in the future.

The aim of the present paper is to provide some insight into the nature of the Bayesian-network and decision-theoretic models developed in the biomedical field. To this end, we will describe some of the network modelling issues involved in developing Bayesian networks in biomedicine. Furthermore, we will discuss some of the additional modelling and formalisation tasks that must be undertaken in solving biomedical problems using Bayesian networks. This will be illustrated by actual Bayesian network models from the biomedical field.

This paper is organised as follows. In the next Section, the formalism of Bayesian networks is introduced and methods for their construction are briefly reviewed. In Section 3, various aspects of modelling biomedical knowledge in Bayesian networks are discussed. Subsequently, in Section 4, we introduce the biomedical problems involving uncertainty for which Bayesian networks are typically employed. Next, in Section 5, some examples of Bayesian networks are discussed. The examples are meant to give the reader an impression of how biomedical problems are mapped to Bayesian network models. Finally, in Section 6, we draw some conclusions.

2 Bayesian Networks

In this section, the formalism of Bayesian networks and the basic methods for their development are reviewed. For a more thorough treatment of the topic, the reader is referred to [11; 45].

2.1 The Formalism

A *Bayesian network*, or probabilistic network, $\mathcal{B} = (G, \Pr)$ is a model of a joint, or multivariate, probability distribution over a set of random variables;

it consists of a graphical structure G and an associated joint probability distribution Pr. The graphical structure takes the form of an acyclic directed graph, or ADG for short, $G = (V(G), A(G))$, with nodes $V(G)$ and arcs $A(G) \subseteq V(G) \times V(G)$. Each node $v \in V(G)$ corresponds to a random variable X_v that in the case Pr is discrete takes one of a finite set of values. The arcs in the ADG model the probabilistic influences between the variables. Informally speaking, an arc $u \to v$ between two nodes u and v indicates that there is an influence between the associated variables X_u and X_v; absence of an arc between u and v means that the corresponding variables do not influence each other directly [11; 26].

Associated with the graphical structure of a Bayesian network is a joint probability distribution Pr, that is taken to be factorised according to the structure of the associated ADG. For each variable X_v in the joint probability distribution, with $v \in V(G)$, is specified a set of conditional probability distributions $\Pr(X_v \mid X_{\pi(v)})$; each of these distributions describes the joint effect of a specific combination of values for the variables $X_{\pi(v)}$ associated to the parents $\pi(v)$ of v, on the probability distribution over X_v's values. These sets of conditional probability distributions define together a unique joint probability distribution that factorises over the ADG's topology through

$$\Pr(X_{V(G)}) = \prod_{v \in V(G)} \Pr(X_v \mid X_{\pi(v)}) \quad (1)$$

by the local Markov property, that says that a random variable X_v is independent of the random variables associated with the nondescendants of v given the parents $\pi(v)$ of v.

By definition, the graphical part of a Bayesian network $\mathcal{B}$ is assumed to represent all dependences encoded in the joint probability distribution Pr, and possibly more. It is said that G is an independence map, or *I-map*, of Pr. As independences and dependences complement each other, the notion of I-map means that every independence in the ADG must be preserved by the joint probability distribution. More about the representation of independence information in Bayesian networks can be found in [11] and in this book.

Figure 1 shows an example Bayesian network; the notations x and $\neg x$ are used to indicate $X = true$ and $X = false$, respectively. The ADG of the network models *Cancer* to be independent of *Heart disease* given a value for their common parent *Smoking*. The conditional probability distributions associated with the variable *Cancer* in the figure further demonstrate that the local Markov property provides for a localised representation of the joint probability distribution. The property, in fact, serves to significantly reduce the amount of probabilistic information that has to be explicitly specified to uniquely describe the joint distribution. The property also allows for the design of efficient algorithms for computing any probability of interest over a network's variables [38; 45].

The ADG of a Bayesian network, apart from being acyclic, can have an arbitrarily complex topology to capture the intricacies of its application do-

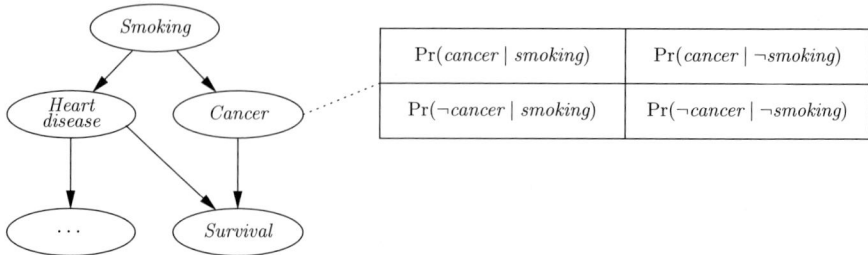

Fig. 1. An example Bayesian network.

main. For classification problems, however, a specific class of networks of limited topology have become popular [8; 16; 17]. In these networks, a distinction is made between a single class variable C and one or more feature variables; the latter variables serve to describe the characteristics of the instances to be classified. The class variable does not have any incoming arcs, but has arcs pointing to every feature variable. Between the feature variables, arcs are allowed under strict topological constraints. In a naive Bayesian network, for example, no arcs are allowed between the feature variables. In a tree-augmented Bayesian network, or TAN, on the other hand, arcs are allowed between the feature variables as long as these constitute a tree. In a forest-augmented network, or FAN, to conclude, the arcs should constitute a forest of trees [40]. The general structure of a naive Bayesian network and of a TAN network are shown in Figure 2.

Fig. 2. A naive Bayesian network, (a), and a tree-augmented Bayesian network, (b); the nodes F_j indicate the feature variables and C is the class variable.

Although the variables in a Bayesian network are often assumed to be discrete, taking a value from a finite set of values, a network may also include continuous variables that adopt a value from a range of real values [37]. Generally Gaussian, or normal, distributions are assumed for the conditional probability distributions for such continuous variables. These distributions then are specified in terms of a limited number of parameters, such as their means and variance. Most Bayesian-network tools nowadays allow for a mixture of discrete and continuous variables to be included in a network, under some topological constraints.

2.2 Manual Construction

Many of the Bayesian networks developed to date for real-life applications in biomedicine and health-care have been constructed by hand [3; 4; 23; 28; 29; 42; 43]. Manual construction of a network involves various development stages. For each of these stages, knowledge is acquired from experts in the domain of application, the relevant medical literature is studied, and available patient data are analysed. The following development stages are generally distinguished:

(1) *Selection of relevant variables.* As a Bayesian network in essence is a graphical model of a joint probability distribution over a set of random variables, the first stage in its construction is the identification of the important variables to be captured, along with the values they may adopt. The selection of the relevant variables is generally based on interviews with experts, descriptions of the domain, and an extensive analysis of the purpose of the network under construction. Often, knowledge about the (patho)physiological processes concerned is used to guide the identification of the relevant variables [34; 39].

(2) *Identification of the relationships among the variables.* Once the variables to be included in the network have been decided upon, the dependence and independence relationships between them have to be analysed and expressed in a graphical structure. For this purpose, generally the notion of causality is employed as a guiding principle: typical questions asked during the interviews with the domain experts are "What could cause this effect?" and "What manifestations could this cause have?". The elicited relationships are then expressed in graphical terms by taking the direction of causality for directing the arcs between the variables. The notion of causality often appears to match the experts' way of thinking about the (patho)physiological processes in their domain [20].

(3) *Identification of qualitative probabilistic and logical constraints.* Knowledge of qualitative probabilistic constraints and of logical constraints among the variables involved can help in the assessment and verification of the probabilities required for the network under construction. Qualitative probabilistic constraints are derived, for example, from properties of stochastic dominance of distributions. These constraints can be expressed as qualitative signs that can be used to study the reasoning behaviour of the projected network prior to its quantification [48]. Logical constraints are derived from functional relationships between the variables and can be used to significantly reduce the number of probabilities that have to be assessed for the network.

(4) *Assessment of probabilities.* In the next development stage, the local conditional probability distributions $\Pr(X_v \mid X_{\pi(v)})$ for each variable X_v are filled in. The required probabilities can be obtained from domain experts. Although the elicitation of judgemental probabilities is generally considered a daunting task, elicitation methods are available that are tailored

to obtaining the large number of probabilities required in reasonable time [22; 23; 47]. Alternatively, the probabilities can be obtained from data. For a network with discrete variables, the conditional probability distributions are often computed as the weighted average of a probability estimate based on the available data and a prior Dirichlet distribution, that is, a multinomial distribution whose parameters can be interpreted as counts on a dataset:

$$\Pr(X_v \mid X_{\pi(v)}, D) = \frac{n}{n+n_0} \widehat{\Pr}_D(X_v \mid X_{\pi(v)}) + \frac{n_0}{n+n_0} \Theta(X_v \mid X_{\pi(v)})$$

where $\widehat{\Pr}_D$ is the probability distribution estimated from a given dataset D, and Θ is the Dirichlet prior over the possible values of X_v; Θ is often taken to be uniform. The parameter n is the size of the dataset D and n_0 is equal to an imaginary or real number of past cases on which the contribution of Θ is based. The resulting probability distribution Pr is again a Dirichlet distribution.

(5) *Sensitivity analysis and evaluation.* With the previous development stage, a fully specified Bayesian network is obtained. Before the network can be used in real-life practice, its quality and clinical value have to be established. One of the techniques for assessing a network's quality is to perform a sensitivity analysis with patient data. Such an analysis serves to provide insight in the robustness of the output of the network to possible inaccuracies in the underlying probability distribution [10; 21]. Evaluation of a Bayesian network can be done in various different ways. Examples include measuring classification performance on a given set of real patient data and measuring similarity of structure or probability distribution to a gold-standard network or other probabilistic model.

As developing a Bayesian network is a creative process, the various stages are iterated in a cyclic fashion where each stage may, on each iteration, induce further refinement of the network under construction. An ontology may be developed to support the process [31].

2.3 Learning

In many fields of biomedicine and health-care, data have been collected and maintained, sometimes over numerous years. Such a data collection usually contains highly valuable information about the relationships between the variables discerned, be it implicitly. If a comprehensive dataset is available, a Bayesian network can be learnt from the data, that is, it can be developed without explicit access to knowledge of human experts.

To be suitable for learning purposes, a dataset has to satisfy various properties. First of all, the data contained in the dataset must have been collected very carefully. Biases that are introduced in the dataset as a result of the data collection strategies used will have impact on the resulting Bayesian network, yet may not be desirable for the purpose for which the network is being developed. Also, the variables and associated values that occur in the dataset should match the variables and values that are to be modelled in the network, or should at least admit easy translation. Moreover, the dataset should comprise enough data to allow for reliable identification of probabilistic relationships among the variables discerned. In addition to these general prerequisites, a dataset should satisfy several properties that are implicitly assumed by most learning algorithms. One of these is the assumption that each case in the dataset specifies a value for every variable discerned, that is, there are no missing values. Unfortunately, for most real-life datasets this property does not hold. To use a dataset with missing values for learning purposes, the missing values have to be filled in, or *imputated*, for example based upon (roughly) estimated probabilities for these values or with the help of domain experts. Most learning algorithms further assume that the cases in the dataset have been generated independently, that is, the values specified for the variables in a case are assumed not to be influenced in any way by the values in previously generated cases. Also, it is assumed that the process of data generation is not time-dependent.

Learning a Bayesian network from data involves the tasks of *structure learning*, that is, identifying the graphical structure of the network, and *parameter learning*, that is, estimating the conditional probability distributions to be associated with the network's ADG. In many learning algorithms, the two tasks are performed simultaneously and, as a consequence, are not easily distinguished.

One of the early algorithms for learning a Bayesian network from data is the *K2* algorithm [9]. Given a dataset D, this algorithm searches, in a greedy heuristic way, for an ADG that, supplemented with maximum likelihood estimates for its probabilities, best explains the data at hand. More formally, it searches for a digraph G^* that maximises the joint probability $\Pr(G, D)$ over all possible digraphs G. Given a topological ordering on the random variables concerned, the algorithm constructs, for every subsequent variable V_i, an optimal set of parents. To this end, it starts by assuming the parental set to be empty and then adds, iteratively, the parent whose addition most increases the probability of the resulting structure and the dataset; it stops adding variables to a parental set as soon as the addition of a single parent cannot increase the probability $\Pr(G, D)$. The K2 algorithm is an example of a *search & scoring method*. These methods search the space of all possible acyclic digraphs by generating various different graphs in a heuristic way and comparing these as to their ability to explain the data at hand. Other search & scoring methods build, for example, upon the use of the minimum description length (MDL) principle [35], or use a genetic algorithm for the search involved [36].

Another approach to learning a Bayesian network from data is to build upon the use of a *dependence analysis* [7]. A Bayesian network in essence models a collection of conditional dependence and independence statements, through its Markov condition. By studying the available dataset, the dependences and independences between the various variables can be extracted, for example by means of statistical tests, and subsequently captured in a graphical structure. The *information-theoretical algorithm* of Cheng and Bell is an example of an algorithm taking this approach [7]. The algorithm has three subsequent phases, termed drafting, thickening and thinning. In the drafting phase, the algorithm establishes, from the data, the mutual information for each pair of variables and constructs a draft digraph from this information. In the thickening phase, the algorithm adds arcs between pairs of nodes if the corresponding variables are not conditionally independent given a certain conditioning set of variables. In the thinning phase, to conclude, each arc of the graph obtained so far is examined using conditional independence tests, and is removed if the two variables connected by the arc prove to be conditionally independent.

Based upon the observation that independence tests quickly become unreliable for larger conditioning sets and the search space of all possible digraphs is unfeasibly large, learning algorithms have been proposed that take a hybrid approach [15; 55]. These algorithms are composed of two phases. In the first phase, a graph is constructed from the data, generally using lower-order dependence tests only. This graph is subsequently used to explicitly restrict the search space of graphical structures for the second phase in which a search algorithm is employed to find an ADG that best explains the data.

To conclude, there is also a great deal of interest in estimating probability distributions from data using maximum likelihood estimation [27]. The *expectation-maximisation (EM) algorithm* is a two-step algorithm used by many researchers for this purpose [13]. It consists of a step of computing the expected value of the relevant parameter and a maximisation step, which are carried out in an interleaved fashion until convergence. In contrast with the learning algorithms reviewed above, the EM algorithm is able to deal with missing values.

2.4 Manual Construction versus Learning

Manual construction of a Bayesian network requires access to knowledge of human experts and, in practice, turns out to be quite time consuming. With the increasing availability of clinical and biological data, learning evidently is the more feasible alternative for developing a Bayesian network. Learning, as a consequence, is attracting considerable interest, both from developers and within the research community. Whether or not building a Bayesian network by hand would result in a network of higher quality when compared to learning it from data, is yet an open question. One would expect that, in many areas of biomedicine, human knowledge of the underlying (patho)physiological processes is more robust than the knowledge embedded in a dataset of limited size. To date there is little evidence, however, to corroborate this expectation. It is an equally open question whether learning a Bayesian network of more complex topology pays off when compared to learning a simple Bayesian classifier. One would expect that the more faithful the ADG of a Bayesian network is in reflecting the dependences and independences embedded in the data, the better its performance. Research by Domingos and Pazzani has shown, however, that, when used for classification problems, naive Bayesian networks tend to outperform more sophisticated networks [16]. This finding has led to the suggestion that more complex network structures do not pay off. Friedman et al. [17], and Cheng and Greiner [8], on the other hand, have shown that tree-augmented networks, which in comparison to naive Bayesian networks incorporate extra dependences among their feature variables, often outperform these naive Bayesian networks. Allowing for even more complex relationships between the feature variables, as in a forest-augmented network, moreover, has been shown to yield still better performance [40].

3 Exploiting Biomedical Knowledge in Bayesian Network Design

So far, we have considered general properties of Bayesian network construction. Developing a Bayesian-network model for a realistic biomedical problem is usually far from easy. Without having detailed knowledge of the problem domain, or without access to people with biomedical domain knowledge, it is impossible to develop Bayesian networks to solve biomedical problems, even in the situation when lots of data are available. However, mapping biomedical knowledge to the formalism of Bayesian networks is greatly facilitated if particular principles of biomedical knowledge are followed.

As mentioned above, modelling in terms of cause-effect relationships is particularly popular in developing biomedical applications. We will explore some of the alternatives in this section, trying to remain as close as possible to what we consider to be the essence of biomedical knowledge, with an emphasis on the representation of clinical knowledge.

3.1 Signs, Symptoms and Tests

Whether concerned with the diagnosis or treatment of a disease, medical management of patients is guided by proper history taking, making observations on the current clinical state of the patient and observing the evolution of the patient's disease.

Figure 3 shows a typical Bayesian network fragment, which includes nodes representing signs, symptoms and tests. Modelled here are a disorder, flu, with one sign (body temperature), one symptom (chills) and one test (erythrocyte sedimentation rate, ESR). As one may see, there is no difference in the way signs, symptoms and tests are modelled in a Bayesian network. In all cases, there is an arc going from the disorder to the associated sign, symptom or test. This graphical representation assumes that the signs, symptoms and tests

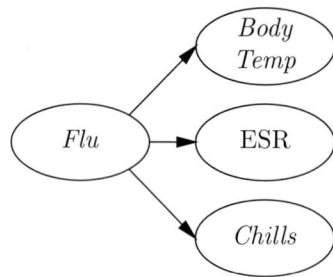

Fig. 3. Fictitious Bayesian network fragment concerning flu representing signs, symptoms and tests.

are conditionally independent of each other given the disorder; this is often not the case. For example, literature on human physiology tells us that body temperature and chills are dependent of one another.

For some signs and symptoms, and for most tests (using T as a random variable representing signs, symptoms or tests), it is known from literature how often they are positive in the presence of a disorder D, i.e., $\Pr(t \mid d)$ is known, and how often they are negative in the absence of the given disorder, i.e., $\Pr(\neg t \mid \neg d)$ is known. The probability $\Pr(t \mid d)$ is known as the *sensitivity* or true positive rate; the probability $\Pr(\neg t \mid \neg d)$ is known as the *specificity* or true negative rate. The sensitivity and specificity of particular tests can often be looked up in standard textbooks.

If particular symptoms or signs T are known to be uniquely associated with a disease D, these symptoms are said to be *pathognomonic*. Probabilistically, pathognomonic signs or symptoms T can be characterised by the fact that $\Pr(t \mid d) = 1$.

3.2 Causal Modelling

Figure 3 offers a typical, abstract clinical view on the clinical interpretation of information that is used in medical decision-making. However, in many cases much more is known about disease processes than is modelled in the naive Bayesian network structure shown in Figure 3. If one wishes to build a Bayesian network model for a biomedical problem, it is necessary to select the right level of abstraction in the building process. Often this is difficult, and much insight into the nature of biomedical knowledge is required in order to do so properly.

Figure 4 depicts a more elaborate version of the Bayesian network structure shown in Figure 3. It now turns out that there is a hidden variable, *Fever*, that is seen as a *common cause* of changes in the likelihood of *Body Temp* and *Chills*. The problem here is that fever is a complex response of the body to an infection, of which much is known, but that can only be characterised by measuring particular molecular substances. Thus, the node *Fever* summarises

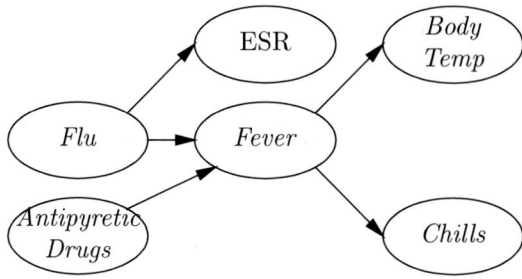

Fig. 4. Fictitious Bayesian network; more elaborate fragment concerning flu including a hidden variable (fever) with associated signs, symptoms and tests.

the entire process, and it may not be easy to quantify the corresponding random variable probabilistically. It is apparent from the figure that the changes in ESR are a direct consequence of flu, and not of fever. Finally, the likelihood of the occurrence of fever is moderated by the fact whether or not the patient uses antipyretic drugs, such as aspirin or acetaminophen. The occurrence of a fever is a *common consequence* of the interaction between the presence or absence of flu and the use of antipyretic drugs, yet the role of both variables is not the same. This is reflected in the probability distribution, not in the structure of the independence relation modelled in the ADG; here we have that

$$\Pr(\textit{fever} \mid \textit{flu, antipyretic drugs}) \ll \Pr(\textit{fever} \mid \textit{flu}, \neg\textit{antipyretic drugs})$$

3.3 Causal Independence

In biomedical problems, often a number of causal factors interact to give rise to an effect. If this is modelled in a Bayesian network, the resulting probability table may be prohibitively large. The theory of *causal independence* has been shown to be useful in this case, as it allows for the specification of the interactions among variables in terms of cause-effect relationships, adopting particular statistical independence assumptions [41]. Causal independence is frequently used in the construction of practical networks for situations where the underlying probability distributions are complex. The theory has also been exploited to increase the efficiency of probabilistic inference in Bayesian networks [56].

The global structure of a causal-independence model is shown in Figure 5; it expresses the idea that causes $C_1, \ldots, C_n$ influence a given common effect E through intermediate variables $I_1, \ldots, I_n$ and a deterministic function f, called the *interaction function*. The influence of each cause C_k on the common effect E is independent of each other cause C_j, $j \neq k$. The function f represents in which way the intermediate effects I_k, and indirectly also the causes C_k, interact to yield a final effect E. Hence, this function f is defined

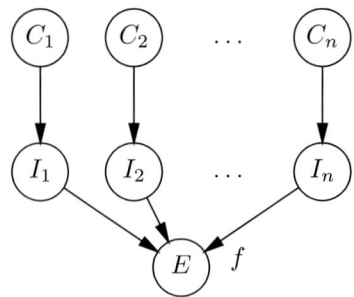

Fig. 5. Causal independence model.

in such way that when a relationship, as modelled by the function f, between I_k, $k = 1, \ldots, n$, and $E = \text{true}$ is satisfied, then it holds that $e = f(I_1, \ldots, I_n)$. In terms of probability theory, the notion of causal independence can be formalised for the occurrence of effect E, i.e. $E = \text{true}$, as follows:

$$\Pr(e \mid C_1, \ldots, C_n) = \sum_{f(I_1, \ldots, I_n) = e} \Pr(e \mid I_1, \ldots, I_n) \Pr(I_1, \ldots, I_n \mid C_1, \ldots, C_n) \tag{2}$$

meaning that the causes $C_1, \ldots, C_n$ influence the common effect E through the intermediate effects $I_1, \ldots, I_n$ only when $e = f(I_1, \ldots, I_n)$ for certain values of I_k, $k = 1, \ldots, n$. Under this condition, it is assumed that $\Pr(e \mid I_1, \ldots, I_n) = 1$; otherwise, when $f(I_1, \ldots, I_n) = \neg e$, it holds that $\Pr(e \mid I_1, \ldots, I_n) = 0$. Note that the effect variable E is conditionally independent of $C_1, \ldots, C_n$ given the intermediate variables $I_1, \ldots, I_n$, and that each variable I_k is only dependent on its associated variable C_k; hence, it holds that

$$\Pr(e \mid I_1, \ldots, I_n, C_1, \ldots, C_n) = \Pr(e \mid I_1, \ldots, I_n)$$

and

$$\Pr(I_1, \ldots, I_n \mid C_1, \ldots, C_n) = \prod_{k=1}^{n} \Pr(I_k \mid C_k)$$

Formula (2) can now be simplified to:

$$\Pr(e \mid C_1, \ldots, C_n) = \sum_{f(I_1, \ldots, I_n) = e} \prod_{k=1}^{n} \Pr(I_k \mid C_k) \tag{3}$$

Based on the assumptions above, it also holds that

$$\Pr(e \mid C_1, \ldots, C_n) = \sum_{I_1, \ldots, I_n} \Pr(e \mid I_1, \ldots, I_n) \prod_{k=1}^{n} \Pr(I_k \mid C_k) \tag{4}$$

Finally, it is assumed that $\Pr(i_k \mid \neg c_k) = 0$ (absent causes do not contribute to the effect); otherwise, the probabilities $\Pr(I_k \mid C_k)$ are assumed to be positive.

Formula (3) is practically speaking not very useful, because the size of the specification of the function f is exponential in the number of its arguments. The resulting probability distribution is therefore in general computationally intractable, both in terms of space and time requirements. An important subclass of causal independence models, however, is formed by models in which the deterministic function f can be defined in terms of separate binary functions g_k, also denoted by $g_k(I_k, I_{k+1})$. Such causal independence models have been called *decomposable* causal independence models [30]; these models are of significant practical importance. Usually, all functions $g_k(I_k, I_{k+1})$ are identical for each k; a function $g_k(I_k, I_{k+1})$ may therefore be simply denoted by $g(I, I')$. Typical examples of decomposable causal independence models are the noisy-OR [14; 32; 45] and noisy-MAX [14; 30] models, where the function g represents a logical OR and a MAX function, respectively.

In [41], the qualitative behaviour of causal independence models is studied when taking f to be a Boolean function. Insight into such qualitative behaviour may be useful when selecting the right Boolean function for a biomedical problem.

As an example, consider the interaction between bactericidal antimicrobial agents, i.e., drugs that kill bacteria by interference with their metabolism, and bacteriostatic antimicrobial agents, i.e., drugs that inhibit the multiplication of bacteria. Penicillin is an example of a bactericidal drug, whereas chlortetracyclin is an example of a bacteriostatic drug. It is well known among medical doctors that the interaction between bactericidal and bacteriostatic drugs can have antagonistic effects; e.g., the drug combination penicillin and chlortetracyclin may be have as little effect against an infection as prescribing no antimicrobial agent at all, even if the bacteria are susceptible to each of these drugs. Note that here we interpret drugs as statistical variables, not as decision variables as in clinical decision making. The depiction of the causal interaction of the relevant variables is shown in Figure 6. The interaction between penicillin and chlortetracyclin as depicted in Figure 6 can be described my means of an exclusive OR, as presence of either of these in the patient's body tissues leads to a decrease in bacterial growth, whereas if both are present or absent, there will be little or no effect on bacterial growth.

3.4 Context and Conditioning

Not all biomedical knowledge that can be modelled in a Bayesian network is causal in nature; another type of knowledge that is employed in biomedicine concerns subgroup knowledge, i.e., knowledge where there is much similarity in the description of a process, with the exception of some of the entries in some probability tables. Random variables that allow taking into account characteristics of subgroups can be looked upon as defining a *context*. Gender and age are typical examples of contextual random variables; they are frequently employed in real-world biomedical Bayesian networks.

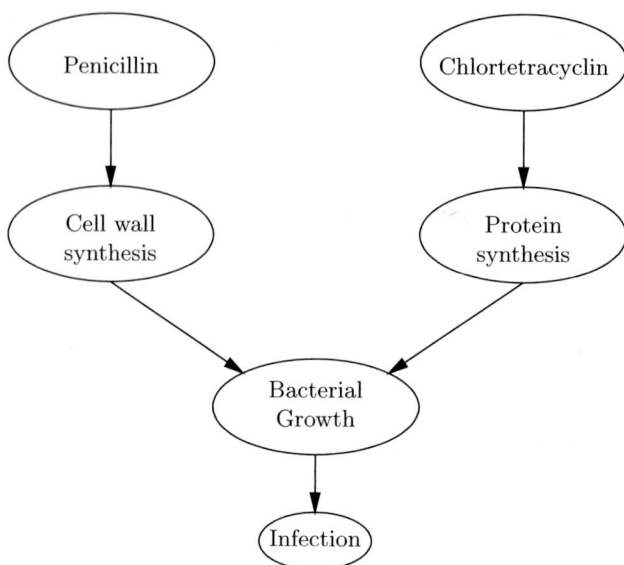

Fig. 6. Example Bayesian network, modelling the interaction between the antimicrobial agents penicillin and chlortetracyclin on infection.

Figure 7 depicts a fragment of a Bayesian network of non-Hodgkin lymphoma of the stomach, where each of the variables different from the variable 'Age' has been modelled to be dependent of the age of the patient. Usually, contextual nodes do not have incoming arcs, and have outgoing arcs to other nodes in the ADG.

4 Problem Solving in Biomedicine and Health-care

Bayesian networks are increasingly used in biomedicine and health-care to support different types of problem solving, four of which are briefly reviewed here.

4.1 Diagnostic Reasoning

Establishing a diagnosis for an individual patient in essence amounts to constructing a hypothesis about the disease the patient is suffering from, based upon a set of indirect observations from diagnostic tests. Diagnostic tests, however, generally do not serve to unambiguously reveal the condition of a patient: the tests typically have true-positive rates and true-negative rates unequal to 100%. To avoid misdiagnosis, the uncertainty in the test results obtained for a patient should be taken into consideration upon constructing a diagnostic hypothesis. Bayesian networks offer a natural basis for this type

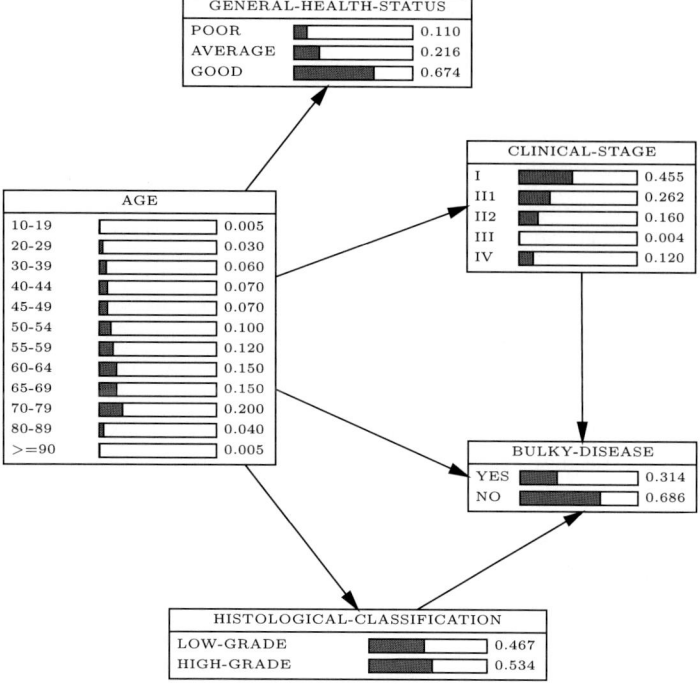

Fig. 7. Bayesian network fragment including some variables concerning non-Hodgkin lymphoma of the stomach.

of reasoning with uncertainty. A significant number of network-based systems for medical diagnosis have in fact been developed in the past and are currently being developed.

Formally, a diagnosis may be defined as a value assignment $\mathcal{D}^*$ to a subset of the random variables concerned, such that

$$\mathcal{D}^* = \arg\max_{\mathcal{D}} \Pr(\mathcal{D} \mid \mathcal{E})$$

where $\mathcal{E}$ is the observed evidence, composed of symptoms, signs and test results. A diagnosis thus is a *maximum a posteriori assignment*, or MPA, to a given subset of variables. Establishing a maximum a posteriori assignment from a Bayesian network, however, is extremely hard from a computational point of view. Since in addition combinations of disease do not occur very often, diagnostic reasoning is generally focused on single diseases. One approach is to assume that all diseases are mutually exclusive. The different possible diseases then are taken as the values of a *single disease variable*. Another approach is to capture each possible disease by a separate variable. Reasoning then amounts to computing the probability distribution for each such variable separately. The combination of the most likely values for these separate disease variables, however, need not be a maximum a posteriori assignment to these variables.

To assist physicians in the complex task of diagnostic reasoning, a Bayesian network is often equipped with a test-selection method that serves to indicate which tests had best be ordered to decrease the uncertainty about the disease present in a specific patient [2]. A test-selection method typically employs an information-theoretic measure for assessing diagnostic uncertainty. Such a measure is defined on a probability distribution over a disease variable and expresses the expected amount of information required to establish the value of this variable with certainty. An example measure often used for this purpose is the Shannon entropy. The measure can be extended to include information about the costs involved in performing a specific test and about the side effects it can have. Since it is computationally hard to look beyond the immediate next diagnostic test, test selection is generally carried out *non-myopically*, that is, in a sequential manner. The method then suggests a test to be performed and awaits the user's input; after taking the test's result into account, the method suggests a subsequent test, and so on.

4.2 Prognostic Reasoning

Prognostic reasoning in biomedicine and health-care amounts to making a prediction about what will happen in the future. As knowledge of the future is inherently uncertain, in prognostic reasoning uncertainty is even more predominant than in diagnostic reasoning. Another prominent feature of prognostic reasoning when compared to diagnostic reasoning is the exploitation of knowledge about the evolution of processes over time. Even if temporal knowledge is not represented explicitly, prognostic Bayesian networks still have a clear general temporal structure, which is depicted schematically in Figure 8. The outcome predicted for a specific patient is generally influenced by the particular sequence of treatment actions to be performed, which in turn may depend on the information that is available about the patient before the treatment is started. The outcome is often also influenced by progress of the underlying disease itself.

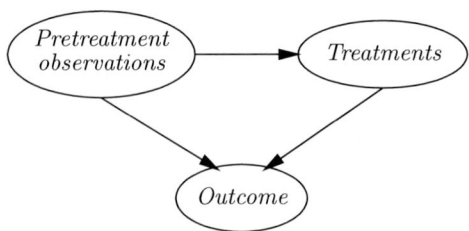

Fig. 8. General structure of a prognostic Bayesian network; each box denotes a part of the network.

Formally, a prognosis may be defined as a probability distribution

$$\Pr(\textit{Outcome} \mid \mathcal{E}, \mathcal{T})$$

where $\mathcal{E}$ again is the available patient data, including symptoms, signs and test results, and $\mathcal{T}$ denotes a selected sequence of treatment actions. The outcome of interest may be expressed by a single variable, for example modelling life expectancy. The outcome of interest, however, may be more complex, modelling not just length of life but also various aspects pertaining to quality of life. A subset of variables may then be used to express the outcome.

Prognostic Bayesian networks are a rather new development in medicine. Only recently have researchers started to develop such networks, for example in the areas of oncology [25; 42] and infectious disease [3; 43]. There is little experience as yet with integrating ideas from, for example, traditional survival analysis into Bayesian networks. Given the importance of prognostication in health-care, it is to be expected, however, that more prognostic networks will be developed in the near future.

4.3 Treatment Selection

The formalism of Bayesian networks provides only for capturing a set of random variables and a joint probability distribution over them. A Bayesian network therefore allows only for probabilistic reasoning, as in establishing a diagnosis for a specific patient and in making a prediction of the effects of treatment. For making decisions, as in deciding upon the most appropriate treatment alternative for a specific patient, the network formalism does not provide. Reasoning about treatment alternatives, however, involves reasoning about the effects to be expected from the different alternatives. It thus involves diagnostic reasoning and, even more prominently, prognostic reasoning. To provide for selecting an optimal treatment, a Bayesian network and its associated reasoning algorithms are therefore often embedded in a decision-support system that offers the necessary constructs from decision theory to select an optimal treatment given the predictions [3; 42]. Alternatively, the Bayesian-network formalism can be extended to include knowledge about decisions and preferences. An example of such an extended formalism is the influence-diagram formalism [51]. Like a Bayesian network, an *influence diagram* includes an acyclic directed graph. In this graph, the set of nodes is partitioned into a set of probabilistic nodes modelling random variables, a set of decision nodes modelling the various different treatment alternatives, and a value node modelling the preferences involved. Influence diagrams for treatment selection once again have a clear general structure, which is depicted schematically in Figure 9.

4.4 Discovering Functional Interactions

So far we have focused on the use of once constructed Bayesian networks for problem solving in medicine and health-care. However, the insight obtained

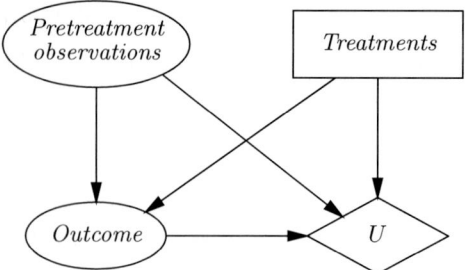

Fig. 9. General structure of an influence diagram, including a prognostic Bayesian network and a utility node U; each ellipse and box denotes a part of the diagram.

by the construction process itself, in particular when done automatically by using one of the learning methods described above, may also be exploited to solve problems. As the topology of a Bayesian network can be interpreted as a representation of the uncertain interactions among variables, there is a growing interest in bioinformatics to use Bayesian network for the unravelling of molecular mechanisms at the cellular level. Such mechanisms are called *metabolic pathways*. For example, finding interactions between genes based on experimentally obtained expression data in microarrays is currently a significant research topic [18]. Biological data are often collected over time; the analysis of the temporal patterns may reveal how the variables interact as a function of time. This is a typical task undertaken in molecular biology. Bayesian networks are now also being used for the analysis of such biological time series data [46].

5 Some Examples

In the previous sections, we have sketched some of the developments in Bayesian-networks research in biomedicine and health-care. We conclude this paper with a brief description of some actual systems, some old and some new. The description of the old systems allows us to draw some new conclusions in light of what we know today. The description of the new systems gives some insight on how far the field has progressed. No attempt has been made to be complete.

5.1 Clinical Diagnostic Systems

One of the earliest examples of a Bayesian network for the diagnosis of disease is MUNIN (MUscle an Nerve Inference Network), a large Bayesian network that is aimed at supporting clinicians in diagnosing muscle and nerve disorders [4]. The 2001 version of MUNIN included descriptions of 22 disorders, 186 diagnostic findings and consisted of 1100 nodes in total [53]. MUNIN is a typical example of a Bayesian network that has been modelled on the basis of clinical expertise and knowledge from medical literature. Similar remarks can be made regarding Pathfinder, the other frequently cited early diagnostic application of Bayesian networks, in this case concerned with the diagnosis of lymphatic disease [28; 29].

The work on MUNIN and Pathfinder is related to the even earlier work on probabilistic diagnosis, as exemplified by the work of De Dombal [12] and Spiegelhalter and Knill-Jones [52]. In the early work, modelling was done in the framework of naive Bayesian networks [12], sometimes by extending this framework by incorporating ways to express dependence information [52]. As Bayesian networks offer a systematic way to model such (possibly subjective) dependence information, the advances in probabilistic graphical models appear to have made this early work outdated. However, on the one hand, one needs to take into account that modelling can be very time consuming (in the case of MUNIN it took more than 10 years), and, as mentioned in Section 2.4, restricted Bayesian network models can be hard to beat when it comes to diagnostic performance. On the other hand, building an accurate diagnostic Bayesian network that includes descriptions of multiple disorders may be very difficult without resorting to the full technology, of Bayesian networks, and this is something that may be concluded from work in the Pathfinder project.

That unrestricted Bayesian networks have certain merits has also become clear when developing a Bayesian network for a situation when biomedical data is scarce. In [5], Antal et al. have explored the potential of the huge collection of information available on the World Wide Web as prior information for learning Bayesian networks for the classification of ovarian tumours. Such information can complement subjective clinical information and collected clinical data. In this research, techniques developed in the area of information retrieval were used as a basis for finding relationships among variables from the Web.

5.2 Treatment Management Systems

Whereas it can be doubted whether unrestricted Bayesian networks offer the best approach in all circumstances in building diagnostic systems, with systems that support treatment selection the trade-off is different. Here it is mandatory that the resulting probabilistic model is as accurate as possible given the resources available, and, thus, the use of restricted Bayesian network topologies is not an obvious choice here.

The literature includes a number of good examples of Bayesian networks for treatment selection, which, as discussed in Section 4.3, are normally extended to obtain a decision-theoretic model, possibly represented as an influence diagram. Decision support in the treatment of infectious disease and various types of cancer are relevant research subjects.

A typical example of a Bayesian network for treatment selection is the model of ventilator-associated pneumonia shown in Figure 10. The Bayesian

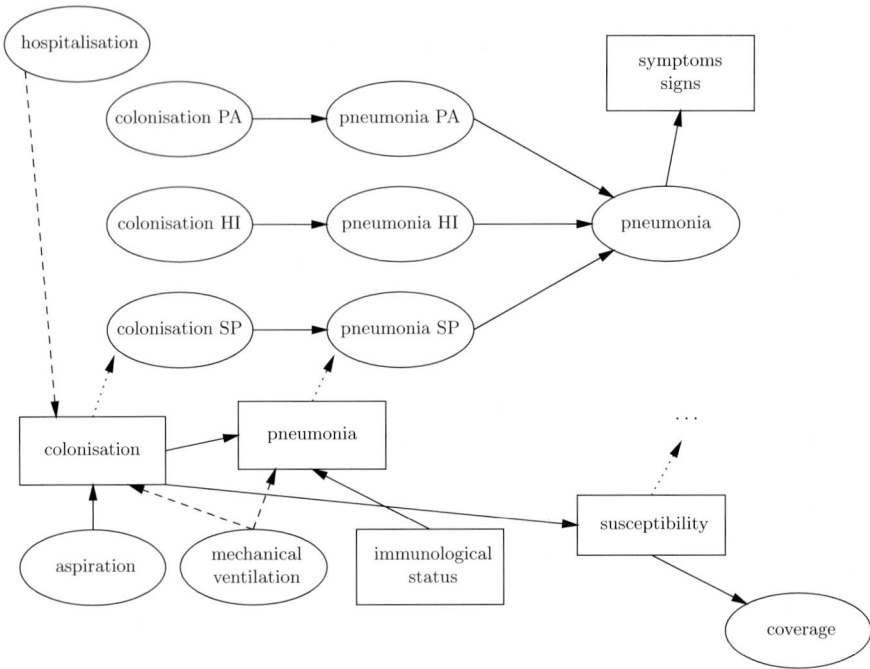

Fig. 10. Detailed structure of part of the probabilistic VAP model. Only three of the seven microorganisms included in the model are shown. Dotted arcs point to the actual topology of the Bayesian network; boxes represent sets of similar nodes. Abbreviations for bacteria: PA = Pseudomonas aeruginosa, HI = Haemophilus influenzae, SP = Streptococcus pneumoniae.

network models the evolution of pneumonia in patients by means of bacterial colonisation. In addition, the signs and symptoms resulting of pneumonia are represented in the network as is the extent to which combinations of antibiotics are able to cover particular bacteria.

The aims of the TREAT project is to develop a decision-support system, based on a Bayesian network, that can assist in the treatment of various causes of bacteremia (presence of bacteria in the blood that can give rise to sepsis). The resulting system has been one of few systems that has been extensively evaluated using prospective data from various hospitals [44]. The Bayesian network underlying this system has been carefully designed, guided by expert microbiological knowledge [3].

5.3 Health Care and Epidemiology

The use of Bayesian networks in the wider scope of health care has recently become a focus of study by some researchers. Acid at al. have carried out a preliminary study of the use of Bayesian networks as tools for the management of health services [1]. This has been done by learning Bayesian network structures from data collected at a hospital.

Getoor et al. have examined the use of Bayesian models to analyse tuberculosis epidemiology [24]. This was done be an extended Bayesian network formalism: statistical relational models. The difference between learning statistical relational models and ordinary Bayesian networks is that in the former it is assumed that data are organised as a collection of tables (relations), so that learning takes place by inspecting tables in a relational dataset that are explicitly linked to each other.

Sebastiani et al. have studied the use of dynamic Bayesian networks, i.e., Bayesian networks with time-stamped variables, to develop a surveillance system for influenza [49]. This is another example of a Bayesian network used in health care, where its structure has been inspired by the wish to model biological principles.

5.4 Discovering Regulatory Processes

Bayesian networks can also be employed to explore regulatory, or control, principles of biological processes. Segal et al. have shown that by learning Bayesian networks from data it is possible to obtain insight into the way genes are regulated [50]. The results suggested regulatory roles for previously uncharacterised proteins. Bulashevka et al. have used Bayesian-network learning to uncover the mechanisms underlying the progression in genetic changes in the development of urothelial cancers of the bladder [6]. Using probabilistic reasoning, taking into account the statistical dependences and independences encoded in the Bayesian network, the authors were able to propose hypotheses regarding the primary and secondary events in tumour pathogenesis.

The use of Bayesian networks in discovering protein-protein interaction has been studied by Jansen et al. [33], using a naïve Bayesian network, as shown in Figure 2(a), and a fully connected network, i.e. where each pair of variables in the network is connected by a link. The first structure was used to encode data where information about interactions between proteins were not available; the fully connected network captures all possible interactions among proteins.

An overview of recent successes in discovering cellular processes by means of Bayesian networks is given in Ref. [19], which is of particular interest to clinical researchers engaged in unravelling the molecular mechanisms underlying disease.

6 Conclusions

In this paper, we have sketched the complete process of the development of Bayesian networks and related probabilistic graphical models in biomedicine and health care. Not only general principles – many of these are identical or at least similar to those in other fields – but also special features of Bayesian network semantics in these areas were described. It was also discussed how these representation formalisms can be deployed in the process of biomedical decision making. Finally, a number of example systems were reviewed.

Many researchers believe that probabilistic graphical models, and in particular Bayesian networks, are *the* appropriate tools for medical decision support and data analysis. Whereas graphical representations seem to have found their way in statistical data analysis, it is as yet too early to conclude anything about whether or not probabilistic graphical models will be the future formalism for building medical decision-support systems, replacing current paper-based medical guidelines and protocols. Many researchers in the field have recently started with such research.

Acknowledgements

Linda van der Gaag is provided significant input on an earlier version of this paper.

References

[1] S. Acid, L.M. de Campos, J.M. Fernández-Luna, S. Rodríguez, J.M. Rodríguez and J.L. Salcedo. A comparison of learning algorithms for Bayesian networks: a case study based on data from an emergency medical service. *Artif Intell Med*, 30(3):215–232, 2004.

[2] S. Andreassen. Planning of therapy and tests in causal probabilistic networks. *Artif Intell Med*, 4:227–241, 1992.

[3] S. Andreassen, C. Riekehr, B. Kristensen, Schønheyder HC, Leibovici L. Using probabilistic and decision-theoretic methods in treatment and prognosis modeling. *Artif Intell Med*, 15:121–134, 1999.

[4] S. Andreassen, M. Woldbye, B. Falck, S.K. Andersen. MUNIN – A causal probabilistic network for interpretation of electromyographic findings. In: McDermott J, editor. *Proceedings of the 10th International Joint Conference on Artificial Intelligence*, pages 366–372, Los Altos, CA, 1987. Morgan Kaufmann.

[5] P. Antal, G. Fannes, D. Timmerman, Y. Moreau and B. De Moor. Using literature and data to learn Bayesian networks as clinical models of ovarian tumors. *Artif Intell Med*, 30(3):257–281, 2004.

[6] S. Bulashevka, O. Skakacs, B. Brors, R. Eils, G. Kovacs. Pathways of urothelial cancer progression suggested by Bayesian network analysis of allelotyping data. *Int J Cancer*, 110:850–856, 2004.

[7] J. Cheng, D. Bell, W. Liu. Learning Bayesian networks from data: an efficient approach based on information theory. In: *Proceeding of the sixth*

ACM International Conference on Information and Knowledge Management, pages 325–331, 1997. ACM.
[8] J. Cheng, R. Greiner. Comparing Bayesian network classifiers. In: *Proceedings of UAI'99*, pages 101–107, San Francisco, CA, 1999. Morgan Kaufmann.
[9] G.F. Cooper, E. Herskovitz. A Bayesian method for the induction of probabilistic networks from data. *Machine Learning*, 9:309–347, 1992.
[10] V.M.H. Coupé, L.C. van der Gaag. Sensitivity analysis: an aid for probability elicitation. *Knowledge Engineering Review*, 15:215–232, 2000.
[11] R.G. Cowell, A.P. Dawid, S.L. Lauritzen, D.J. Spiegelhalter. *Probabilistic Networks and Expert Systems*. Springer, New York, 1999.
[12] F.T. de Dombal, D.J. Leaper, J.R. Staniland, A.P. McAnn and J.C. Horrocks. Computer-aided diagnosis of acute abdominal pain. *British Medical Journal*, ii:9–13, 1972.
[13] A. Dempster, N. Laird, D. Rubin. Maximisation likelihood from incomplete data via the EM algorithm. *Royal Statistical Society (Series B)*, 39:1–38, 1977.
[14] F.J. Díez. Parameter adjustment in Bayes networks: the generalized noisy OR-gate. In: *Proc of the 9th Conference on Uncertainty in Artificial Intelligence*, 1993, pp. 99–105.
[15] S. van Dijk, D. Thierens, L.C. van der Gaag. Building a GA from design principles for learning Bayesian networks. In: Cantú-Paz E, Foster JA, Deb K, et al, editors. *Proceedings of the Genetic and Evolutionary Computation Conference.* pages 886 – 897, San Francisco, CA, 2003. Morgan Kaufmann.
[16] P. Domingos, M. Pazzani. On the optimality of the simple Bayesian classifier under zero–one loss. *Machine Learning*, 29:103–130, 1997.
[17] N.I.R. Friedman, D. Geiger, M. Pazzani. On the optimality of the simple Bayesian network classifier. *Machine Learning*, 29:131–163, 1997.
[18] N.I.R. Friedman, M. Linial, I. Nachman, D. Peér. Using Bayesian network to analyze expression data. *J Computational Biology*, 7:601–620, 2000.
[19] N.I.R. Friedman. Inferring cellular networks using probabilistic graphical models. *Science*, 3003:799–805, 2004.
[20] L.C. van der Gaag and E.M. Helsper. Experiences with modelling issues in building probabilistic networks. In: Gómez-Pérez A, Benjamins VR, editors. Knowledge Engineering and Knowledge Management: Ontologies and the Semantic Web. *Proceedings of EKAW 2002*. Lecture Notes in Artificial Intelligence, LNAI 2473, pages 21–26, 2002, Berlin. Springer Verlag.
[21] L.C. van der Gaag, S. Renooij. Analysing sensitivity data. In: Breese J, Koller D, editors. *Proceedings of the 17th Conference on Uncertainty in Artificial Intelligence*, pages 530–537, 2001, San Francisco, CA. Morgan Kaufmann.
[22] L.C. van der Gaag, S. Renooij, C.L.M. Witteman, B. Aleman, B.G. Taal. How to elicit many probabilities. In: *Proceedings of the 15th Conference*

on Uncertainty in Artificial Intelligence, pages 647–654, San Francisco, CA, 1999. Morgan Kaufmann.

[23] L.C. van der Gaag, S. Renooij, C.L.M. Witteman, B.M.P. Aleman, B.G. Taal. Probabilities for a probabilistic network: a case study in oesophageal cancer. *Artif Intell Med*, 25:123–148, 2002.

[24] L. Getoor, J.T. Rhee, D. Koller and P. Small. Understanding tuberculosis epidemiology using structured statistical models. *Artif Intell Med*, 30(3):233–256, 2004.

[25] S.F. Galán, F. Aguado, F.J. Díez, J. Mira. NasoNet: joining Bayesian networks and time to model nasopharyngeal cancer spread. In: *Proceedings of the 8th Conference on Artificial Intelligence in Medicine in Europe (AIME2001)*, LNAI2101, pages 207–216, 2001, Berlin. Springer-Verlag.

[26] C. Glymour and G.F. Cooper. *Computation, Causation & Discovery*. The MIT Press, Menlo Park, CA, 1999.

[27] T. Hastie, R. Tibshirani, J. Friedman. *The Elements of Statistical Learning: Data Mining, Inference, and Prediction*. Springer, New York, 2001.

[28] D.E. Heckerman, E.J. Horvitz and B.N. Nathwani. Towards normative expert systems: part I – The Pathfinder project. *Methods of Information in Medicine*, 31:90–105, 1992.

[29] D.E. Heckerman and B.N. Nathwani. Towards normative expert systems: part II – probability-based representations for efficient knowledge acquisition and inference. *Methods of Information in Medicine*, 31:106–116, 1992.

[30] D. Heckerman and J.S. Breese. Causal independence for probabilistic assessment and inference using Bayesian networks. *IEEE Transactions on Systems, Man and Cybernetics*, 26(6):826–831, 1996.

[31] E.M. Helsper and L.C. van der Gaag. Building Bayesian networks through ontologies. In: F. van Harmelen, editor. *Proceedings of ECAI-2002*, pages 680–684, Amsterdam, 2002. IOI Press.

[32] M. Henrion. Some practical issues in constructing belief networks. In: J.F. Lemmer and L.N. Kanal, editors, *Uncertainty in Artificial Intelligence*, 3, pages 161–173, Amsterdam, 1989. Elsevier.

[33] R. Jansen, H. Yu, D. Greenbaum, Y. Kluger, N.J. Krogan, S. Chung, A. Emili, M. Snyder, J.F. Greenblatt and M. Gerstein. A Bayesian network approach for predicting protein-protein interaction from genomic data. *Science* 302:449–253, 2003.

[34] M. Korver, P.J.F. Lucas. Converting a rule-based expert system into a belief network. *Medical Informatics*, 18(3):219–241, 1993.

[35] W. Lam and F. Bacchus. Learning Bayesian belief networks: an approach based on the MDL principle. *Computational Intelligence*, 10:269–293, 1994.

[36] P. Larrañaga, M. Poza, Y. Yurramendi, R. Murga, C. Kuijpers. Structure learning of Bayesian networks by genetic algorithms: A performance analysis of control parameters. *IEEE Transactions on Pattern Analysis and Machine Intelligence*, 18(9):912–926, 1996.

[37] S.L. Lauritzen. Propagation of probabilities, means and variances in mixed graphical models. *J American Statistical Association*, 87:1098–1108, 1992.
[38] S.L. Lauritzen and D.J. Spiegelhalter. Local computations with probabilities on graphical structures and their application to expert systems. *J Royal Statistical Society (Series B)*, 50:157–224, 1987.
[39] P.J.F. Lucas. Knowledge acquisition for decision-theoretic expert systems. *AISB Quaterly*, 94:23–33, 1996.
[40] P.J.F. Lucas. Restricted Bayesian-network structure learning. In: Gámez JA, Moral S, Salmerón A, editors. *Advances in Bayesian Networks*, Studies in Fuzziness and Soft Computing, volume 146, pages 217–234, Berlin, 2004. Springer-Verlag.
[41] P.J.F. Lucas. Bayesian network modelling through qualitative patterns. *Artificial Intelligence*, 163:233–263, 2005.
[42] P.J.F. Lucas, H. Boot and B.G. Taal. Computer-based decision-support in the management of primary gastric non-Hodgkin lymphoma. *Methods of Information in Medicine*, 37:206–219, 1998.
[43] P.J.F. Lucas, De Bruijn N.C., Schurink K., Hoepelman I.M.. A Probabilistic and decision-theoretic approach to the management of infectious disease at the ICU. *Artif Intell Med*, 19(3):251–279, 2000.
[44] M. Paul, S. Andreassen, A.D. Nielsen, et al. Prediction of bacteremia using TREAT, a computerized decision-support system. *Clinical Infectious Diseases*, 42:1274–1282, 2006.
[45] J. Pearl. *Probabilistic Reasoning in Intelligent Systems*. Morgan Kaufman, San Mateo, CA, 1988.
[46] M. Ramoni, P. Sebastiani and P. Cohen. Bayesian clustering by dynamics. *Machine Learning* 47:91–121, 2002.
[47] S. Renooij. Probability elicitation for belief networks: issues to consider. *Knowledge Engineering Review*, 16(3):255–269, 2001.
[48] S. Renooij, L.C. van der Gaag. From qualitative to quantitative probabilistic networks. In: Darwiche A., Friedman N., editors. *Proceedings of the 18th Conference on Uncertainty in Artificial Intelligence*, pages 422–429, San Francisco, CA, 2002. Morgan Kaufmann.
[49] P. Sebastiani, K.D. Mandl, P. Szolovits, I.S. Kohane and M.F. Ramoni. A Bayesian dynamic model for influence surveillance. *Statist Med* 25:18-3–1816, 2006.
[50] E. Segal, M. Shapira, A. Regev, D. Peér, D. Botstein, D. Koller and N. Friedman. Module networks: identifying regulatory modules and their condition-specific regulators from gene expression data. *Nat Genet*, 34(2):166–76, 2003.
[51] R.D. Shachter RD. Evaluating influence diagrams. *Operation Research*, 34(6):871–882, 1986.
[52] D.J. Spiegelhalter and R.P. Knill-Jones. Statistical and knowledge-based approaches to clinical decision-support systems, with an application in gastroenterology. *Journal of the Royal Statistical Society*, 147:35–77, 1984.

[53] M. Suojanen. *Construction of large Medical Decision Support Systems: a causal probabilistic network for diagnosis of neuromuscular disorders.* PhD Thesis, University of Aalborg, 2001.

[54] M. Studený. *Probabilistic Conditional Independence Structures.* Springer-Verlag, London, 2005.

[55] M.L. Wong, S.Y. Lee, K.S. Leung. A hybrid data mining approach to discover Bayesian networks using evolutionary programming. In: Langdon WB, et al., editors. *Proceedings of the Genetic and Evolutionary Computation Conference,* pages 214–222, San Francisco, CA, 2002. Morgan Kaufmann.

[56] N.L. Zhang and L. Yan. Independence of Causal Influence and Clique Tree Propagation. *International Journal of Approximate Reasoning,* 19:335–349, 1997.

Learning and Validating Bayesian Network Models of Gene Networks

Jose M. Peña[1], Johan Björkegren[2], and Jesper Tegnér[1,2]

[1] IFM, Linköping University, SE-58183 Linköping, Sweden
{jmp, jespert}@ifm.liu.se
[2] CGB, Karolinska Institute, SE-17177 Stockholm, Sweden
johan.bjorkegren@cgb.ki.se

Summary. We propose a framework for learning from data and validating Bayesian network models of gene networks. The learning phase selects multiple locally optimal models of the data and reports the best of them. The validation phase assesses the confidence in the model reported by studying the different locally optimal models obtained in the learning phase. We prove that our framework is asymptotically correct under the faithfulness assumption. Experiments with real data (320 samples of the expression levels of 32 genes involved in *Saccharomyces cerevisiae*, i.e. baker's yeast, pheromone response) show that our framework is reliable.

1 Introduction

The cell is the functional unit or building block of all the organisms. The cell is self-contained, as it includes the information necessary for regulating its function. This information is encoded in the DNA of the cell, which is divided into a set of genes, each coding for one or more proteins. Proteins are required for practically all the functions in the cell, and they are produced through the expression of the corresponding genes. The amount of protein produced is determined by the expression level of the gene, which may be regulated by the protein produced by another gene. As a matter of fact, much of the complex behavior of the cell can be explained through the concerted activity of genes. This concerted activity is typically represented as a network of interacting genes. Identifying this network, which we call gene network (GN), is crucial for understanding the behavior of the cell which, in turn, can lead to better diagnosis and treatment of diseases. This is one of the most exciting challenges in bioinformatics. For the last few years, there has been an increasing interest in learning Bayesian network (BN) models of GNs [1; 9; 12; 14; 17; 20; 21; 23], mainly owing to the following two reasons. First, there exist principled algorithms for learning BN models from data [3; 5; 19; 21; 27]. Second, BN models can represent stochastic relations between genes. This is particularly important when inferring models of GNs from gene expression data, because gene expression has a stochastic component [2; 18], and because gene expression data typically include measurement noise [9; 25].

Following the papers cited above, we view a GN as a probability distribution $p(\mathbf{U})$, where $\mathbf{U}$ is a set of random variables such that each of them represents the expression level of a gene in the GN. And we aim to learn about the (in)dependencies in $p(\mathbf{U})$ by learning a BN model from some given gene expression data sampled from $p(\mathbf{U})$. Specifically, we define a BN model $M(G)$ as the set of independencies in G, where G is an acyclic directed graph (DAG) whose set of nodes is $\mathbf{U}$. The independencies in G correspond to the d-separation statements in G: $\mathbf{X}$ and $\mathbf{Y}$ are d-separated given $\mathbf{Z}$ in G if for every undirected path in G between a node in $\mathbf{X}$ and a node in $\mathbf{Y}$ there exists a node W in the path such that either (i) W does not have two parents in the path and $W \in \mathbf{Z}$, or (ii) W has two parents in the path and neither W nor any of its descendants in G is in $\mathbf{Z}$. The probability distributions that do not satisfy any other independence than those in G are called faithful to G.

In this paper, we follow the so-called model selection approach to learning a BN model from some given data: Given a scoring criterion that evaluates the quality of a model with respect to the data, model selection searches the space of models for the highest scoring model. Unfortunately, model selection is NP-complete [4]. For this reason, most algorithms for model selection are heuristic and they only guarantee convergence to a locally optimal model. Validating this model is crucial, as the number of locally optimal models can be large [19]. When inferring a BN model of a GN from gene expression data, validation becomes even more important: Gene expression data are typically scarce and noisy [9; 25] and, thus, they may not have enough power to discriminate between those locally optimal models that are close to the set of independencies in the probability distribution of the GN and those that are not.

In this paper, we propose a framework for learning from data and validating BN models of GNs. The learning phase consists in running repeatedly a stochastic algorithm for model selection in order to discover multiple locally optimal models of the learning data and, then, reporting the best of them. The validation phase assesses the confidence in some features of the model reported by studying the different locally optimal models obtained in the learning phase. The higher the confidence in the features of the model reported, the more believable or valid it is. We prove that our framework is asymptotically, i.e. in the large sample limit, correct under the faithfulness assumption. We show with experiments on real data that our framework is reliable.

In the sections below, we describe the learning and validation phases of our framework (Sects. 2 and 3, respectively) and, then, we evaluate it on synthetic and real data (Sects. 4 and 5, respectively). We conclude in Sect. 6 with a discussion on this and related works.

2 Learning Phase

As mentioned in the previous section, the learning phase runs repeatedly a stochastic algorithm for model selection in order to obtain multiple locally optimal models of the learning data and, then, reports the best of them. We use the k-greedy equivalence search algorithm (KES) [19] for this purpose. Like most algorithms for model selection, KES consists of three components: A neighborhood, a scoring criterion, and a search strategy. The neighborhood of a model restricts the search to a small part of the search space around the model, and it is usually defined by means of local transformations of the model. The scoring criterion evaluates the quality of a model with respect to the learning data. The search strategy selects a new model, based on the scoring criterion, from those in the neighborhood of the current best model. The paragraphs below describe these components in the case of KES.

KES uses the inclusion boundary of a model as the neighborhood of the model. The inclusion boundary of a model $M(G_1)$, $IB(M(G_1))$, is the union of the upper and lower inclusion boundaries, $UIB(M(G_1))$ and $LIB(M(G_1))$, respectively. $UIB(M(G_1))$ is the set of models $M(G_2)$ that are strictly included in $M(G_1)$ and such that no model strictly included in $M(G_1)$ strictly includes $M(G_2)$. Likewise, $LIB(M(G_1))$ is the set of models $M(G_2)$ that strictly include $M(G_1)$ and such that no model strictly including $M(G_1)$ is strictly included in $M(G_2)$. $IB(M(G_1))$ is characterized using DAGs as the set of models represented by all the DAGs that can be obtained by adding or removing a single edge from any representative DAG of $M(G_1)$, where a DAG G_2 is representative of $M(G_1)$ if $M(G_1) = M(G_2)$ [5]. Any representative DAG G_2 of a model can be obtained from any other representative DAG G_1 of the model through a sequence of covered edge reversals in G_1, where the edge $X \rightarrow Y$ is covered in G_1 if X and Y share all their parents but X in G_1 [5].[3]

KES scores a model by scoring any representative DAG of the model. Thus, KES requires that all the representative DAGs of a model receive the same score. Furthermore, KES also requires that the scoring criterion is locally consistent: Given an i.i.d sample from a probability distribution $p(\mathbf{U})$, the scoring criterion is locally consistent if the score assigned to a DAG G asymptotically increases (resp. decreases) with each edge removal that adds independencies to $M(G)$ that hold (resp. does not hold) in $p(\mathbf{U})$. The two most commonly used scoring criteria, the Bayesian Dirichlet metric with uniform prior (BDeu) [15] and the Bayesian information criterion (BIC) [24], satisfy the two requirements above and can be used with KES [5]. BDeu scores the exact marginal likelihood of the learning data for a given DAG, whereas BIC scores an asymptotic approximation to it. Finally, KES uses the following search strategy:

---

[3] A more efficient, though more complex, characterization of $IB(M(G))$ using completed acyclic partially directed graphs is reported in [28; 29].

```
KES (k∈[0,1])
M = model of the DAG without any edge
repeat
   B = set of models in IB(M) with higher score than M
   if |B| > 0 then
      C = random subset of B with size max(1,|B|·k)
      M = the highest scoring model in C
   else return M
```

where $|B|$ denotes the cardinality of the set B. For the sake of simplicity, KES represents each model in the search space by one of its representative DAGs. Thus, B and C are sets of DAGs. The input parameter $k \in [0,1]$ allows to trade off greediness for randomness. This makes KES ($k \neq 1$) able to reach different locally optimal models when run repeatedly. KES ($k = 1$) corresponds to the greedy equivalence search algorithm (GES) proposed in [5].[4] We refer the reader to [19] for a thorough study of KES, including the proof of the following property.

Theorem 1. *Given a fully observed i.i.d sample from a probability distribution faithful to a DAG G, KES asymptotically returns $M(G)$.*

3 Validation Phase

In the light of the experiments in [19], the learning phase described in the previous section is very competitive. However, when the learning data are as scarce, noisy and complex as gene expression data are, the best locally optimal model discovered in the learning phase may not be reliable, because the learning data may lack the power to discriminate between those locally optimal models that are close to the set of independencies in the sampled probability distribution and those that are not. Therefore, validating the model learnt is of much importance. Our proposal for validating it consists of two main steps. First, extraction of relevant features from the model. Second, assessment of the confidence in the features extracted. The higher the confidence in these features, the more believable or valid the model is. The following sections describe these two steps.

3.1 Feature Extraction

First of all, we need to adopt a model representation scheme that allows interesting features to be extracted. Representing a model by a DAG does not

---

[4] To be exact, GES is a two-phase algorithm that first uses only $UIB(M(G))$ and, then, only $LIB(M(G))$. KES ($k = 1$) corresponds to a variant of GES described in [5] that uses $IB(M(G))$ in each step.

seem appropriate here, because there may be many representative DAGs of the model. A completed acyclic partially directed graph (CPDAG) provides, on the other hand, a canonical representation of a model. A CPDAG represents a model by summarizing all its representative DAGs: The CPDAG contains the directed edge $X \to Y$ if $X \to Y$ exists in all the representative DAGs, while it contains the undirected edge X–Y if $X \to Y$ exists in some representative DAGs and $Y \to X$ in some others. See [5] for an efficient procedure to transform a DAG into its corresponding CPDAG.

We pay attention to four types of features in a CPDAG: Directed edges, undirected edges, directed paths, and Markov blanket neighbors. Two nodes are Markov blanket neighbors if there is an edge between them or if they have a child in common. We focus on these types of features because they suggest relevant features of the probability distribution of the learning data. A directed or undirected edge suggests an unmediated dependence. A directed path suggests a causal pathway because it appears in all the representative DAGs of the model. Finally, the Markov blanket neighborhood of a random variable suggests the minimal set of predictors of the probability distribution of the random variable, because the Markov blanket neighborhood is the minimal set conditioned on which the random variable is independent of the rest of random variables.

3.2 Confidence Assessment

Despite the fact that the different locally optimal models discovered in the learning phase disagree in some features, we expect them to share some others. In fact, the more strongly the learning data support a feature, the more frequently it should appear in the different locally optimal models found. Likewise, the more strongly the learning data support a feature, the higher the likelihood of the feature being true in the probability distribution that generated the learning data. This leads us to assess the confidence in a feature as the fraction of models containing the feature out of the different locally optimal models obtained in the learning phase. Note that we give equal weight to all the models available, no matter their scores. Alternatively, we could weight each model by its score. We prove below that this approach to confidence estimation is asymptotically correct under the faithfulness assumption. We show in Sect. 4 that it is accurate for finite samples as well.

Theorem 2. *Given a fully observed i.i.d sample from a probability distribution faithful to a DAG G, the features in $M(G)$ asymptotically receive confidence equal to one and the rest equal to zero.*

Proof. Under the conditions of the theorem, KES asymptotically returns $M(G)$ owing to Theorem 1. □

3.3 Validity Assessment

Let M^* denote the best locally optimal model found in the learning phase. Deciding on the validity of M^* on the basis of the confidence values scored by its features may be difficult. We suggest a sensible way to ease making this decision. We call true positives (TPs) to the features in M^* with confidence value equal or above a given threshold value t. Likewise, we call false positives (FPs) to the features not in M^* with confidence value equal or above t, and false negatives (FNs) to the features in M^* with confidence value below t. In order to decide on the validity of M^*, we propose studying the trade-off between the number of FPs and FNs for each type of features under study as a function of t. The fewer FPs and FNs for high values of t, the more believable or valid M^* is. In other words, we trust M^* as a valid model of the probability distribution of the learning data if the features in M^* receive high confidence values, while the features not in M^* score low confidence values. Note that we treat on equal basis FPs and FNs. Alternatively, we can attach different costs to them according to our preferences, e.g. we may be less willing to accept FPs than FNs. The following property follows directly from Theorem 2.

Theorem 3. *Given a fully observed i.i.d sample from a probability distribution faithful to a DAG G, the number of FPs and FNs is asymptotically zero for any $t > 0$.*

Therefore, our framework for learning from data and validating BN models of GNs is asymptotically correct under the faithfulness assumption, i.e. the learning phase always returns the true model (Theorem 1) and the validation phase always confirms its validity (Theorem 3). We note that, although the faithfulness assumption may not hold in practice, the theorems above are desirable properties for any work on BN model validation to have.

4 Evaluation on Synthetic Data

We have proven in Theorems 2 and 3 that our approach to confidence estimation is asymptotically correct under the faithfulness assumption. We now show that it is also accurate for finite samples under the faithfulness assumption. The database used in the evaluation is the Alarm database [16]. This database consists of 20000 cases sampled from a BN model representing potential anesthesia problems in the operating room. The CPDAG of the BN model sampled has 37 nodes and 46 edges, and each node has from two to four states. We perform experiments with samples of sizes 1 %, 2 %, 5 %, 10 %, 25 %, 50 % and 100 % of the Alarm database. The results reported are averages over five random samples of the corresponding size.

The setting for the evaluation is as follows. We consider KES ($k = 0.6, 0.8, 0.9$) with BIC as the scoring criterion. We avoid values of k close to 0 so as to prevent convergence to poor locally optimal models [19]. For each sample in the evaluation, we first run KES 1000 independent times and use the different locally optimal models discovered to estimate the confidence in the features of interest, i.e. directed edges, undirected edges, directed paths, and Markov blanket neighbors. We give equal weight to all the models used for confidence estimation. Then, we compute the trade-off between the number of FPs and FNs for each type of features under study as a function of the threshold value t. We treat equally FPs and FNs when computing the trade-off. Unlike in Sect. 3.3, FPs and FNs are calculated with respect to the true model so as to assess the accuracy of our method for confidence estimation.

We report results for $k = 0.8$ and omit the rest because they all lead to the same conclusions. Out of the 1000 independent runs of KES performed for each of the samples in the evaluation, we obtained an average of 203 different locally optimal models for the sample size 1 %, 233 for 2 %, 161 for 5 %, 115 for 10 %, 119 for 25 %, 85 for 50 %, and 70 for 100 %. We note that the number of different locally optimal models obtained decreases as the size of the learning data increases, which is expected because Theorem 1 applies. Figure 1 shows the trade-off curves between the number of FPs and FNs as a function of the threshold value t. We note that the CPDAG of the true model has 42 directed edges, 4 undirected edges, 196 directed paths, and 65 Markov blanket neighbors. We do not report trade-off curves for undirected edges because they are difficult to visualize as there are only four undirected edges in the true model. Instead, the trade-off curves in Fig. 1 (top) summarize the number of FPs and FNs for both directed and undirected edges. The shape of the trade-off curves for the three types of features, concave down and closer to the horizontal axis (FNs) than to the vertical axis (FPs), indicates that our method for confidence estimation is reliable: For all the sample sizes except 1 %, there is a wide range of values of t such that (i) the number of TPs is higher than the number of FNs, and (ii) the number of FNs is higher than the number of FPs. For the sample size 1 %, these observations are true only for Markov blanket neighbors, which indicates that these features are easier to learn. This makes sense as Markov blanket neighbors are less sensitive than the other types of features to whether the edge between two nodes is directed or undirected. The trade-off curves in the figure also show that the number of FPs and FNs decreases as the size of the learning data increases, which is expected because Theorem 3 applies. In particular, when setting t to the value that minimizes the sum of FPs and FNs for the sample size 100 %, there are 1 FP and 1 FN (45 TPs) for edges ($t = 0.45$), 1 FP and 10 FNs (186 TPs) for directed paths ($t = 0.6$), and 0 FPs and 3 FNs (62 TPs) for Markov blanket neighbors ($t = 0.7$). Figure 2 depicts the TP and FP edges for the sample size 100 % when $t = 0.45, 0.95$. Recall that $t = 0.45$ is the threshold value that minimizes the sum of FPs and FNs for edges and that it implies 1 FP and 1 FN (45 TPs). The FN edge $12 \rightarrow 32$ is reported in [6] to be not supported by the data. When $t = 0.95$, there are 0 FPs and 17 FNs (29 TPs). Therefore, our method for confidence assessment assigns to a considerable amount of

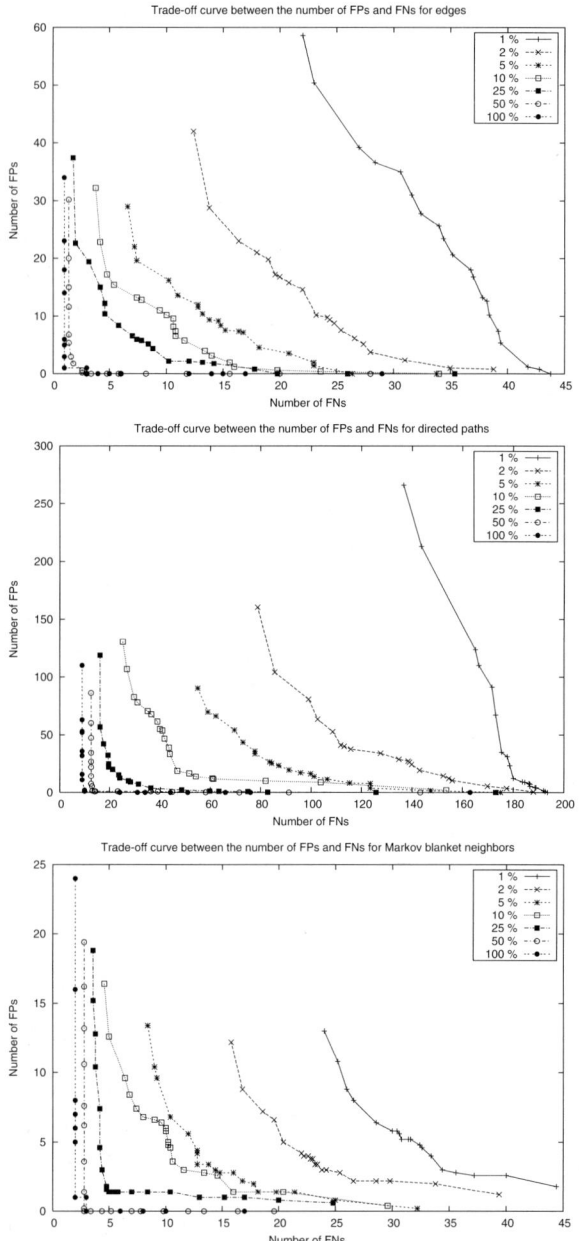

Fig. 1. Trade-off between the number of FPs and FNs for the Alarm databases ($k = 0.8$) at threshold values $t = 0.05 \cdot r$, $r = 1, \ldots, 20$. Top, directed and undirected edges. Middle, directed paths. Bottom, Markov blanket neighbors.

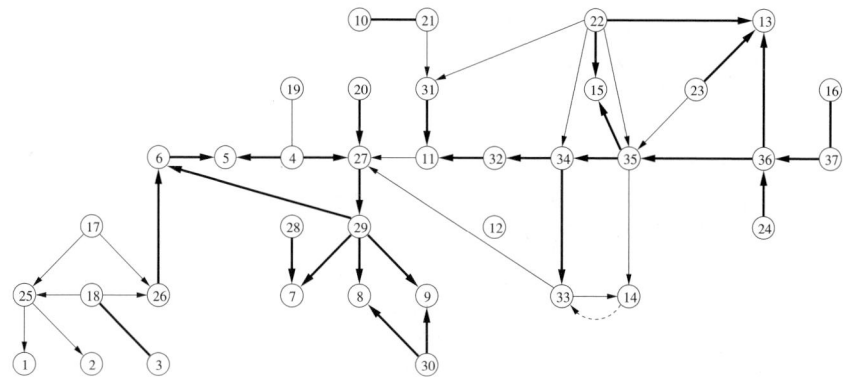

Fig. 2. Directed and undirected edges for the Alarm database of size 100 % ($k = 0.8$) when $t = 0.45$ (plain and bold edges) and when $t = 0.95$ (bold edges). Solid edges are TPs and dashed edges are FPs.

TPs higher confidence than to any FP. This is also true for directed paths and Markov blanket neighbors as can be seen in Fig. 1 (middle and bottom).

5 Evaluation on Real Data

In this section, we evaluate the framework for learning from data and validating BN models of GNs that we have proposed in Sects. 2 and 3. The experimental setting is the same as in the previous section with the only exception that FPs and FNs are now calculated with respect to the best locally optimal model found in the learning phase (recall Sect. 3.3). The data used in the evaluation are the data in [14], which we call the Yeast database hereinafter. This database consists of 320 records characterized by 33 attributes. The records correspond to 320 samples of unsynchronized *Saccharomyces cerevisiae* (baker's yeast) populations observed under different experimental conditions.[5] The first 32 attributes of each record represent the expression levels of 32 genes involved in yeast pheromone response. This pathway plays an essential role in the sexual reproduction of yeast. The last attribute of each record, named MATING_TYPE, indicates the mating type of the strain of yeast in the corresponding sample, either MATa or MATα, as some of the 32 genes measured express only in strains of a specific mating type. Gene expression levels are discretized into four states. We refer the reader to [14] for details on the data collection and preparation process. Table 1 reproduces the description of the 32 genes in the database that is given in [14]. The

---

[5] Yeast is extensively studied in molecular biology and bioinformatics because it is considered an ideal organism: It is quick and easy to grow, and it provides insight into the workings of other organisms, including humans.

Table 1. Top, description of the 32 genes in the Yeast database. The genes are divided into functional groups according to the current knowledge of yeast pheromone response. Each group has a different color assigned. Bottom, description of the groups of genes.

Gene	Group	Function of the protein encoded by the gene
STE2	Magenta	Transmembrane receptor peptide
MFA1	Magenta	a-factor mating pheromone
MFA2	Magenta	a-factor mating pheromone
STE6	Magenta	Responsible for the export of a-factor from MATa cells
AGA2	Magenta	Binding subunit of a-agglutinin complex, involved in cell-cell adhesion during mating by binding Sag1
BAR1	Magenta	Protease degrading α-factor
STE3	Red	Transmembrane receptor peptide
MFALPHA1	Red	α-factor mating pheromone
MFALPHA2	Red	α-factor mating pheromone
SAG1	Red	Binding subunit of α-agglutinin complex, involved in cell-cell adhesion during mating by binding Aga2 (also known as Agα1)
FUS3	Blue	Mitogen-activated protein kinase (MAPK)
STE12	Blue	Transcriptional activator
FAR1	Blue	Substrate of Fus3 that leads to G1 arrest, known to bind to STE4 as part of complex of proteins necessary for establishing cell polarity required for shmoo formation after mating signal has been received
FUS1	Blue	Required for cell fusion during mating
AGA1	Blue	Anchor subunit of a-agglutinin complex, mediates attachment of Aga2 to cell surface
GPA1	Green	Component of the heterotrimeric G-protein (Gα)
STE4	Green	Component of the heterotrimeric G-protein (Gβ)
STE18	Green	Component of the heterotrimeric G-protein (Gγ)
STE7	Yellow	MAPK kinase (MAPKK)
STE11	Yellow	MAPKK kinase (MAPKKK)
STE5	Yellow	Scaffolding peptide holding together Fus3, Ste7 and Ste11 in a large complex
KSS1	Orange	Alternative MAPK for pheromone response (in some dispute)
STE20	Orange	p21-activated protein kinase (PAK)
STE50	Orange	Unknown function but necessary for proper function of Ste11
SNF2	Brown	Implicated in induction of numerous genes in pheromone response pathway (component of SWI-SNF global transcription activator complex)
SWI1	Brown	Implicated in induction of numerous genes in pheromone response pathway (component of SWI-SNF global transcription activator complex)
SST2	White	Involved in desensitization to mating pheromone exposure
KAR3	White	Essential for nuclear migration step of karyogamy
TEC1	White	Transcriptional activator believed to bind cooperatively with Ste12 (more active during induction of filamentous or invasive growth response)
MCM1	White	Transcription factor believed to bind cooperatively with Ste12 (more active during induction of pheromone response)
SIN3	White	Implicated in induction or repression of numerous genes in pheromone response pathway
TUP1	White	Implicated in repression of numerous genes in pheromone response pathway

Group	Description of the group
Magenta	Genes expressed only in MATa cells
Red	Genes expressed only in MATα cells
Blue	Genes whose promoters are bound by Ste12
Green	Genes coding for components of the heterotrimeric G-protein complex
Yellow	Genes coding for core components of the signaling cascade (except FUS3 which is in the group Blue)
Orange	Genes coding for auxiliary components of the signaling cascade
Brown	Genes coding for components of the SWI-SNF complex
White	Others

description is based on [7; 8; 22]. The table also divides the genes into groups according to their function in the domain under study.

We first report the results of the learning phase. Out of the 1000 independent runs of KES performed for each value of k considered in the evaluation, we obtained 967 different locally optimal models for $k = 0.6$, 330 for $k = 0.8$, and 159 for $k = 0.9$. In the three cases, the best model found was the same. Figure 3 (top) shows its CPDAG. We remark that the graph in the figure does not intend to represent the biological or physical GN, but the (in)dependencies in it. We note that all the edges in the CPDAG are undirected, meaning that each edge appears in opposite directions in at least two representative DAGs of the model. As a matter of fact, none of the CPDAGs of the different locally optimal models obtained in the 3000 runs of KES performed has directed edges. This reduces the types of features to study hereinafter to only two: Undirected edges and Markov blanket neighbors. However, when there is not any directed edge in a CPDAG, two nodes are Markov blanket neighbors if and only if they are connected by an undirected edge. Therefore, we only pay attention to undirected edges hereinafter.

We now discuss the results of the validation phase. Figure 3 (bottom left) shows the trade-off between the number of FPs and FNs for undirected edges as a function of the threshold value t. We note that the CPDAG of the best model found in the learning phase has 32 undirected edges. As can be appreciated from the figure for each value of k considered in the evaluation, FNs only happen for high values of t, while FPs only occur for low values of t. Therefore, TPs receive substantially higher confidence values than FPs. For $k = 0.8$, for instance, no TP scores lower than 0.60, while no FP scores higher than 0.25. These observations support the validity and meaningfulness of the best model discovered in the learning phase. Figure 3 (bottom right) depicts the undirected edges for $k = 0.8$ when $t = 0.60, 0.90$. We note that all the edges in the figure are TPs. As a matter of fact, there are 0 FPs and 0 FNs (32 TPs) for $t = 0.60$, and 0 FPs and 11 FNs (21 TPs) for $t = 0.90$. The figures for $k = 0.6, 0.9$ are similar to the one shown. We omit them for the sake of readability.

It is worth mentioning that we repeated the experiments in this section with a random database created by randomly reshuffling the entries of each attribute in the Yeast database. In such a database, we did not expect to find features scoring high confidence values. As a matter of fact, no edge was added in any of the 3000 runs of KES performed. This leads us to believe that the results presented above are not artifacts of the learning and validation phases but reliable findings. We give below further evidence that the (in)dependencies in the best model induced in the learning phase are consistent with the existing knowledge of yeast pheromone response. This somehow confirms the results of the validation phase, namely that the best model obtained in the learning phase is reliable.

We first discuss consistency with respect to the knowledge in Table 1. Magenta-colored genes are marginally dependent one on another as well as on

MATING_TYPE. Moreover, no gene from other group mediates these dependencies. Likewise, red-colored genes are marginally dependent one on another as well as on MATING_TYPE, and no gene from other group mediates these dependencies. These observations are consistent with the fact that magenta-colored genes express only in MATa cells, while red-colored genes express only in MATα cells. This also supports the fact that MATING_TYPE is the only node that mediates between magenta and red-colored genes. Green-colored genes are marginally dependent one on another, and no gene from other group mediates these dependencies. These observations are consistent with the fact that green-colored genes code for components of the heterotrimeric G-protein complex. Yellow-colored genes are marginally dependent one on another, which is consistent with these genes coding for core components of the signaling cascade. While STE5 and STE7 are adjacent, two green-colored genes (GPA1 and STE18) mediate between them and STE11. A similar result is reported in [14]. The authors conjecture that this finding may indicate common or serial regulatory control between green and yellow-colored genes. Orange-colored genes are marginally dependent one on another, which is consistent with the fact that they code for auxiliary components of the signaling cascade. Only TUP1 mediates these dependencies, specifically orange-colored genes are independent one of another given TUP1. As a matter of fact, TUP1 has the highest number of adjacencies in the model, which is consistent with its role as repressor of numerous genes in pheromone response pathway. We note that several nodes mediate between the core (yellow-colored) and the auxiliary (orange-colored) components of the signaling cascade. This agrees with [14]. The authors suggest that this finding may indicate that these two groups of genes have different regulatory mechanisms. Brown-colored genes are marginally dependent one on another, which is consistent with these genes coding for components of the SWI-SNF complex. However, TUP1 and STE20 mediate this dependency. A similar result is reported in [14]. Blue-colored genes are marginally dependent one on another, which is consistent with the promoters of these genes being bound by Ste12. However, several other genes mediate these dependencies.

We now discuss further evidence that does not appear in Table 1. The edges STE2–STE6, STE3–SAG1, and SST2–AGA1 are consistent with the genes connected by each edge being expressed similarly and being cell cycle-regulated [26]:[6] STE2 and STE6 peak at the M phase, while the rest of the genes peak at the M/G1 transition. Likewise, the genes connected by each of the edges MFALPHA2–STE3, MFA1–AGA2, and FAR1–TEC1 are also substantially correlated as well as cell cycle-regulated [26], though they do

---

[6] The cell cycle is the sequence of events by which the cell divides into two daughter cells and, thus, it is the biological basis of life. The cell cycle is divided into four main phases: G1, S, G2 and M. In G1 and G2, the cell grows and prepares to enter the next phase, either S or M. In S, the DNA is duplicated. In M, the actual cell division happens.

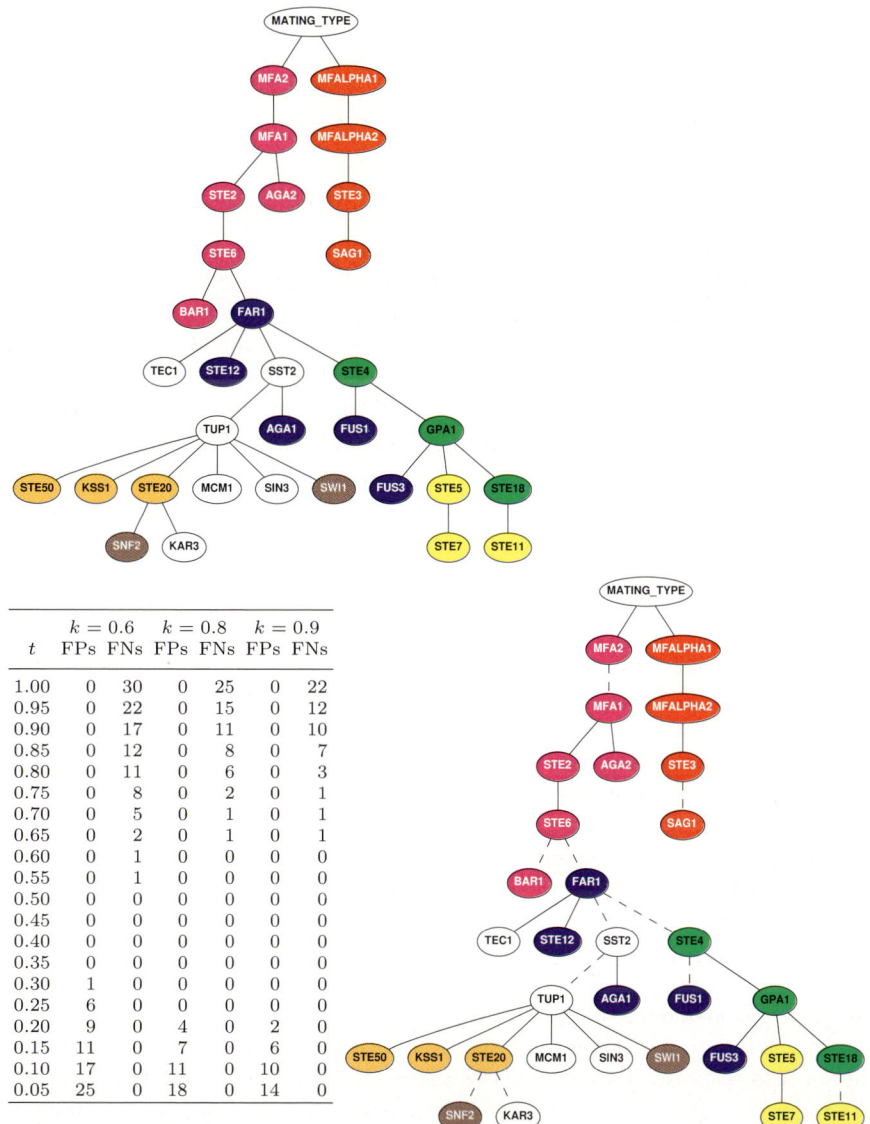

Fig. 3. Top, CPDAG of the best model learnt from the Yeast database ($k = 0.8$). Bottom left, trade-off between the number of FPs and FNs for undirected edges for the Yeast database at threshold values $t = 0.05 \cdot r$, $r = 1, \ldots, 20$. Bottom right, undirected edges for the Yeast database ($k = 0.8$) when $t = 0.60$ (solid and dashed edges) and when $t = 0.90$ (solid edges). Nodes are colored with the color of the functional group they belong to in Table 1.

not peak at the same phase of the cell cycle (MFALPHA2 and MFA1 peak at the G1 phase, FAR1 at the M phase, and STE3, AGA2 and TEC1 at the M/G1 transition). The edge STE6–FAR1 is consistent with these genes being cell cycle-regulated, both peaking at the M phase [26]. The edge TUP1–MCM1 is consistent with the fact that these genes interact in the cell [13].

Finally, it is worth mentioning that most of the edges scoring high confidence values in the validation phase are supported by the existing knowledge of yeast pheromone response. For instance, most edges in Fig. 3 (bottom right) with confidence values equal of above 0.90 have been discussed in the paragraphs above. Therefore, we can conclude that the framework proposed in this paper for learning from data and validating BN models of GNs is accurate and reliable: The learning phase has produced a model that is consistent with the existing knowledge of the domain under study, and the validation phase has confirmed, independently of the existing knowledge, that the model is indeed meaningful.

6 Discussion

There exist numerous works showing that a BN model induced from gene expression data can provide accurate biological insight into the GN underlying the data [1; 9; 12; 14; 17; 20; 21; 23]. This work is yet another example. However, learning BN models from data is a challenging problem (NP-complete and highly multimodal), specially if the learning data are as scarce and noisy as gene expression data are. For these reasons, any BN model of a GN obtained from gene expression data must be biologically validated before being accepted. Validating the model through biological experiments is expensive and, thus, the validation step typically reduces to checking whether the model agrees with the existing biological knowledge of the domain under study. Unfortunately, this way of proceeding condemns models providing true but new biological insight to be rejected. In this paper, we suggest a solution to this problem: We propose a method for checking whether the model learnt is statistically reliable, independently of the existence of biological knowledge. If the model fails to be reliable as a whole, we can instead report the features that are reliable, which are usually very informative. As a matter of fact, some of the works cited above focus on learning features with confidence value above a given threshold rather than on model selection [12; 14; 20]. A major limitation of this approach is that, in general, a set of features does not represent a (global) model of the probability distribution of the learning data but a collection of (local) patterns, because each feature corresponds to a piece of local information. Therefore, the reasoning about the (in)dependencies of the probability distribution of the learning data that a set of features allows is much less powerful than that of a model, e.g. a model can be queried about any (in)dependence statement but a set of features cannot. For this reason, we prefer our framework for model selection and validation and, only if the model selected does not pass the validation phase, we report features.

The works on learning features cited above use the methods in [10; 14] to estimate the confidence in a feature. See [11] for yet another interesting method. Like our method, these methods assess the confidence in a feature as the fraction of models containing the feature out of a set of models. However, they differ from our method in how this set of models is obtained. In [10] it is obtained by running a greedy hill-climbing search on a series of bootstrap samples of the learning data, in [11] by Markov chain Monte Carlo simulation, and in [14] by selecting the highest scoring models visited during a simulated annealing search. No proof of asymptotic correctness is reported for any of these methods. We have proven that our method is asymptotically correct under the faithfulness assumption. The key in the proof is that our algorithm for model selection uses the inclusion boundary neighborhood, which takes into account all the representative DAGs of the current best model to produce the neighboring models. This is a major difference with the works on learning BN models of GNs cited at the beginning of this section, which use classical neighborhoods based on local transformations (single edge additions, removals and reversals) of a single representative DAG of the current best model. The inclusion boundary neighborhood outperforms the classical neighborhoods in practice without compromising the runtime, because it reduces the risk of getting stuck in a locally but not globally optimal model [3]. Moreover, unlike the classical neighborhoods, the inclusion boundary neighborhood allows to develop asymptotically optimal algorithms for model selection [3; 5; 19].

We are currently engaged in two lines of research. First, we are interested in replacing the faithfulness assumption by a weaker assumption such as the composition property assumption. Second, we would like to use the results of the validation phase to design informative gene perturbations, gather new data, and refine the models obtained in the learning phase accordingly. We hope that by influencing the data collection process we will reduce the amount of data required for learning a reliable model. This is important given the high cost of gathering gene expression data. Moreover, combining observational and interventional data will also provide insight into the causal relations in the GN under study.

Acknowledgements

We thank Alexander J. Hartemink for providing us with the Yeast database. This work is funded by the Swedish Foundation for Strategic Research (SSF) and Linköping Institute of Technology.

References

[1] Bernard, A. and Hartemink, A. J. (2005) Informative Structure Priors: Joint Learning of Dynamic Regulatory Networks from Multiple Types of Data. In Pacific Symposium on Biocomputing 10.

[2] Blake, W. J., Kærn, M., Cantor, C. R. and Collins, J. J. (2003) Noise in Eukaryotic Gene Expression. Nature 422:633-637.

[3] Castelo, R. and Kočka, T. (2003) On Inclusion-Driven Learning of Bayesian Networks. Journal of Machine Learning Research 4:527-574.

[4] Chickering, D. M. (1996) Learning Bayesian Networks is NP-Complete. In Learning from Data: Artificial Intelligence and Statistics V:121-130.

[5] Chickering, D. M. (2002) Optimal Structure Identification with Greedy Search. Journal of Machine Learning Research 3:507-554.

[6] Cooper, G. and Herskovits, E. H. (1992) A Bayesian Method for the Induction of Probabilistic Networks from Data. Machine Learning 9:309-347.

[7] Costanzo, M. C., Crawford, M. E., Hirschman, J. E., Kranz, J. E., Olsen, P., Robertson, L. S., Skrzypek, M. S., Braun, B. R., Hopkins, K. L., Kondu, P., Lengieza, C., Lew-Smith, J. E., Tillberg, M. and Garrels, J. I. (2001) YPD™, PombePD™ and WormPD™: Model Organism Volumes of the BioKnowledge™ Library, an Integrated Resource for Protein Information. Nucleic Acids Research 29:75-79.

[8] Elion, E. A. (2000) Pheromone Response, Mating and Cell Biology. Current Opinion in Microbiology 3:573-581.

[9] Friedman, N. (2004) Inferring Cellular Networks Using Probabilistic Graphical Models. Science 303:799-805.

[10] Friedman, N., Goldszmidt, M. and Wyner, A. (1999) Data Analysis with Bayesian Networks: A Bootstrap Approach. In Proceedings of the Fifteenth Conference on Uncertainty in Artificial Intelligence 196-205.

[11] Friedman, N. and Koller, D. (2003) Being Bayesian About Network Structure. A Bayesian Approach to Structure Discovery in Bayesian Networks. Machine Learning 50:95-125.

[12] Friedman, N., Linial, M., Nachman, I. and Pe'er, D. (2000) Using Bayesian Networks to Analyze Expression Data. Journal of Computational Biology 7:601-620.

[13] Gavin, I. M., Kladde, M. P. and Simpson, R. T. (2000) Tup1p Represses Mcm1p Transcriptional Activation and Chromatin Remodeling of an a-Cell-Specific Gene. The EMBO Journal 19:5875-5883.

[14] Hartemink, A. J., Gifford, D. K., Jaakkola, T. S. and Young, R. A. (2002) Combining Location and Expression Data for Principled Discovery of Genetic Regulatory Network Models. In Pacific Symposium on Biocomputing 7:437-449.

[15] Heckerman, D., Geiger, D. and Chickering, D. M. (1995) Learning Bayesian Networks: The Combination of Knowledge and Statistical Data. Machine Learning 20:197-243.

[16] Herskovits, E. H. (1991) Computer-Based Probabilistic-Network Construction. PhD Thesis, Stanford University.
[17] Imoto, S., Goto, T. and Miyano, S. (2002) Estimation of Genetic Networks and Functional Structures Between Genes by Using Bayesian Network and Nonparametric Regression. In Pacific Symposium on Biocomputing 7:175-186.
[18] McAdams, H. H. and Arkin, A. (1997) Stochastic Mechanisms in Gene Expression. In Proceedings of the National Academy of Science of the USA 94:814-819.
[19] Nielsen, J. D., Kočka, T. and Peña, J. M. (2003) On Local Optima in Learning Bayesian Networks. In Proceedings of the Nineteenth Conference on Uncertainty in Artificial Intelligence 435-442.
[20] Pe'er, D., Regev, A., Elidan, G. and Friedman, N. (2001) Inferring Subnetworks from Perturbed Expression Profiles. Bioinformatics 17:S215-S224.
[21] Peña, J. M., Björkegren, J. and Tegnér, J. (2005) Growing Bayesian Network Models of Gene Networks from Seed Genes. Bioinformatics 21:ii224-ii229.
[22] Ren, B., Robert, F., Wyrick, J. J., Aparicio, O., Jennings, E. G., Simon, I., Zeitlinger, J., Schreiber, J., Hannett, N., Kanin, E., Volkert, T. L., Wilson, C. J., Bell, S. P. and Young, R. A. (2000) Genome-Wide Location and Function of DNA Binding Proteins. Science 290:2306-2309.
[23] Sachs, K., Perez, O., Pe'er, D., Lauffenburger, D. A. and Nolan, G. P. (2005) Causal Protein-Signaling Networks Derived from Multiparameter Single-Cell Data. Science 308:523-529.
[24] Schwarz, G. (1978) Estimating the Dimension of a Model. Annals of Statistics 6:461-464.
[25] Sebastiani, P., Gussoni, E., Kohane, I. S. and Ramoni, M. (2003) Statistical Challenges in Functional Genomics (with Discussion). Statistical Science 18:33-60.
[26] Spellman, P. T., Sherlock, G., Zhang, M. Q., Iyer, V. R., Anders, K., Eisen, M. B., Brown, P. O., Botstein, D. and Futcher, B. (1998) Comprehensive Identification of Cell Cycle-Regulated Genes of the Yeast *Saccharomyces cerevisiae* by Microarray Hybridization. Molecular Biology of the Cell 9:3273-3297.
[27] Spirtes, P., Glymour, C. and Scheines, R. (1993) Causation, Prediction, and Search. Springer-Verlag, New York.
[28] Studený, M. (2003) Characterization of Inclusion Neighbourhood in Terms of the Essential Graph: Upper Neighbours. In Proceedings of the Seventh European Conference on Symbolic and Quantitative Approaches to Reasoning with Uncertainty 161-172.
[29] Studený, M. (2003) Characterization of Inclusion Neighbourhood in Terms of the Essential Graph: Lower Neighbours. In Proceedings of the Sixth Workshop on Uncertainty Processing 243-262.

The Role of Background Knowledge in Bayesian Classification

Marcel van Gerven and Peter J.F. Lucas

Institute for Computing and Information Sciences, Radboud University Nijmegen,
Toernooiveld 1, 6525 ED Nijmegen, The Netherlands
{marcelge,peterl}@cs.ru.nl

Summary. The development of Bayesian classifiers is frequently accomplished by means of algorithms that are highly data-driven. However, for many domains data-availability is scarce such that the resulting classifiers show poor performance. Even if performance is acceptable, Bayesian classifier structures are highly restricted and may therefore be unintelligable to the user. In this paper we address both issues. In the first part, we explore the trade-offs between classifiers constructed from clinical background knowledge and classifiers learned from a small clinical dataset. It is shown that the construction of classifiers from (partial) background knowledge is a feasible approach. In the second part, we introduce a construction algorithm that allows for a less restricted classifier structure, allowing easier clinical interpretation.

1 Introduction

The problem of representing and reasoning with expert knowledge has attracted considerable attention during the last three decades; in particular, ways of dealing with the *uncertainty* involved in decision making has been identified again and again as one of the key issues in this area. *Bayesian networks* are nowadays considered as standard tools for representing and reasoning with uncertain knowledge [13]. A Bayesian network consists of a structural part, representing the statistical (in)dependencies that hold among domain variables, and a probabilistic part specifying a joint probability distribution over these variables [16].

Learning a Bayesian network structure is NP hard [2] and manually constructing a Bayesian network for a realistic domain is a very laborious and time-consuming task. When one is more interested in the classification performance of a Bayesian network rather than in the accurate modeling of the independence structure that holds in the domain, we may use Bayesian classifiers. A Bayesian classifier can be identified as a Bayesian network with a fixed or severely constrained structural part, dedicated to classification. Here, the task is to assign the correct value to a *class variable* based on the available evidence. Examples of such Bayesian classifiers are the naive Bayesian classifier

[6], where evidence variables $E = \{E_1, \ldots, E_n\}$ are assumed to be conditionally independent given the class variable C and the tree-augmented Bayesian classifier [7], where correlations between evidence variables are represented as arcs between evidence variables in the form of a tree.

In medicine, Bayesian classifiers have proven to be a valuable tool for automated diagnosis and prognosis, but are lacking in some respects. Although Bayesian classifiers are robust in the presence of sufficient amounts of data [7; 15; 11], the heavy reliance of classifier construction algorithms on available data is not always justified, as there are many domains in which this availability is limited. For instance, in the medical domain, a substantial amount of medical disorders has only a sporadic occurrence and, therefore, even clinical research datasets may only include data of a hundred to a few hundred patients. Clearly, in such cases there is a role for human domain knowledge to compensate for the limited availability of data, which then may act as background knowledge to a learning algorithm.

A second perceived difficulty with the use of Bayesian classifiers is the fact that the restrictions on classifier structure disallow many statements of conditional (in)dependence. This may severely degrade the performance of particular classification algorithms on certain datasets. Such restrictions also lead to classifier structures that may be totally unintelligible. We feel that intelligible classifier structures will increase the acceptance of the use of Bayesian classifiers. Especially in medicine, a physician would like to see a correspondence between his knowledge and the (in)dependence assumptions that are captured by the structure of the Bayesian classifier.

In this paper we address the two difficulties identified above. In section 3 we examine the role of background knowledge with respect to the construction of Bayesian classifiers for the domain of prognosis in oncology where we have only a small amount of data at our disposal. In section 4 we address the restrictive nature of Bayesian classifiers by the introduction of a new algorithm to construct Bayesian classifiers which relaxes the structural assumptions and may therefore yield a network structure which is more intuitive from a medical point of view. In order to determine the performance and interpretability of this algorithm we make use of a clinical dataset of hepatobiliary (liver and biliary) disorders whose reputation has been firmly established. Performance of the algorithm is compared with an existing system for diagnosis of hepatobiliary disorders and other Bayesian classifiers such as the naive Bayesian classifier and the tree-augmented Bayesian classifier. We round off with some conclusions concerning the role of background knowledge in Bayesian classification.

2 Preliminaries

Prior to addressing the issues that have been mentioned in the previous section, we discuss the preliminaries necessary for the remainder of this paper.

A *Bayesian network* is defined as a pair (G, P), where G is an *acyclic directed graph* (ADG) with *vertices* $V(G)$ and *directed edges* or *arcs* $E(G)$. P is a joint probability distribution defined over a set of random variables, assuming a one-to-one correspondence between vertices $V(G)$ and random variables $X_{V(G)}$. We use X_U with $U \subseteq V(G)$ to refer to a subset of random variables. If U is a singleton set $\{i\}$ then we will write X_i to refer to the random variable that corresponds to the vertex i.

The ADG G is assumed to be a *(directed) independence map* (I-map), representing the independence structure between a set of random variables [16]. Let $\pi_G(v) = \{v' \mid (v', v) \in E(G)\}$ be the set of *parents* of v in G. The I-map property of G admits the following recursive factorization of the joint probability distribution:

$$P(X_1, \ldots, X_n) = \prod_{i \in V(G)} P(X_i \mid X_{\pi_G(i)}). \qquad (1)$$

As there is a one-to-one correspondence between $V(G)$ and a set X of random variables, we often use vertices and random variables interchangeably. In that case, we just write $P(X_i \mid \pi(X_i))$ to denote $P(X_i \mid X_{\pi_G(i)})$ for given G. Let S_X denote the *sample space* of a discrete random variable or set of random variables X. We use x as shorthand notation for a tuple $(x_1, \ldots, x_n) \in S_X$.

We call a Bayesian network that offers a task-neutral representation of the knowledge concerning the independence structure between variables in a domain a *declarative model*. In contrast, Bayesian classifiers are Bayesian networks with a fixed or severely constrained structural part that are specifically suited to the classification task.

In order to systematically assess the performance of Bayesian classifiers with structures of varying complexity we utilize the *forest-augmented naive classifier*, or FAN classifier for short (Fig. 1). A FAN classifier is a modification of the *tree-augmented naive* (TAN) classifier, where the topology of the resulting graph over evidence variables is restricted to a forest of trees [17; 11]. The algorithm to construct FAN classifiers used in this paper is based on a modification of the algorithm to construct TAN classifiers [7], where the *conditional mutual information*

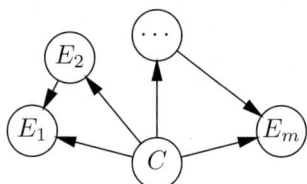

Fig. 1. A forest-augmented naive (FAN) classifier. For each evidence variable E_i there is at most one incoming arc allowed from $E \setminus \{E_i\}$ and exactly one incoming arc from the class variable C. Both the naive classifier and the tree-augmented naive classifier are extreme cases of the forest-augmented naive classifier.

$$I(E_i, E_j \mid C) = \sum_{e_i, e_j, c} P(e_i, e_j, c) \log \frac{P(e_i, e_j \mid c)}{P(e_i \mid c) P(e_j \mid c)}, \quad (2)$$

computed from a dataset D is used to build a maximum cost spanning tree between evidence variables $E = \{E_1, \ldots, E_m\}$.

The performance of FAN classifiers may be determined by computing *zero-one loss*. Let D be a dataset consisting of p cases and let c^k be the value of the class variable C given the k-th example $e^k \in S_E$. Then, $L(e^k) = 1$ if $\arg\max_c P(C = c \mid E = e^k) = c^k$ and 0 otherwise. The *classification accuracy* is defined as the percentage of correctly classified cases:

$$\frac{1}{p} \sum_{k=1}^{p} \left(1 - L(e^k)\right) \times 100\%.$$

A disadvantage of this straightforward method of comparing the quality of the classifiers is that the actual posterior probabilities are ignored. A more precise indication of the behavior of Bayesian classifiers can be obtained with the *logarithmic scoring rule* [3]. With each prediction generated by a Bayesian network we associate a score $s_k = -\log P(c^k \mid e^k)$, which can be interpreted formally as the entropy and has the informal meaning of a penalty. When the probability $P(c^k \mid e^k) = 1$, then $s_k = 0$ (actually observing c^k generates no information); otherwise, $s_k > 0$. The total score for dataset D is now defined as the average of the individual scores:

$$S = \frac{1}{p} \sum_{k=1}^{p} s_k.$$

The logarithmic scoring rule is a rule which measures differences in probabilities for a class c_k given evidence E. A global measure of the difference between two distributions P and Q is the Kullback-Leibler (KL) distance [8]:

$$D(P, Q) = \sum_x P(x) \log \frac{P(x)}{Q(x)},$$

where $D(P, Q) \geq 0$, with equality iff $P = Q$.

When classifier parameters are learned from data, we compute the conditional probability distribution for a variable X_i as the weighted average of a probability estimate and the Dirichlet prior, as follows:

$$P(X_i \mid \pi(X_i)) = \frac{N}{N + N_0} P_D(X_i \mid \pi(X_i)) + \frac{N_0}{N + N_0} \Theta_i \quad (3)$$

where P_D is the probability distribution estimate based on a given dataset D, and Θ_i is a uniform Dirichlet prior $\frac{1}{|S_{X_i}|}$. N is the size of the dataset and N_0 accounts for the contribution of the prior to the posterior and can be thought of as representing a number of pseudo-counts.

2.1 Estimating Classifiers from Background Knowledge

Learning FAN classifiers directly from a declarative model is accomplished as follows. If we have a joint probability distribution $P(C, E, X)$ with class-variable C, evidence variables $E = \{E_1, \ldots, E_m\}$ and remaining variables $X = \{X_1, \ldots, X_n\}$ underlying the declarative Bayesian network $\mathcal{B} = (G, P)$, then the following decomposition is associated with the Bayesian network:

$$P(C, E, X) = P(C \mid \pi(C)) \prod_{i=1}^{m} P(E_i \mid \pi(E_i)) \prod_{j=1}^{n} P(X_j \mid \pi(X_j)).$$

The joint probability distribution underlying the FAN classifier $\mathcal{B}' = (G', P')$ with $V(G') = V(G)$ is defined as $P'(C, E)$. The probability distribution P is used as a basis for the estimation of P', as follows:

$$P'(E_i \mid \rho(E_i), C) = \sum_{x \in S_{X \cup E \setminus \{E_i\} \cup \rho(E_i)}} P(E_i, x \mid \rho(E_i), C) \qquad (4)$$

with $\rho(E_i) = \pi_{G'}(E_i) \setminus \{C\}$. Building FAN classifiers based on the declarative model amounts to estimating three-vertex networks of the form shown in Fig. 2 using Eqn. (4) that act as input to the FAN construction algorithm.

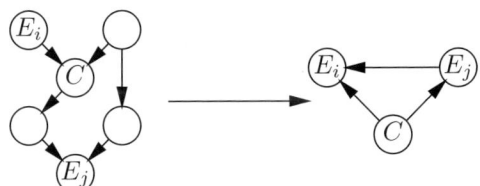

Fig. 2. Declarative model, used in computing the joint probability distributions for a three-vertex network, where $P(E_i, E_j, C) = P(E_i \mid E_j, C) P(E_j \mid C) P(C)$ and $P(E_i, E_j \mid C) = P(E_i \mid E_j, C) P(E_j \mid C)$.

The construction of declarative models is a difficult undertaking since experts need to state perfectly the (in)dependence structure and conditional probability distributions associated with a given domain. Since this is a very time-consuming task and an instantiation of the infamous *knowledge acquisition bottleneck* [4], we will examine how background knowledge of different degrees of completeness influences the quality of the resulting classifiers built from this knowledge. More formally, let $\mathcal{B} = (G, P)$ be a declarative model with joint probability distribution $P(X)$, representing full knowledge of a domain. Let $\mathcal{B}' = (G', P')$ with $V(G') = V(G)$ be a Bayesian network with $P'(X)$. $\mathcal{B}'$ is said to represent *partial* background knowledge if $0 < D(P, P') < \epsilon$ for $\epsilon > 0$, where ϵ is the least upper-bound of $D(P, P')$ for an uninformed prior P'. We focus on the incomplete specification of dependencies as our operationalization of partial background knowledge, such that

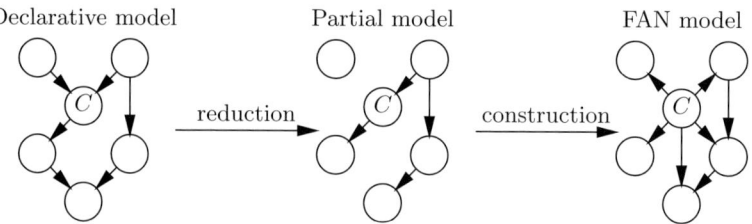

Fig. 3. A declarative model is reduced to a partial model. Subsequently, FAN models are constructed from the partial model. The structure of the FAN model typically does not correspond to that of the declarative model due to the strong restrictions placed on the resulting network structure.

for a *partial model* $\mathcal{B}'$, $E(G') \subseteq E(G)$. The probability distribution P is used to estimate P', as follows:

$$P'(X_i \mid X_{\pi_{G'}(i)}) = \sum_{x' \in S_{X'}} P(X_i \mid X_{\pi_{G'}(i)}, x')P(x' \mid X_{\pi_{G'}(i)}), \quad (5)$$

with $X' = X_{\pi_G(i)} \setminus X_{\pi_{G'}(i)}$. Figure 3 shows how a partial model is estimated from a declarative model using equation (5) and employed to estimate the probabilities for a FAN classifier.

2.2 The Maximum Mutual Information Algorithm

The maximum mutual information algorithm is a classifier construction algorithm that is less restrictive than the discussed FAN algorithm. It uses both the computed mutual information between evidence variables and the class-variable and the computed conditional mutual information between evidence-variables as a basis for constructing a Bayesian classifier.

Mutual information (MI) between an evidence variable E_i and the class-variable C can be estimated from a network with structure $C \to E_i$ by:

$$I(E_i, C) = \sum_{e_i, c} P(e_i \mid c)P(c) \log \frac{P(e_i \mid c)}{P(e_i)}. \quad (6)$$

Conditional mutual information between multiple evidence variables is defined similarly to mutual information, where the conditional may be an arbitrary set of variables $A = \{A_1, \ldots, A_n\}$. It is computed as follows:

$$I(E_i, E_j \mid A) = \sum_{e_i, e_j, a} P(e_i \mid e_j, a)P(e_j \mid a)P(a) \log \frac{P(e_i \mid e_j, a)}{P(e_i \mid a)} \quad (7)$$

where $P(e_i \mid a) = \sum_{e_j} P(e_i \mid e_j, a)P(e_j \mid a)$ and $a \in S_A$.

Contrary to naive and TAN classifiers, the MMI algorithm makes no assumptions whatsoever about the initial network structure. It starts from a

Algorithm 1: MMI construction algorithm

input: G {empty Bayesian network structure},
D {database}, C {class variable},
E {evidence-variables}, N {number of arcs}

$\mathcal{C} \leftarrow$ a sequence $\{C_1, \ldots, C_k\}$ of elements (C,E) for $E \in E$, sorted by $I(C, E)$
$\mathcal{A} \leftarrow$ an initially empty sequence $\{A_1, \ldots, A_m\}$ of pairs of evidence variables
$\mathcal{O} \leftarrow \varnothing$ {ordering on the attributes}
for $i = 0$ to N **do**
  if $\mathcal{A} = \varnothing$ or $I(C_1) > I(A_1 \mid \pi(E))$ with $A_1 = (E', E)$ **then**
    Let E be the evidence variable in C_1
    remove C_1 from $\mathcal{C}$
    add E to the ordering $\mathcal{O}$
    add (C, E) to the arcs of G
    for all $E' \in E \setminus \mathcal{O}$ **do**
      add candidate (E', E) to $\mathcal{A}$
    end for
    sort$(\mathcal{A})$ by $I(E', E \mid \pi(E))$
  else
    add A_1 to the arcs of G
    remove A_1 from $\mathcal{A}$
    for all pairs $(E', E) \in \mathcal{A}$ **do**
      recompute $I(E', E \mid \pi(E))$
    end for
    sort$(\mathcal{A})$
  end if
end for
return G

---

fully disconnected graph, whereas the FAN algorithm starts with a naive classifier structure such that $(C, E_i) \in E(G)$ for all evidence variables E_i. Since redundant attributes are not encoded, network structures are sparser, at the same time indicating important information on the independence between class and evidence variables. In this sense, the MMI algorithm can be said to resemble *selective Bayesian classifiers* [9].

The algorithm iteratively selects the arc with highest (conditional) mutual information from the set of candidates and adds it to the Bayesian network $\mathcal{B}$ with classifier structure G (algorithm 1). It starts by computing $I(C, E_i)$ for a sequence $\mathcal{C}$ of pairs (C, E_i). From this sequence, the candidate having highest mutual information, say (C, E_i) is selected. This candidate is removed from $\mathcal{C}$ and added to the classifier structure. Subsequently, it will construct all candidates of the form (E_j, E_i) where (C, E_j) is not yet part of the classifier structure G and add them to $\mathcal{A}$. The conditional mutual information $I(E_i, E_j \mid \pi(E_i))$ is computed for these candidates. Now, the algorithm

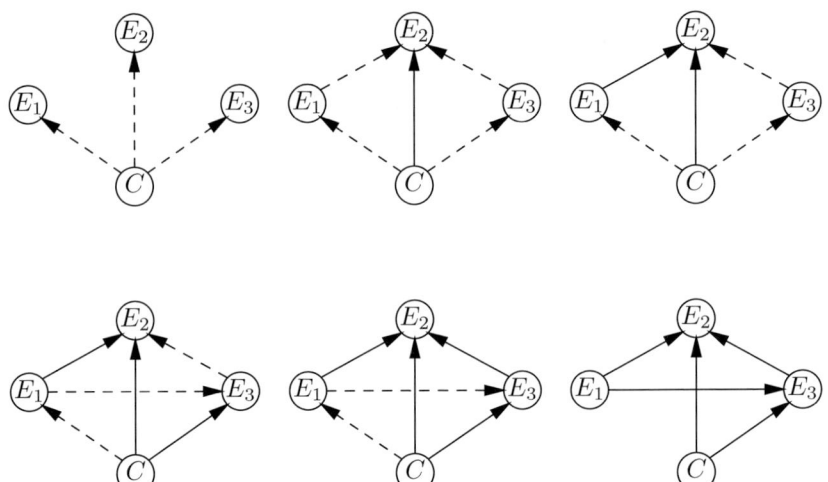

Fig. 4. An example of the MMI algorithm building a Bayesian classifier structure from the top left to the bottom right. Dashed arrows represent candidate dependencies. The final structure incorporates feature selection, orientational preference of dependencies and the encoding of a third-order dependency $P(E_2 \mid C, E_1, E_3)$.

iteratively selects the candidate of list $\mathcal{C}$ or $\mathcal{A}$ having the highest (conditional) mutual information. If a candidate E_i from $\mathcal{A}$ is chosen, then $I(E_i, E_j \mid \pi(E_i))$ for all pairs $(E_i, E_j) \in \mathcal{A}$ is recomputed since the parent set of E_i has changed. By directing evidence arcs to attributes that show high mutual information with the class variable, we enforce that the resulting structure remains an acyclic digraph. Figure 4 shows an example of how the algorithm builds a Bayesian classifier structure.

Looking back at Eqn. (7) a possible complication is identified. Since the set $\pi(E_i)$ of an evidence variable E_i may grow indefinitely and the number of parent configurations grows exponentially with n, the network may become victim of its own unrestrictedness in terms of structure. Note also that since one has a finite (and often small) database at ones disposal, this means that the actual conditional probability $P(E_i \mid \pi(E_i))$ will become increasingly inaccurate when the number of parents grows; configurations associated with large parent-sets cannot be reliably estimated from moderate size databases, introducing what may be termed *spurious dependencies*. When we compute conditional information over a database consisting of k records, the average number of records providing information about a particular configuration of a parent set of size n containing binary variables will only be $k2^{-n}$ on average. So even for moderate size databases such inaccuracies will arise rather quickly.

In order to prevent the occurrence of spurious dependencies, we make use of the following heuristic. The probability $P(E_i, E_j \mid a)$ for $a \in S_A$ is estimated to be equal to

$$\frac{N_a}{N_a + N_0^c} P(E_i, E_j \mid a) + \frac{N_0^c}{N_a + N_0^c} P(E_i \mid a) P(E_j \mid a), \qquad (8)$$

where P is computed according to Eqn. 3, N_a is the number of times the configuration a occurs in D and N_0^c is a parameter that is used during computation of the conditional mutual information. In this manner, the conditional mutual information computed according to Eqn. 7 will be small if the number of occurrences of the conditioning case is small. In the following we will use $N_0^c = 500$ throughout our experiments, unless indicated otherwise.

3 Data-driven versus Model-driven Classifiers

The aim in this section is to gain insight into the quality of Bayesian classifiers when learned from either data or background knowledge. Such a comparison is fairly uncommon since most machine learning research is either based on the availability of large amounts of data or on a model from which the data is generated. These models and data are often explicitly designated for benchmarking purposes, but it is not known and even doubted whether they properly represent the real-world situation [11]. Therefore, we have chosen to use both a dataset taken directly from clinical practice and a Bayesian network constructed by expert clinicians. We refer to classifiers that are learned from data as *data-driven* classifiers (denoted by F_d) and to classifiers that are derived from a declarative model as *model-driven* classifiers (denoted by F_m). We use F_k^n to refer to a type k FAN classifier containing n arcs of the sort (E_i, E_j) with $i \neq j$. Note that F_k^n is equivalent to a naive classifier when $n = 0$ and equivalent to a TAN classifier when n is equal to $|E| - 1$, forming a spanning tree over the evidence variables.

3.1 Non-Hodgkin Lymphoma Model and Data

In this research, we used a Bayesian network incorporating most factors relevant for the management of the uncommon disease *gastric non-Hodgkin lymphoma* (NHL for short), referred to as the *declarative model*, which is shown in Fig. 5. It is fully based on expert knowledge and has been developed in collaboration with clinical experts from the Netherlands Cancer Institute (NKI) [12]. The model has been shown to contain a significant amount of high quality knowledge [1]. Furthermore, we are in possession of a database containing 137 patients which have been diagnosed with gastric NHL.

We excluded post-treatment variables and have built FAN classifiers as depicted in Fig. 6, where the structure and underlying probability distributions are either learned from the available patient data or estimated directly from the (partial) declarative model using equation (4).

Classifiers were evaluated by computing classification accuracy and logarithmic score for 137 patient cases for the class-variable 5-YEAR-RESULT.

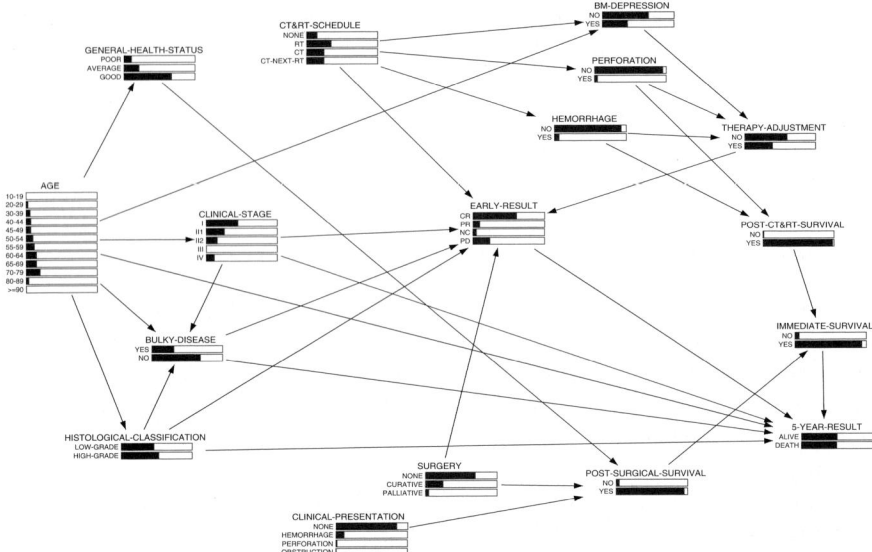

Fig. 5. Declarative Bayesian network as designed with the help of expert clinical oncologists.

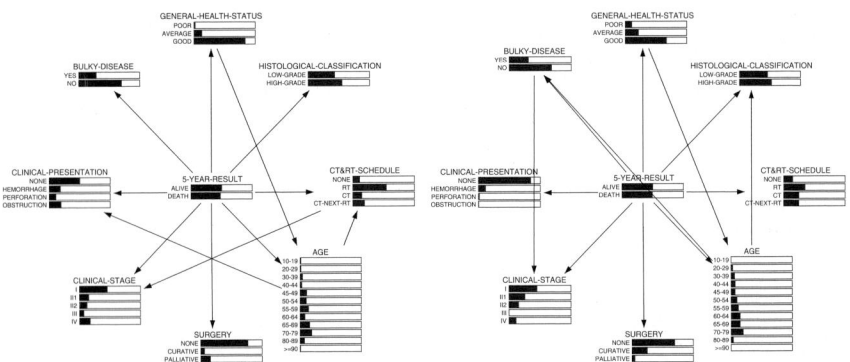

Fig. 6. Differing resulting structures for data-driven FAN classifiers (left) and model-driven FAN classifiers (right) for the class-variable 5-YEAR-RESULT. The bars represent prior probability estimates.

This variable represents whether a patient has died from NHL (DEATH) or lives (ALIVE) five years after therapy. For the classifiers learned from patient data leave-one-out cross-validation was carried out such that test cases were excluded during estimation of the joint probability distribution of the resulting classifiers. Probability distributions of the classifiers were compared with that of the declarative model by means of the KL distance.

3.2 Data-driven and Model-driven Classification

The results for both classification accuracy and logarithmic score (Fig. 7) show that performance was consistently better for model-driven classifiers than for data-driven classifiers. Construction of a classifier from a database with a limited number of cases obviously leads to a performance degradation and the use of background knowledge considerably enhances classifier quality. Although this is not surprising, it does show that for a realistic domain such as prognosis for non-Hodgkin lymphoma it can be a better strategy to construct a classifier based on expert opinion rather than on available data.

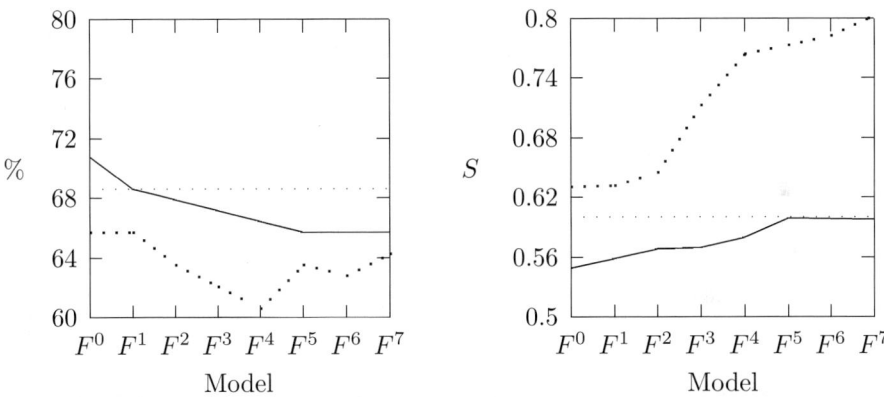

Fig. 7. Classification accuracy (left) and logarithmic score (right) for Bayesian classifiers with a varying number of arcs learned from either patient data (data-driven, dotted line) or the declarative model (model-driven, solid line). Classification accuracy and logarithmic score for the declarative model are shown for reference (straight line).

Although model-driven classifiers perform better than data-driven classifiers for this domain, a decrease in classification accuracy and logarithmic score can be observed for classifiers with more complex structures. This is in agreement with previous research, where it has been shown that a classifier with high bias and low variance tends to produce higher classification accuracy than one with low bias and high variance, due to insensitivity to

Table 1. KL distances for model-driven and data-driven FAN classifiers.

	F^0	F^1	F^2	F^3	F^4	F^5	F^6	F^7
Model-driven	0.52	0.27	0.22	0.18	0.15	0.14	0.13	0.13
Data-driven	6.56	6.58	8.40	9.24	11.55	11.56	12.36	13.77

bias in the classifier's probability estimates [5]. On the other hand, more complex structures allow a more accurate representation of the joint probability distribution underlying the declarative model, as can be seen in Table 1.

When comparing the model-driven and data-driven classifiers we can distinguish both qualitative differences in terms of classifier structure and quantitative differences in terms of probability estimates. In the following, we will discuss these differences. When structures are compared, it is found that entirely different dependencies were added due to large differences in CMI when computed either from patient data or background knowledge. The strongest dependency computed from patient data is the dependency between CT&RT-SCHEDULE (chemotherapy and radiotherapy schedule) and CLINICAL-STAGE having a CMI of 0.212. An indirect dependency with a CMI of 0.0112 indeed exists between these variables, since the two post-treatment variables EARLY-RESULT and 5-YEAR-RESULT are mutual descendants (Fig. 5). Because post-treatment information is unknown at the time of therapy administration, clinicians tend to base therapy selection directly on the clinical stage of the tumor. This is an example of a discrepancy between expert opinion and clinical practice, which must be taken into account when validating a model based on patient data. In Ref. [12] more such discrepancies are identified, which are due to evolution in treatment policy and due to the use of indirect and inaccurate measurements of clinically relevant variables. One should always be aware of such discrepancies when constructing a classifier from background knowledge.

As has been remarked, classification performance decreases for both model-driven and data-driven classifiers of increasing structural complexity. For model-driven classifiers this cannot be attributed to a small sample size since the declarative model is a generative model representing an infinite-size dataset. However, this is under the assumption of a perfect estimate of the independence structure and conditional probability distributions by the expert physician. This is an unrealistic assumption in practice and it is to be expected that the accurate estimation of conditional probabilities tends to become more difficult when the size of the conditioning set grows.

For data-driven classifiers a decreasing performance for more complex structures in case of a small sample size is to be expected and can be understood as follows. Recall that a classifier structure is built by selecting those dependencies between evidence variables with highest conditional mutual information. The probabilities that are used for computing the CMI are based on a probability estimate based on a dataset D and on a Dirichlet prior Θ_i. For small datasets the probability estimates may deviate considerably from the actual probability and the contribution of the prior due to the N_0 parameter may lead to incorrect values for the CMI such that incorrect arcs are added to the classifier structure. Secondly, (conditional) probabilities estimated from such a small dataset will also be inaccurate. These effects can be observed in Fig. 6 where the estimated structures and prior probabilities differ considerably between model-driven and data-driven classifiers.

3.3 Classification using Partial Models

Although the benefit of using background knowledge has been demonstrated in previous sections, it will not usually be the case that full knowledge of the domain is available. Instead, one expects the expert to deliver partial knowledge about the structure and underlying probabilities of the domain. In this section we investigate how partial specifications influence the quality of Bayesian classifiers. To this end, we created partial models retaining 0, 5, 10, 15, 20, 25 and all 32 arcs of the original declarative model. In total 77 different partial models were generated and the KL distances between the declarative and partial models were computed. From these models we have generated model-driven FAN classifiers F_m^0 and F_m^7. Linear regressions on classification accuracy and logarithmic score are shown in Fig. 8.

The outliers at the bottom right and top right of the figure were identified to be partial models where the class-variable 5-YEAR-RESULT is a disconnected vertex and were not included in the regression. Such a model encodes just the class-variable's prior probabilities and can be regarded as the model with baseline performance. Superimposed + and ⋄ symbols represent models

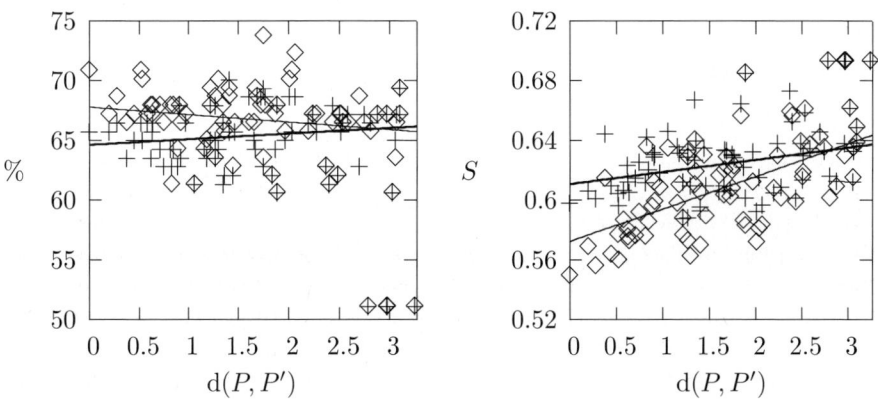

Fig. 8. Regression results on classification accuracy and logarithmic score for the naive classifier F_m^0 (⋄, thin line) and TAN classifier F_m^7 (+, thick line) for partial models containing varying amounts of partial background knowledge as measured by the KL distance between the declarative model $\mathcal{B} = (G, P)$ and partial models $\mathcal{B}' = (G', P')$.

whose relevant dependencies can be fully represented within the conditional probability tables of the naive classifier.

It is hard to discern a pattern in the left part of Fig. 8 and little value can be assigned to the regression results. On average, the naive classifier does show better classification accuracy than the TAN model with a best performance of 73.72% for a model containing ten arcs with a KL distance of 1.75. The large variance in classification accuracy for partial models with equal relative entropies confirms previous results reported in Ref. [11] where it was indicated that the relationship between the quality of a probability distribution, as measured here more precisely by means of KL distance, and classification accuracy is not straightforward. In the right part of Fig. 8 one can observe, on average, an increase in logarithmic score with increasing KL distance, which is more pronounced for the naive classifier. In general, the results indicate a positive effect on classification performance for increasingly detailed background knowledge although the relationship is not straightforward. On average, partial models containing 10 arcs attain performances similar to that of the model which was learned from data, which demonstrates that the use of partial background knowledge is indeed a feasible alternative to the use of data for the construction of Bayesian classifiers.

4 The MMI Classifier

In the previous section we have examined the performance of classifiers learned from background knowledge when there is only a small amount of data. In this section we focus more on the interpretability of classifier structures by weakening the restrictions of the FAN classifier.

4.1 The COMIK Dataset

In order to validate classifier performance we made use of the COMIK dataset, which was collected by the Copenhagen Computer Icterus (COMIK) group and consists of data on 1002 jaundiced patients. The COMIK group has been working for over a decade on the development of a system for diagnosing liver and biliary disease which is known as the Copenhagen Pocket Diagnostic Chart [14]. Using a set E of 21 evidence variables, the system classifies patients into one of four diagnostic categories: *acute non-obstructive*, *chronic non-obstructive*, *benign obstructive* and *malignant obstructive*. The chart offers a compact representation of three logistic regression equations, where the probability of *acute obstructive jaundice*, for instance, is computed as follows: $P(acute\ obstructive\ jaundice \mid E) = P(acute \mid E) \cdot P(obstructive \mid E)$. The performance of the system has been studied using retrospective patient data and it has been found that the system is able to produce a correct diagnostic conclusion (in accordance with the diagnostic conclusion of expert clinicians) in about $75 - 77\%$ of jaundiced patients [10].

4.2 Classification Results and Classifier Interpretability

In this section we will demonstrate the usefulness of the N_0^c parameter that was introduced in Eqn. 8, compare the classification performance of both the FAN and MMI classifiers on the COMIK dataset and give a medical interpretation of the resulting structures.

First we present the results of varying the parameter N_0^c in order to determine whether this has an effect on the classification performance and network structure of our classifiers. To this end, we have determined the classification accuracy and summed squared fan-in of the nodes in the classifier for a network of 30 arcs. Let $\mid \pi(i) \mid$ denote the cardinality of the parent set of a vertex i. The summed squared fan-in $F(\mathcal{B})$ of a Bayesian network $\mathcal{B} = (G, P)$ is defined as $F(\mathcal{B}) = \sum_{i \in V(G)} \mid \pi(i) \mid^2$. Table 2 clearly shows that the summed squared fan-in decreases when N_0^c increases; indicating that spurious dependencies are removed. This removal also has a beneficial effect on the classification accuracy of the classifier, which rises from 74.75% for $N_0^c = 1$ to 76.25% for $N_0^c = 660$. A setting of $N_0^c = 500$ seems reasonable, for which classification accuracy is high and the influence on structural complexity is considerable, but not totally restrictive.

Table 2. Effects of varying parameter N_0^c for a model consisting of 30 arcs.

N_0^c	%	$F(\mathcal{B})$	N_0^c	%	$F(\mathcal{B})$	N_0^c	%	$F(\mathcal{B})$
1	74.75	87	102	75.95	65	800	76.25	59
4	74.75	77	290	75.95	63	900	76.25	59
36	74.85	71	610	75.95	61	2000	76.25	57
56	75.15	67	660	76.25	61			

We have compared the performance of the MMI algorithm with that of the FAN algorithm. Figure 9 shows that both algorithms perform comparably and within the bounds of the Copenhagen Pocket Diagnostic Chart. Both the MMI and FAN algorithm show a small performance decrease for very complex network structures, which may be explained in terms of overfitting artifacts. The last arcs added will be arcs having very small mutual information, which can be a database artifact instead of a real dependency within the domain, thus leading to the encoding of spurious dependencies. Best classifier accuracy for the MMI algorithm is 76.65% for a network of 19 arcs versus 76.45% for a network of 27 arcs for the FAN algorithm.

In terms of classifier structure, one can observe that both algorithms represent similar dependencies, with the difference that those of the MMI algorithm form a subset of those of the FAN algorithm. The best FAN classifier has a structure with an arc from the class variable to every evidence variable and the following arcs between evidence variables: *biliary-colics-gallstones → upper-abdominal-pain → leukemia-lymphoma → gall-bladder, history-ge-2-weeks →*

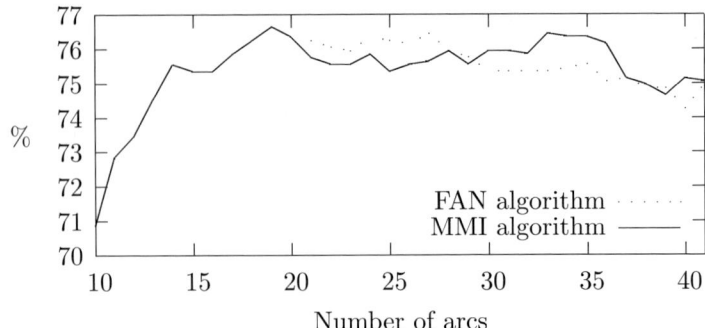

Fig. 9. Classification accuracy for Bayesian classifiers with a varying number of arcs learned using the FAN algorithm or the MMI algorithm for the COMIK dataset.

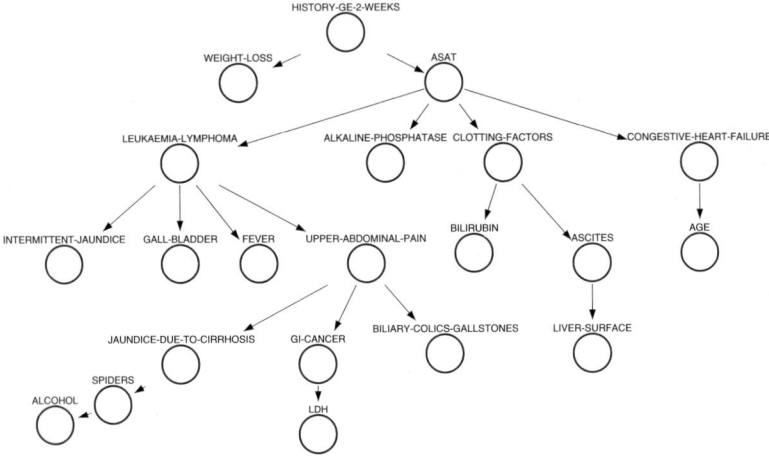

Fig. 10. Dependencies between evidence variables for the COMIK dataset using a FAN classifier containing 41 arcs. The class-variable was fully connected with all evidence variables (not shown).

weight-loss, ascites → *liver-surface* and *ASAT* → *clotting-factors*. The MMI algorithm has left *leukemia-lymphoma, congestive-heart-failure* and *LDH* independent of the class-variable and shows just the dependency *liver-surface* → *ascites* between evidence variables.

Given our aim of learning Bayesian classifiers that not only display good classification performance, but are comprehensible to medical doctors as well, we have carried out a qualitative comparison between two of the Bayesian networks learned from the COMIK data: Figure 10 shows a FAN classifier

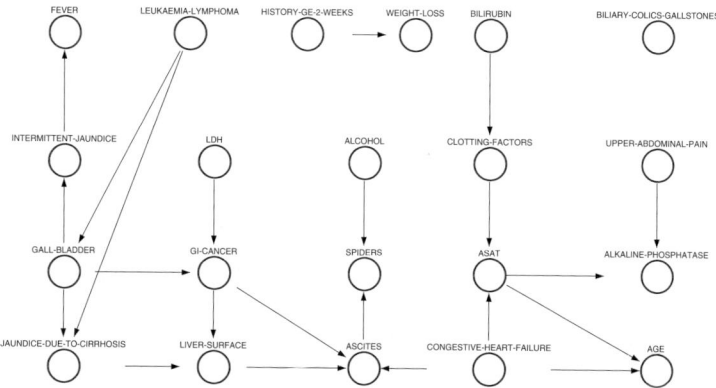

Fig. 11. Dependencies between evidence variables for the COMIK dataset using an MMI classifier containing 41 arcs. The class-variable was fully connected with all evidence variables (not shown).

which was learned using the FAN algorithm described previously [11], whereas Figure 11 shows an MMI network with the same number of arcs. Clearly, the restriction imposed by the FAN algorithm that the arcs between evidence variables form a forest of trees does have implications with regard to the understandability of the resulting networks. Yet, parts of the Bayesian network shown in Figure 10 can be given a clinical interpretation. Similar remarks can be made for the MMI network, although one would hope that giving an interpretation is at least somewhat easier.

If we ignore the arcs between the class vertex and the evidence vertices, there are 20 arcs between evidence vertices in the FAN and 22 arcs between evidence vertices in the MMI network. Ignoring arc orientation, 9 of the arcs in the MMI network are shared by the FAN classifier. As the choice of the direction of arcs in the FAN network is arbitrary, it is worth noting that in 4 of these arcs the direction is different; in 2 of these arcs it is medically speaking impossible to establish the right direction of the arcs, as hidden variables are involved, in 1 the arc direction is correct (*congestive-heart-failure* → *ASAT*), whereas in the remaining arc (*GI-cancer* → *LDH*) the direction is incorrect.

Some of the 13 non-shared arcs of the MMI network have a clear clinical interpretation. For example, the arcs *GI-cancer* → *ascites*, *congestive-heart-failure* → *ascites* and *GI-cancer* → *liver-surface* are arcs that can be given a causal interpretation, as gastrointestinal (GI) cancer and right-heart failure do give rise to the accumulation of fluid in the abdomen (i.e. ascites), and there are often liver metastases in that case that may change the liver surface. Observe that the multiple causes of ascites cannot be represented in the FAN network due to its structural restrictions. The path *gallbladder* → *intermittent-jaundice* → *fever* in the MMI network offers a reasonably accurate picture of the course of events of the process giving rise to fever; in contrast, the situation depicted in the FAN, where *leukemia-lymphoma* acts as a common cause, does not reflect clinical reality. However, the arc from *upper-abdominal-pain* to *biliary-colics-gallstones* in the FAN, which is correct, is missing in the MMI network. Overall, the MMI network seems to reflect clinical reality somewhat better than the FAN, although not perfectly.

5 Conclusion

We have investigated the role of background knowledge with respect to Bayesian classification. In this research we have addressed the issues of learning classifier structures in data-poor domains and the construction of a minimally restrictive classifier structure.

The construction of classifiers for data-poor domains is an important issue since many real-world problems are characterized by the absence of sufficient statistical data and most algorithms for constructing Bayesian classifiers are highly data-driven. In this research, we presented a method for constructing model-driven classifiers from partial background knowledge and showed that they outperform data-driven classifiers for a realistic clinical domain. We have shown that in this domain, the classification performance of both data- and model-driven classifiers decreases when the structural complexity of the classifiers increases. This is in agreement with previous research which demonstrated that for small sample sizes, classifiers with high bias perform best. Although structurally simple classifiers show good classification performance, it may still be desirable to use classifiers of larger structural complexity. The large KL distance between the declarative model and structurally simple classifiers demonstrated that high bias comes at the price of limitations on how well the joint probability distribution underlying a declarative model can be approximated.

The MMI algorithm makes few structural assumptions and iteratively builds classifier structures that reflect higher-order dependencies between evidence variables. In this sense, the MMI algorithm resembles Sahami's k-dependence classifiers [18] with the difference that we do not require the addition of an arc between the class variable and each evidence variable. Furthermore, our use of Dirichlet priors during the estimation of conditional mutual information prevents the construction of overly complex network structures and the introduction of spurious dependencies. As is shown, the number of higher-order dependencies will only increase if this is warranted by sufficient data. The experimental results show that classification performance of the resulting classifiers is good while the less restrictive assumptions allow for a network structure that is less ad-hoc and somewhat better to interpret from a medical point of view.

We hope that this research conveys the message that there are other ways of constructing Bayesian classifiers rather than employing large amounts of data and that classification accuracy need not be the only criterion of interest.

References

[1] C. Bielza, J.A. Fernández del Pozo, and P.J.F. Lucas. Finding and explaining optimal treatments. In *AIME 2003*, pages 299–303, 2003.
[2] D.M. Chickering, D. Geiger, and D. Heckerman. Learning Bayesian networks is NP-hard. Technical report, Microsoft Research, November 1994.
[3] R.G. Cowell, A.P. Dawid, and D. Spiegelhalter. Sequential model criticism in probabilistic expert systems. *PAMI*, 15(3):209–219, 1993.
[4] J. Cullen and A. Bryman. The knowledge acquisition bottleneck: Time for reassessment. *Expert Systems*, 5(3):216–225, 1988.
[5] P. Domingos and M. Pazzani. On the optimality of the simple Bayesian classifier under zero-one loss. *Machine Learning*, 29:103–130, 1997.
[6] R.O. Duda and P.E. Hart. *Pattern Classification and Scene Analysis*. Wiley, 1973.
[7] N. Friedman, D. Geiger, and M. Goldszmidt. Bayesian network classifiers. *Machine Learning*, 29:131–163, 1997.
[8] S. Kullback and S. Leibler. On information and sufficiency. *Annals of Mathematical Statistics*, 22:29–86, 1951.
[9] P. Langley and S. Sage. Induction of selective Bayesian classifiers. In *Proceedings of UAI-94*, 1994.
[10] G. Lindberg, C. Thomson, A. Malchow-Møller, P. Matzen, and J. Hilden. Differential diagnosis of jaundice: applicability of the Copenhagen pocket diagnostic chart proven in Stockholm patients. *Liver*, 7:43–49, 1987.
[11] P.J.F. Lucas. Restricted Bayesian network structure learning. In J.A. Gámez, S. Moral, and A. Salmeron, editors, *Advances in Bayesian Networks, Studies in Fuzziness and Soft Computing*, volume 146, pages 217–232. Springer-Verlag, Berlin, 2004.
[12] P.J.F. Lucas, H. Boot, and B.G. Taal. Computer-based decision support in the management of primary gastric non-Hodgkin lymphoma. *Methods of Information in Medicine*, 37:206–219, 1998.
[13] P.J.F. Lucas, L.C. van der Gaag, and A. Abu-Hanna. Bayesian networks in biomedicine and health-care. *Artificial Intelligence in Medicine*, 30:201–214, 2004.
[14] A. Malchow-Møller, C. Thomson, P. Matzen, and et al. Computer diagnosis in jaundice: Bayes' rule founded on 1002 consecutive cases. *Journal of Hepatology*, 3:154–163, 1986.
[15] M. Pazzani. Searching for dependencies in Bayesian classifiers. In *Learning from data: Artificial intelligence and statistics V*, pages 239–248. New York, NY: Springer-Verlag, 1996.

[16] J. Pearl. *Probabilistic Reasoning in Intelligent Systems: Networks of Plausible Inference*. Morgan Kaufmann Publishers, 1988.
[17] J.P. Sacha, L. Goodenday, and K.J. Cios. Bayesian learning for cardiac SPECT image interpretation. *Artificial Intelligence in Medicine*, 26:109–143, 2002.
[18] M. Sahami. Learning limited dependence Bayesian classifiers. In *Second International Conference on Knowledge Discovery in Databases*, 1996.

Printing: Krips bv, Meppel
Binding: Stürtz, Würzburg